Functionality, Advancements and Industrial Applications of Heat Pipes

Functionality, Advancements and Industrial Applications of Heat Pipes

Bahman Zohuri
Associate Research Professor
Department of Electrical and Computer Engineering
University of New Mexico
Albuquerque, New Mexico, United States

ELSEVIER
ACADEMIC PRESS
An imprint of Elsevier

Academic Press is an imprint of Elsevier
125 London Wall, London EC2Y 5AS, United Kingdom
525 B Street, Suite 1650, San Diego, CA 92101, United States
50 Hampshire Street, 5th Floor, Cambridge, MA 02139, United States
The Boulevard, Langford Lane, Kidlington, Oxford OX5 1GB, United Kingdom

Library of Congress Cataloging-in-Publication Data
A catalog record for this book is available from the Library of Congress

British Library Cataloguing-in-Publication Data
A catalogue record for this book is available from the British Library

ISBN: 978-0-12-819819-3

For information on all Academic Press publications visit our website at https://www.elsevier.com/books-and-journals

Publisher: Brian Romer
Acquisitions Editor: Maria Convey
Editorial Project Manager: Andrae Akeh
Production Project Manager: Sruthi Satheesh
Cover Designer: Matthew Limbert

Typeset by TNQ Technologies

This book is dedicated to my son Sasha.

Contents

About the author xiii
Preface xvii
Acknowledgements xxiii

1. Heat pipe infrastructure 1

1.1 Introduction 1
1.2 Basic principles of heat pipes and history 2
1.3 History 4
1.4 Description and types of heat pipes 5
1.5 Principles of operation 18
1.5.1 Container 18
1.5.2 Working fluid 19
1.5.3 Wick or capillary structure 20
1.5.4 Sintered powder 21
1.5.5 Grooved tube 21
1.5.6 Screen mesh 21
1.5.7 How the heat pipe is working 23
1.5.8 Heat pipe assemblies design guidelines 24
1.5.9 Orientation with respect to gravity 24
1.5.10 Temperature limits 25
1.5.11 Heat removal 25
1.5.12 Reliability 25
1.5.13 Forming or shaping 26
1.5.14 Effects of length and pipe diameter 26
1.5.15 Wick structures 26
1.6 Heat pipe operating ranges 27
1.7 Constraints 29
1.8 Lessons learned 35
1.9 Applications 36
1.10 Summary 41
References 44

2. Application of heat pipe in industry 47
2.1 Introduction 48
2.2 Overview industrial application of heat pipes 51
2.2.1 Cooling of electronic components 53
2.2.2 Spacecraft 53
2.2.3 Energy conservation 54
2.2.4 Heat pipe driven heat exchanger (HPHX) 54
2.2.5 Preservation of permafrost 55
2.2.6 Snow melting and deicing 55
2.2.7 Heat pipe inserts for thermometer calibration 56
2.2.8 High-temperature heat pipe furnace 56
2.2.9 Miscellaneous heat pipe applications 57
2.3 Energy-dependent boundary equations 58
2.4 Heat pipe in space 61
2.4.1 Radioisotope systems 62
2.4.2 Fission systems: heat 68
2.4.3 Fission systems: propulsion 69
2.4.4 Nuclear thermionic technology development 69
2.4.5 Potential space nuclear thermionic missions 74
2.4.6 Heat pipe power system 77
2.4.7 Space reactor power systems 78
2.4.8 Stirling engine system 85
2.4.9 Heat pipe design 86
2.4.10 Nuclear reactor power system 87
2.4.11 Material choices 92
2.4.12 Safety considerations 92
2.4.13 Reactor control 92
2.4.14 Neutron shielding 93
2.4.15 Reactor sitting 94
2.4.16 Nuclear energy propulsion of aircraft (NEPA) 95
2.4.17 Project prometheus 2003 97
2.4.18 Mars one mission 99
2.4.19 Kilopower reactor using stirling technology (KRUSTY) experiment 101
2.5 Space shuttle orbiter heat pipe applications 104
2.6 Heat pipe in electronics 107
2.6.1 Electronic and electrical equipment cooling 109
2.7 Heat pipe in defense and avionics 113
2.7.1 On the ground application 114
2.7.2 In the sea application 115
2.7.3 In the air application 116
2.7.4 In the space application 116

2.8 Heat pipe as heat exchanger 118
2.9 Heat pipe in residential building 121
2.10 Heat pipe applications in thermal energy storage systems 123
2.10.1 Energy storage methods 124
2.10.2 Latent heat thermal energy storage 126
2.10.3 Latent heat thermal storage materials 127
2.10.4 Phase Change Material (PCM) classification 128
2.10.5 Latent heat thermal energy storage systems assisted by heat pipes 134
2.11 Passive thermal technical discipline lead (TDL) for luna lander 153
2.12 Heat pipe driving home energy system 159
2.13 Heat pipe driving heat exchangers and heat pumps 165
2.14 Gas turbine engines and the automotive industry 166
2.15 Heat pipes driving production tools 169
2.16 Medicine and human body temperature control via heat pipe 169
2.17 Heat pipes driving ovens and furnaces 171
2.18 Heat pipes driving Permafrost Stabilization 172
2.19 Heat pipes driving transportation systems and deicing 173
References 174

3. Different types of heat pipes 183

3.1 Introduction 183
3.2 Compatible fluids and materials 189
3.2.1 Freeze – thaw and thermal cycling 192
3.3 Other types of heat pipes 195
3.4 Thermosyphon 195
3.5 Loop heat pipes/capillary pumped loop 196
3.5.1 Loop heat pipe advantages 196
3.6 Pulsating heat pipes 196
3.7 Micro heat pipes (MHP) 197
3.8 Constant-condenser heat pipes (CCHP) 202
3.9 Constant-condenser heat pipes (CCHP) 203
3.9.1 Variable conductance with gas-loaded heat pipes 206
3.10 Rotating and revolving heat pipes 208
3.11 High-temperature heat pipes (liquid metal heat pipes) 211
3.12 Cryogenic heat pipes 213
3.13 Wrap-around heat pipe (WAHP) in air conditioning systems 214
3.14 Oscillating Heat Pipes 215
3.15 Liquid trap diode heat pipes 217
3.16 Vapor trap diode heat pipes 218

3.17 Diode heat pipes for Venus Landers concept 221
3.17.1 Function 1 – collecting heat 222
3.17.2 Function 2 – rejecting heat 222
3.17.3 The role of the diode heat pipe 223
3.17.4 Background – diode heat pipe 224
3.18 Annular heat pipes concept 225
3.19 HiK™ heat pipe plates 227
3.19.1 HiK™ plates CFD analysis 230
3.19.2 Cooling embedded VME and VPX systems 231
3.20 Pressure controlled heat pipes (PCHPs) 233
References 236

4. Heat pipe manufacturing 239

4.1 Introduction 239
4.2 Manufacturing of heat pipes 240
4.2.1 Envelope 241
4.2.2 End cap 242
4.2.3 Fill tube 243
4.2.4 Wick 244
4.2.5 Working fluid 245
4.3 Heat pipe manufacturing procedures 245
4.4 Cleaning of parts 245
4.5 Assembly of heat pipes 253
4.5.1 Wick forming and insertion 253
4.5.2 End cap installation 253
4.5.3 End closure and welding 255
4.5.4 Summary of assembly of heat pipes procedures 255
4.6 Evacuation and charging 259
4.6.1 Fluid charging 262
4.6.2 Fluid purity and inventory 262
4.6.3 Analysis of gas blockage 264
4.6.4 Effect of heat pipe design and operating conditions on gas blockage 266
4.7 Full tube closure 267
4.8 Heat pipe testing techniques 267
4.8.1 Mechanical sounders 270
4.8.2 Wick wetting 271
4.8.3 Performance versification 271
References 273

5. Heat pipe heat exchanger opportunities and industrial applications 275

5.1 Introduction 275
5.2 General theory of heat pipe design 277
5.2.1 Capillary limitation 277
5.2.2 Sonic limitation 277

5.2.3 Entrainment limitation 278
5.2.4 Boiling limitation 278
5.3 Holistic approach to heat pipe application 281
5.3.1 Merit number derivation 288
5.3.2 Lowest heat pipe limit driven temperature 289
5.3.3 Heat pipe working fluids 292
5.4 Heat pipe heat exchangers, an innovation for heat transfer management 293
5.4.1 Direct contact heat exchangers highly efficient HVAC systems 294
5.4.2 Variable conductance heat pipe (VCHP) heat exchanger 298
5.4.3 Innovative heat exchanger designs 300
5.5 An overview of the heat pipe technology summary 303
References 304

6. Thermosyphon and heat pipe applications 307

6.1 Introduction 308
6.2 Historical development and background of thermosyphon and heat pipe 309
6.3 Heat pipes and thermosyphon 314
6.4 Application of heat pipes and thermosyphon 316
6.4.1 Application in space systems 317
6.4.2 Application in cold regions 317
6.4.3 Application in automobile industry 319
6.4.4 Application in railroad industry 319
6.4.5 Application in electrical, electronics, and nuclear industries 319
6.4.6 Summary statement 320
6.5 Thermosyphon design 321
6.5.1 Geometry 322
6.5.2 Working fluids 323
6.6 Mass flow rate and sonic velocity analysis 326
6.7 Heat transport limitations 327
6.7.1 Sonic limit (choking) of vapor flow 327
6.7.2 Viscous limit 328
6.8 Comparison of alkaline metals thermosyphon with convective loop 329
6.9 Thermosyphon startup 332
6.10 Two-phase instabilities in thermosyphon 333
6.10.1 Surging (chugging) and geysering instability 333
6.10.2 Thermosyphon evaporator instability 334
6.10.3 Fluid superheating (alkaline metals) 334
6.11 Nucleation sites 334
6.12 Inclination effects on a thermosyphon performance 337
6.13 Summary 341
References 342

7. Thermodynamic analysis of thermosyphon 345

7.1 Introduction 345
7.2 General model (vertical thermosyphon) and flooding 347
7.3 Two-phase thermosyphon thermodynamic analysis with spiral heat exchanger 350
7.4 Summary 352
References 352

8. Thermosyphon & heat pipe dimensionless numbers in boiling fluid flow 355

8.1 Introduction 355
8.2 Thermosyphon 355
8.3 Heat pipe 358
8.4 Results and discussion 359
8.5 Summary 363
References 363

Index 365

About the author

Dr. Bahman Zohuri is currently at the Galaxy Advanced Engineering, Inc. a consulting company that he stared himself in 1991 when he left both semiconductor and defense industries after many years working as a chief scientist. After graduating from University of Illinois in field of Physics and Applied Mathematics, as well as University of New Mexico from Nuclear Engineering Department, he joined Westinghouse Electric Corporation where he performed thermal hydraulic analysis and natural circulation for Inherent Shutdown Heat Removal System (ISHRS) in the core of a Liquid Metal Fast Breeder Reactor (LMFBR) as a secondary fully inherent shut system for secondary loop heat exchange. All these designs were used for Nuclear Safety and Reliability Engineering for Self-Actuated Shutdown System. He designed the Mercury Heat Pipe and Electromagnetic Pumps for Large Pool Concepts of LMFBR for heat rejection purpose for this reactor around 1978 where he received a patent for it. He later on was transferred to defense division of Westinghouse where he was responsible for the dynamic analysis and method of launch and handling of MX missile out of canister. The results are applied to MX launch seal performance and muzzle blast phenomena analysis (i.e., missile vibration and hydrodynamic shock formation). He also was involved in analytical calculation and computation in the study of Nonlinear Ion Wave in Rarefying Plasma. The results are applied to the propagation of "Soliton Wave" and the resulting charge collector traces, in the rarefactions characteristic of the corona of a laser irradiated target pellet. As part of his graduate research work at Argonne National Laboratory, he performed computation and programming of multi-exchange integral in surface physics and solid state physics. He holds different patent in areas such as diffusion processes and design of diffusion furnace while he was senior process engineer working for different semiconductor industries such as Intel, Varian, and National Semiconductor corporations. Later on, he joined Lockheed Missile and Aerospace Corporation as Senior Chief Scientist and was responsible Research and Development (R&D) and the study of vulnerability, survivability and both radiation and laser hardening of different components Strategic Defense Initiative known as Star Wars. This included of payload (i.e., IR Sensor) for Defense Support Program (DSP), Boost Surveillance and Tracking Satellite (BSTS) and Space Surveillance and Tracking Satellite (SSTS) against laser or nuclear threat. While in there, he also studied and performed the analysis of characteristics of laser beam and nuclear radiation interaction with materials, Transient Radiation

Effects in Electronics (TREE), Electromagnetic Pulse (EMP), System Generated Electromagnetic Pulse (SGEMP), Single-Event Upset (SEU), Blast and, Thermo-mechanical, hardness assurance, maintenance, device technology.

He did few years of consulting under his company Galaxy Advanced Engineering with Sandia National Laboratories (SNL), where he was supporting development of operational hazard assessments for the Air Force Safety Center (AFSC) in connection with other interest parties. Intended use of the results was their eventual inclusion in Air Force Instructions (AFIs) specifically issued for Directed Energy Weapons (DEW) operational safety. He completed the first version of a comprehensive library of detailed laser tools for Airborne Laser (ABL), Advanced Tactical Laser (ATL), Tactical High Energy Laser (THEL), Mobile/Tactical High Energy Laser (M-THEL), etc.

He also was responsible on SDI computer programs involved with Battle Management C^3 and artificial Intelligent, and autonomous system. He is authoring few publications and holds various patents such as Laser Activated Radioactive Decay and Results of Thru-Bulkhead Initiation.

Recently he has published five books with CRC and Francis Taylor and Springer, Nova, and Elsevier publishing companies in subject of nuclear engineering, plasma physics, thermodynamics, heat transfer and dimensional analysis and few of them are named here:

1. Heat Pipe Design and Technology: A Practical Approach, Published by CRC Publishing Company.
2. Dimensional Analysis and Self-Similarity Methods for Engineering and Scientist Published by Springer Publishing Company.
3. High Energy Laser (HEL): Tomorrow's Weapon in Directed Energy Weapons Volume I, Published by Trafford Publishing Company.
4. Thermodynamics In Nuclear Power Plant Systems, Published by Springer Publishing Company.
5. Thermal-Hydraulic Analysis of Nuclear Reactors, Published by Springer Publishing Company.
6. Application of Compact Heat Exchangers for Combined Cycle Driven Efficiency in Next Generation Nuclear Power Plants: A Novel Approach, Springer Publishing Company.
7. Next Generation Nuclear Plants Driven Hydrogen Production Plants via Intermediate Heat Exchanger a Renewable Source of Energy, Springer Publishing Company.
8. Neural Network Driven Artificial Intelligence: Decision Making Based on Fuzzy Logic, Nova Publishing Company.
9. Physics of Cryogenics: An Ultralow Temperature Phenomenon, Elsevier Publishing Company.
10. Directed Energy Weapons: Physics of High Energy Lasers (HEL), Springer Publishing Company.

The above is the list of few of his books that are published and can be found on Amazon.com under his name along with others by him.

Recently he has been involved with Cloud Computation, Data warehousing, and Data Mining using Fuzzy and Boolean logic and has few books published in these subjects as well as numerous other books that can be found under his name on Amazon or Internet.

Preface

Heat pipes can be designed to operate over a very broad range of temperatures from cryogenic (less than 30K) applications to high temperature systems (more than 2000K). Until recently, the using of heat pipes has been mainly limited to space technology due to cost effectiveness and complex wick construction. There are several applications of heat pipes in this field, such as spacecraft temperature equalization component cooling, temperature control and radiator design in satellites. Currently heat pipe technology has been integrated into modern thermal engineering designs, such as terrestrial thermal control systems, solar energetic, etc.

High temperature heat pipes have been proposed for use in the manufacturing of glass bottles. The glass bottle forming procedure starts by periodically dipping a steel piston into a steel form filled with molten glass. This forms a hollow glass tube, which is later blown into its final shape. The initial glass temperature is around 1100°C and the surface temperature of the piston needs to be kept around 600°C. At higher piston temperatures, the glass will stick to the piston, and at lower temperatures, the glass viscosity increases, causing insufficient deformation during the forming process. Insufficient deformation is the cause of thin-walled bottles which contribute to the waste rate. A stainless-steel/potassium heat pipe was proposed and tested, and it was found that the heat pipe could be kept nearly isothermal. This resulted in a higher dipping frequency and a reduced amount of glass bottle waste.

This book also describes a Variable-Conductance Heat Pipe System (VCHPS) where it has been designed to provide thermal control for a Transmitter Experiment Package (TEP) to be flown on the Communications Technology Satellite. The VCHPS provides for heat rejection during TEP operation and minimizes the heat leak during power down operations. The VCHPS described features a unique method of aiding priming of arterial heat pipes and a novel approach to balancing heat pipe loads by staggering their control ranges.

The heat pipe is compact and efficient because i) the finned-tube bundle is inherently a good configuration for convective heat transfer in both ducts, and ii) the evaporative-condensing cycle within the heat pipes is a highly efficient method of transferring heat internally.

The effects of different factors on the performance of the heat pipe: compatibility of materials, operating temperature range, diameter, power limitations, thermal resistances, and operating orientation, will be considered in the lecture.

Heat pipes can be designed to operate over a very broad range of temperatures from cryogenic (less than 30K) applications to high temperature systems (more than 2000K). Until recently, the using of heat pipes has been mainly limited to space technology due to cost effectiveness and complex wick construction. There are several applications of heat pipes in this field, such as spacecraft temperature equalization component cooling, temperature control and radiator design in satellites. Currently heat pipe technology has been integrated into modern thermal engineering designs, such as terrestrial thermal control systems, solar energetic, etc. The increasing power and shrinking size of electronics components presents growing thermal management challenges. While solid metal conductors such as aluminum extrusions may provide acceptable cooling for individual components in certain situations, board level solutions with more advanced cooling technologies are needed in a growing number of applications. Heat pipes have emerged as an effective and established thermal solution, particularly in high heat flux applications and in situations where there is any combination of non-uniform heat loading, limited airflow over the heat generating components, and space or weight constraints.

Heat pipes have been applied in many ways since their introduction in 1964. Depending on their intended use, heat pipes can operate over a temperature range from 4.0 to 3000K. In all cases, their applications can be divided into three main categories: separation of heat source and sink, temperature equalization, and temperature control. Due to their extremely high thermal conductivity, heat pipes can efficiently transport heat from a concentrated source to a remotely mounted sink. This property can enable dense packing of electronics, for example, without undue regard for heat sink space requirements. Another benefit of the high thermal conductivity is the ability to provide an accurate method of temperature equalization. For example, a heat pipe mounted between two opposing faces of an orbiting platform will enable both faces to maintain constant with equal temperatures, thus minimizing thermal stresses. The temperature control is a result of the capability of heat pipes to transport large quantities of heat very rapidly. This feature enables a source of varying flux to be kept at a constant temperature as long as the heat flux extremes are within the operating range of the heat pipe.

In Chapter One, we describe heat pipe, where a heat pipe is a passive energy recovery heat exchanger that has the appearance of a common plate-finned water coil except the tubes are not interconnected. Additionally, it is divided into two sections by a sealed partition. Hot air passes through one side (evaporator) and is cooled while cooler air passes through the other side (condenser). While heat pipes are sensible heat transfer exchangers, if the air conditions are such that condensation forms on the fins there can be some latent heat transfer and improved efficiency.

In Chapter Two, we are discussing applications of heat pipes in industry such as energy system and others. We also expand on its applications in Space Program as well as Nuclear Industry. We also touch base up heat pipe applications in Electronic Manufacturing where the fast Central Processing Units (CPUs) needs to be cooled down and how heat pipe can be used as a Heat Exchanger(HE). Recently heat pipes have shown very promising results in applications such as thermal energy storage systems and concentrated solar power to produce electricity as part of renewable energy systems as well.

Chapter Three, talks about various types of heat pipe and its market. In today's market there are variety type of heat pipes in term of their geometry structure and their function of operations and/or the methods they are used to transport heat from source to sink or in other hand from evaporator to condenser and bring back the liquid from condenser to evaporator. Reay and Kew[1] are presenting a good description of different types of heat pipes along with their application and reader can obtain a good knowledge from their book[1]. In Chapter One of the book both Constant Condenser Heat Pipe (CCHP) and Variable Condenser Heat Pipe (VCHP type heat pipe are discussed along with their general applications and later of mathematical modeling heat pipe was presented, following Chapter Two and Three. Briefly here both CCHP and VCHP are touched upon again. Readers will be exposed briefly to different types of heat pipes in this chapter although their own research of the market for what type of heat pipe they are looking for and what is their demanding application is highly recommended.

Chapter Four goes over manufacturing of heat pipe and provides different flow charts of the manufacturing process. In this chapter Heat pipe manufacturing methods are examined with the goal of establishing cost effective procedures that will ultimately result in cheaper more reliable heat pipes. Those methods which are commonly used by all heat pipe manufacturers have been considered, including: envelope and wick cleaning, end closure and welding, mechanical verification, evacuation and charging, working fluid purity, and charge tube pinch off. Review and evaluation of available manufacturer's techniques and procedures together with the results of specific manufacturing oriented tests have yielded a set of recommended cost-effective specifications which can be used by all manufacturers.

Chapter Five is describing the heat pipe heat exchanger opportunities and industrial applications in some details. This chapter talks about heat pipes and thermosyphons that are widely recognized as being excellent passive thermal transport devices that can have effective thermal conductivities orders of magnitude higher than similarly-dimensioned solid materials. Thermosyphons and heat pipe application is described in Chapter Six of this book with more details. The integration of heat pipes into heat exchangers (HXs) and heat sinks (HPHXs and HPHSs, respectively) have been shown to have strong potential for energy savings, especially in response to the significant reduction in the manufacturing costs of heat pipes in recent years. This review documents

HPHXs applications, general design procedures, and analysis tools based on the thermal network approach. The thermal network approach is a robust engineering tool that is easy to implement and program, is user friendly, straightforward, computationally efficient, and serves as a baseline methodology to produce results of reasonable accuracy.

Chapter Six is going over thermosyphon and heat pipe application in three hydrogen production processes plant that is powered by a Next Generation Nuclear Plant (NGNP), that are currently under investigation at Idaho National Laboratory. The first is high-temperature steam electrolysis, which uses both heat and electricity; the second is thermo-chemical production through the sulfur iodine process primarily using heat; the third is a hybrid sulfur process (not part of this study) which incorporates sulfur acid decomposition and sulfur dioxide depolarized electrolysis, all processes require a high temperature (>850 °C) for enhanced efficiency; temperatures indicative of the NGNP. Safety and licensing mandates prudently dictate that the NGNP and the hydrogen production facility be physically isolated, perhaps requiring separation of over 100 m.

Chapter Seven lays down the basic understanding of thermodynamic analysis of thermosyphon and shows very high-level approach to this thermodynamic analysis approach. Thermosyphons are devices with high thermal conductivity that can transfer high quantities of heat. In its most simple form, a thermosyphon is a hollow evacuated metal pipe, charged by a pre-determined amount of an appropriate working fluid. It can be divided into three main sections: evaporator, where the heat is delivered to the device, an adiabatic section (which can or cannot exist) and a condenser, where the heat is released. The working fluid located in the evaporator evaporates and, by means of pressure gradients, go toward the condenser region, where it condenses, returning to the evaporator by means of gravity. Complicated mathematical expressions and numerical schemes are helpful but sometimes may mask the real physics from a design engineer's point of view. Usually it is not desirable or necessary to get into such detail for most applications. Therefore, a simple and understandable engineering method for analysis of thermosyphon performance becomes attractive in practice. This chapter provides a different view into the physics behind thermosyphon performance, based on thermodynamics.

Last but not least, Chapter Eight is presenting dimensionless numbers in boiling fluid flow for thermosyphon and heat pipe as well. This chapter introduces the procedure of dimensional analysis at very high-level and describes Buckingham's p-theorem, which follows from it. Dimensional Analysis also called Factor-Label Method, or the Unit Factor Method is a problem-solving method that uses the fact that any number or expression can be multiplied by one without changing its value. It is a useful technique. The only danger is that you may end up thinking that chemistry is simply a math problem - which it definitely is not.

Dimensional analysis is a very powerful tool, not just in fluid mechanics, but in many disciplines. It provides a way to plan and carry out experiments and enables one to scale up results from model to prototype. Consider, for example, the design of an airplane wing. Dimensional analysis is a means of simplifying a physical problem by appealing to dimensional homogeneity to reduce the number of relevant variables.

Bahman Zohuri
Albuquerque, New Mexico, United States

Acknowledgements

I am indebted to the many people who aided me, encouraged me and the people whom supported me beyond the expectation. Some of those who are not around to see end result of their encouragement in production of this book, yet I hope they can see this acknowledgment. My thank goes to Joe Rogers of NASA, one of my best friends who helped with most of the computer codes that are presented in this book to bring them to its present status from their legacy stages.

To Hal Brand of Lawrence Livermore Laboratory who also was there to support me with my computer programming needs. To other friends such as Dr. Patrick Burs of Colorado State University and Dr. David Glass of NASA Langley Research Center provided me with their research papers and computers codes. To Leonardo Tower best person whom I got the honor of knowing him during write up of this book that provided me with his newly developed computer code around the subject. My many thanks to Darryl Johnson and David Antoniuk and Dr. Bruce Marcus of North Grumman (TRW), as well as Mark North of Thermacore Incorporation seamless support is most appreciated by me, since they all had their own share of effort to publish this book and brings to this end.

Another best friend William Kemp of Air Force Weapon Laboratory at Albuquerque New Mexico who is really a true friend and remains to be one. Finally, my many thanks to Senior Editor - Mechanical, Aerospace, and Nuclear & Energy Engineering of Springer Publishing Company Mrs. Tiffany Gasbarrini who made all this to happen. Finally, I am indebted to many people and the individuals and organizations that granted permission to reproduce copyrights materials and published figures.

I am also thankful to David Saunders where he works at NASA Ames in the Aerothermodynamics Branch at ARC under contract to ELORET Corporation. He is always very supportive of me. I am also grateful to Miss Kimberly Hoffman of Catholic University of America for her endless support to obtain few documents from the works of Chi that I needed them so much.

I am also indebted to Dr. Piyush Sabharwall (from Idaho National Laboratory) for creation of Chapters 6–8 of this book. He has been a true friend and advocate of me and very supportive all along. Thank you Piyush for all the support. He would like to dedicate these chapter in the text to his mother Mrs. Vijay Sabharwall, who recently passed away. She has been a true inspiration for him throughout his life, a great teacher and a best mother one could have asked for.

Above all, I offer very special thanks to my mother, father while he was alive, wife and children. They provided constant interest and encouragement, without which this book would not have been written.

Their patience with my many absences from home and long hours in front of computer during preparation of the manuscript is especially appreciated.

Above all, I offer very special thanks to my late mother and father, and to my children, in particular, my son Sasha Zohuri and my daughters Dr. Natasha and her husband Dr. Nikpour as well as my second daughter Natalie Zohuri as well as my Grand Son Darrius Nikpour. They have provided constant interest and encouragement, without which this book would not have been written. Their patience with my many absences from home and long hours in front of the computer to prepare the manuscript are especially appreciated.

Chapter 1

Heat pipe infrastructure

Chapter outline

1.1 Introduction 1

1.2 Basic principles of heat pipes and history 2

1.3 History 4

1.4 Description and types of heat pipes 5

1.5 Principles of operation 18

1.5.1 Container 18

1.5.2 Working fluid 19

1.5.3 Wick or capillary structure 20

1.5.4 Sintered powder 21

1.5.5 Grooved tube 21

1.5.6 Screen mesh 21

1.5.7 How the heat pipe is working 23

1.5.8 Heat pipe assemblies design guidelines 24

1.5.9 Orientation with respect to gravity 24

1.5.10 Temperature limits 25

1.5.11 Heat removal 25

1.5.12 Reliability 25

1.5.13 Forming or shaping 26

1.5.14 Effects of length and pipe diameter 26

1.5.15 Wick structures 26

1.6 Heat pipe operating ranges 27

1.7 Constraints 29

1.8 Lessons learned 35

1.9 Applications 36

1.10 Summary 41

References 44

1.1 Introduction

Heat Pipe Technology (HPT) was founded in 1983 with a grant from the Department of Energy for a project to begin research on new uses for heat pipe technology. Heat pipes are passive-heat-transfer devices that were previously used in various applications ranging from orbiting satellites to the Alaskan Pipeline ground spikes. By applying the principle of heat pipes to air conditioning systems, efficiency was greatly enhanced in both dehumidification performance and energy utilization, with moisture removal increased by 30%–50%. A typical yet a simple heat pipe is illustrated in Fig. 1.1.

Additional research and development followed to determine how to lower fabrication costs. This task was performed by HPT under a three-year, $500,000 contract with NASA's Kennedy Space Center. The outcome of that effort was a new generation of heat pipe technology, costing one-third the price of the existing aerospace heat pipes, while still offering the same level of performance.

Functionality, Advancements and Industrial Applications of Heat Pipes. https://doi.org/10.1016/B978-0-12-819819-3.00001-8

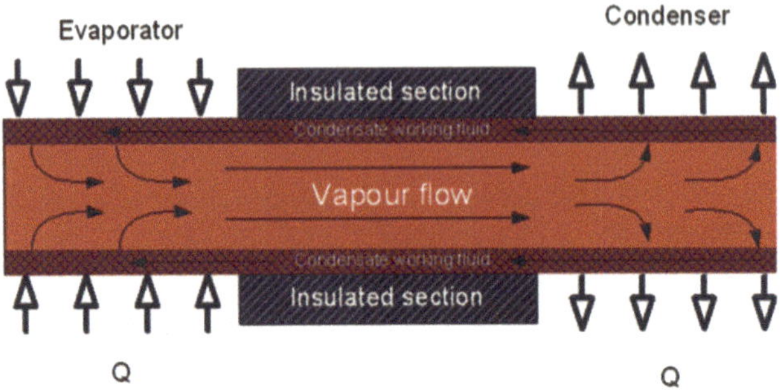

FIG. 1.1 A typical heat configuration heat pipe.

This revolution shattered the price barrier that restricted the widespread commercial implementation of heat pipes, and it yielded a rapid return on investment for building owners, often in as little as a year. Projects soon proved the technology could be applied on virtually any scale, while being commercially viable and practical to implement.

Heat pipes are tubes that have a capillary wick inside running the length of the tube, are evacuated and then filled with a refrigerant as the working fluid and are permanently sealed. The working fluid is selected to meet the desired temperature conditions and is usually a Class I refrigerant. Fins are similar to conventional coils - corrugated plate, plain plate, spiral design. Tube and fin spacing are selected for appropriate pressure drop at design face velocity. Heating, Ventilation, and Air Conditioning (HVAC) systems typically use copper heat pipes with aluminum fins; other materials are available.

The heat pipe is one of the remarkable achievements of thermal physics and heat transfer engineering in this century because of its unique ability to transfer heat over large distances without considerable losses. The main applications of heat pipes deal with the problems of environmental protection and energy and fuel savings.

Heat pipes have emerged as an effective and established thermal solution, particularly in high heat flux applications and in situations where there is any combination of non-uniform heat loading, limited airflow over the heat generating components, and space or weight constraints. This chapter will briefly introduce heat pipe technology and then highlight its basic applications as a passive thermal control device.

1.2 Basic principles of heat pipes and history

The original idea of Heat Pipe was considered in 1944 by Gaugler [1] and in 1962 by Trefethen. Although Gaugler patented a very light weight heat transfer device, that was essentially very basic presentation of Heat Pipe. During that time period the technology did not require a need for such sophisticated yet constructively simple two-phase and passive heat transfer device and there was not much attention that was paid to it. As suggest by Trefeten [2] first 1962 and then inform of a patent application Heat Pipe resurfaced again in 1963 by Wyatt [3]. It was not widely considered and publicized until1964 when George Grove [4] and his co-worker at the Los Alamos National Laboratory independently reinvented the same concept for their existing space program and its application. He is the one who named this most satisfactory and simplistic heat transmission device "Heat Pipe" and developed its applications.

Heat pipes are two-phase flow heat transfer devices where process of liquid to vapor and vice versa circulate between evaporator to condenser with high effective thermal conductivity. Due to the high heat transport capacity, heat exchanger with heat pipes has become much smaller than traditional heat exchangers in handling high heat fluxes. With the working fluid in a heat pipe, heat can be absorbed on the evaporator region and transported to the condenser region where the vapor condenses releasing the heat to the cooling media. Heat pipe technology has found increasing applications in enhancing the thermal performance of heat exchangers in microelectronics, energy saving in Classical Heating, Ventilating, and Air Conditioning (HVAC) systems for operating rooms, surgery centers, hotels, clean rooms etc., temperature regulation systems for the human body and other industrial sectors including space craft and various types of nuclear reactors technologies as a fully inherent cooling apparatus. The Heat Pipe is a self-contained structure which achieves very high thermal energy conductance by means of two-phase fluid flow with capillary circulation. A heat pipe operates within a two phase flow regime as an evaporation-condensation device for transferring heat in which the latent heat of vaporization is exploited to transport heat over long distances with a corresponding small temperature difference. Heat added to the evaporator is transferred to the working fluid by conduction and causes vaporization of the working fluid at the surface of the capillary structure. Vaporization causes the local vapor pressure in the evaporator to increase and vapor to flow toward the condenser thereby transporting the latent heat of vaporization. Since energy is extracted at the condenser, the vapor transported through the vapor space is condensed at the surface of the capillary structure, releasing the latent heat. Closed circulation of the working fluid is maintained by capillary action and/or bulk forces. An advantage of a heat pipe over other conventional methods to transfer heat such as a finned heat sink is that a heat pipe can have an extremely high thermal conductance in steady state operation. Hence, a heat pipe can transfer a high amount of heat over a relatively long length with a comparatively small temperature differential. Heat pipe with liquid metal working fluids can have a thermal conductance of a thousand or even tens of thousands folds better than the best solid metallic conductors, silver or copper. In a heat pipe energy is transported by utilizing phase change of the working substance instead of a large temperature gradient and without external power. Also, the amount of energy transferred through a small cross-section is much larger than that by conduction or convection. Heat pipes may be operated over a broad range of temperatures by choosing an appropriate working fluid. See Fig. 1.2.

However, this useful device has some operating limitations such as the sonic, the capillary, the entrainment, and finally the boiling limit. When any of these limitations is encountered, the capillary structure may dry out leading to failure of the heat pipe. In addition to these limitations, when liquid metal is used as the working fluid, startup difficulty may take place due to possible solid state of the working fluid and extremely low vapor density.

1.3 History

Early research in heat pipes conducted at Los Alamos was directed to applications in space-based thermionic energy conversion systems operating in excess of 1500K. Heat pipes were considered for heating thermionic emitters, for cooling thermionic collectors, and for the ultimate radiation of heat to space fluids and materials were tailored to this temperature regime. Experiments with a Nb-1 %Zr heat pipe, with lithium operating at 1573K, 207 W/cm^2 evaporator radial heat flux; a 1.95 kW/cm^2 axial heat flux, and an Ag-Ta operating at 2273K, 410 W/cm^2 evaporator radial heat flux; and a 4 kW/cm^2 axial heat flux are reported in Deverall and Kemme [5]. The results of early thermionic-related heat pipe fluid-wall compatibility and life test studies with systems of In-W at 2173K for 75 h, Ag-Ta at 2173K for 100 h, Cs-Ti at 673K in excess of 2000 h, Na-stainless steel at 1073K for 500 h, and Li-Nb-1%Zr at 1373K for 4300 h are summarized in Grover et al. [6], Deverall and Kemme [5], Grover et al. [7], Cotter et al. [8], and Ranken and Kemme [9]. A study characterizing both potassium and sodium heat pipes with various wick

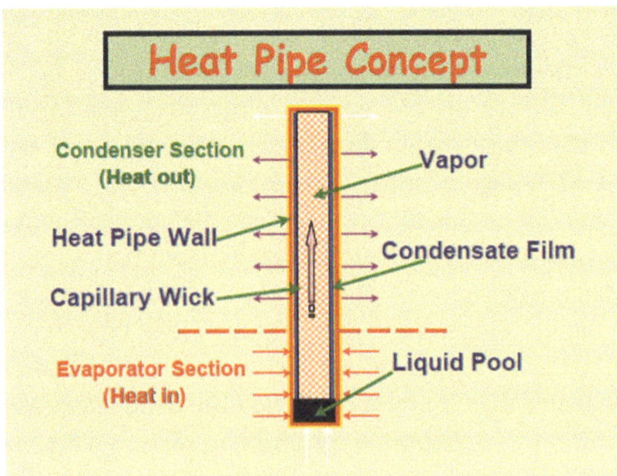

FIG. 1.2 Heat Pipe concept.

structures and a treatment of the limitations to heat pipe start-up and operation is contained in Kemme [10].

On July 24, 1963, George Grover made the following entry into his laboratory notebook: "Heat transfer via capillary movement of fluids. The 'pumping' action of surface tension forces may be sufficient to move liquids from a cold temperature zone to a high temperature zone (with subsequent return in vapor form using as the driving force, the difference in vapor pressure at the two temperatures) to be of interest in transferring heat from the hot to the cold zone. Such a closed system, requiring no external pumps, may be of particular interest in space reactors in moving heat from the reactor core to a radiating system. In the absence of gravity, the forces must only be such as to overcome the capillary and the drag of the returning vapor through its channels."

Thus, began heat pipe research at Los Alamos. Later that year Grover submitted the results of "heat pipe" experiments with water and sodium as working fluids to the *Journal of Applied Physics* [6]. The sodium heat pipe, 90 cm long with a 1.9 cm O.D (Outer Diameter), operated at 1100K with 1-kW heat input. This paper reviews 28 years of space power-related liquid metal heat pipe research that has been conducted at Los Alamos since the invention of the heat pipe.

1.4 Description and types of heat pipes

A heat pipe is essentially a passive heat transfer device with an extremely high effective thermal conductivity. It is a simple closed loop device that can quickly transfer heat from one point to another using a two phase flow schema. It is a high thermal-conductance device as well in which transfers heat by two-phase fluid circulation. The operating temperature range of a heat pipe is determined by the type of working fluid used and its optimum design envelope. They are often referred to as the "super-conductors" of heat as they possess an extra ordinary heat transfer capacity and rate with almost no heat loss. Of the various means of transmitting heat, the heat pipes are known as one of the most satisfactory devices to carry on such a task. In a simple form of its structure this device is transporting heat from one point to another via evaporation and condensation, and the heat transport fluid is re-circulated by capillary forces which automatically develop as induction of the heat transport process.

This closed loop of Heat Pipe is consisting of a sealed hollow tube with two zones namely Evaporator and Condenser in a very simple case of such device, whose inside walls are lined with a capillary structure known as Wick. A thermodynamic working fluid having a substantial vapor pressure at the desired operating temperature saturates the pores of the wick. When heat is applied to any portion of the heat pipe evaporator, this fluid is heated, and it evaporates, readily filling the hollow center of the pipe. The vapor then diffuses throughout the heat pipe. Condensation of the vapor occurs on the pipe wall whenever the temperature is even slightly below that of the evaporation area. As it condenses, the liquid gives up the heat it acquired and returns to the evaporator section or heat source by means of capillary action within the wick. This tends to produce isothermal operation and a high effective thermal conductance. When a heat sink is attached to a portion of the heat pipe, condensation takes place preferentially at this point of heat loss and a vapor flow pattern is then established.

The system, proven in aerospace application, transmits thermal energy at rates hundreds of times greater than the most efficient solid conductor does and at a far superior energy-to-weight ration.

In terms of thermal conduction, a heat pipe is designed to have very high thermal conductance.

Heat is transported from the heat source (evaporator section of the heat pipe) to the heat sink (condenser section of the heat pipe) by means of a condensable fluid contained in a sealed chamber.

Liquid is vaporized, absorbing heat in the evaporator section. Then the vapor flows to the condenser section, where it condenses and releases its latent heat. The liquid is drawn back to the evaporator section by capillary action, where it is re-vaporized to continue the cycle. The temperature gradient along the length of pipe is minimized by designing for a very small vapor pressure drop as the vapor flows from the evaporator section to the condenser section. Thus, the saturation temperatures (temperatures at which evaporation and condensation takes place) are very nearly the same in both sections.

The idea of heat pipes was first suggested by R.S. Gaugler in 1942. However, it was not until 1962, when G.M. Grover invented it that its remarkable properties were appreciated & serious development began. It consists of a sealed aluminum or copper container whose inner surfaces have a capillary wicking material. A heat pipe is similar to a thermosyphons. It differs from a thermosyphons by virtue of its ability to transport heat against gravity by an evaporation-condensation cycle with the help of porous capillaries that form the wick. The wick provides the capillary driving force to return the condensate to the evaporator. The quality and type of wick usually determines the performance of the heat pipe, for this is the heart of the product. Different types of wicks are used depending on the application for which the heat pipe is being used.

The spectrum of heat pipe working fluids extends from cryogens to liquid metals, the choice of fluid being such that its saturation temperature, at the heat pipe operating pressure, be compatible with the heat pipe's application. Also, the fluid is chosen to be chemically inert when wetting the pipe and capillary wick. Ideally, the fluid would have a high thermal conductivity and latent heat. It should have a high surface tension and low viscosity.

Heat transfer in a heat pipe is limited by: the rate at which liquid can flow through the wick; "choking" (the inability to increase vapor flow with increasing pressure differential, also called "sonic limit"); entrainment of liquid in the vapor stream, such that liquid flow to the evaporator is reduced; the rate at which evaporation can take place without excessive temperature differentials in the evaporator section.

Isothermaliser heat pipes will transport heat in either direction, and for a given configuration, the heat flow will depend entirely on the temperature difference between the heat source and heat sink. The isothermaliser heat pipe is therefore a basically passive device with a fixed conductance, provided that none of its limiting conditions are exceeded (sonic, entrainment, capillary, and boiling limits. See Figs. 1.3 and 1.4). The isothermalization function of a heat pipe can be modified to produce active devices in two ways: diode heat pipes, where the pipe operates as an isothermaliser in the forward mode and shuts off in the reverse mode; or variable conductance heat pipes, where the conductance in the forward mode can be actively controlled and again shuts down in the reverse mode.

But today's heat pipes can work both vertically as well as horizontally as well as arbitrary angle of installation and operation of its application. Recent applications of them in zero gravity in particular in satellite are enhanced and proven by NASA and Air Force in collaboration with Los Alamos National Laboratory and other contractors such as TRW and Honeywell and others. Figs. 1.2–1.4 are some examples such applications.

Fig. 1.6 [11], shows application of Loop heat pipe containing two parallel evaporators and two parallel condensers with, Passive and self-regulating as well as Heat load sharing between evaporators. This configuration was implemented to NASA's New Millennium Program, The Space Technology 8 (ST8) mission. Part of new space graft has been illustrated in Fig. 1.7 where the loop heat pipe was used.

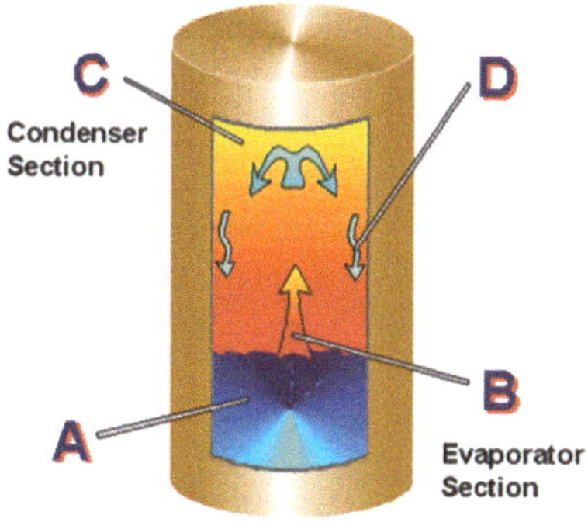

FIG. 1.3 A traditional heat pipe schematic. A traditional heat pipe is a hollow cylinder filled with a vaporizable liquid. (A) Heat is absorbed in the evaporating section. (B) Fluid boils to vapor phase. (C) Heat is released from the upper part of cylinder to the environment; vapor condenses to liquid phase. (D) Liquid returns by gravity to the lower part of cylinder (evaporating section).

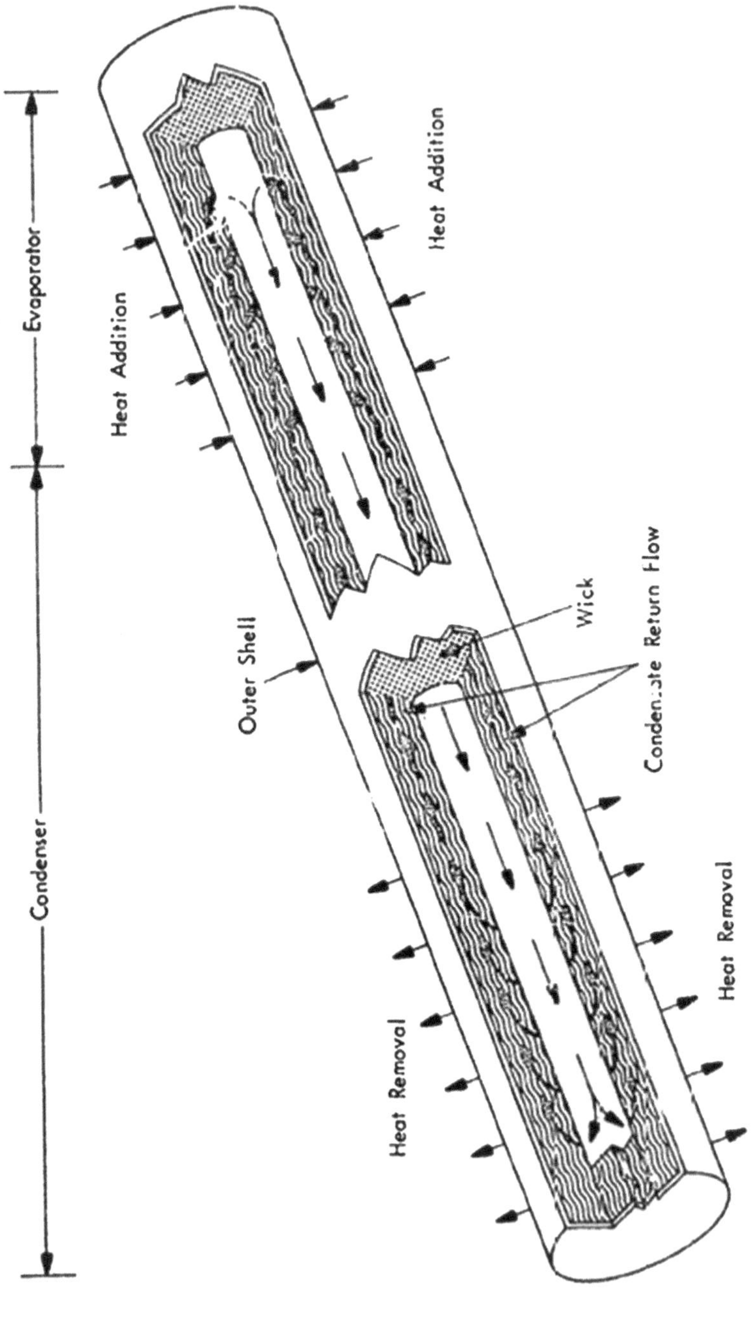

FIG. 1.4 Parts and functions of basic heat pipe.

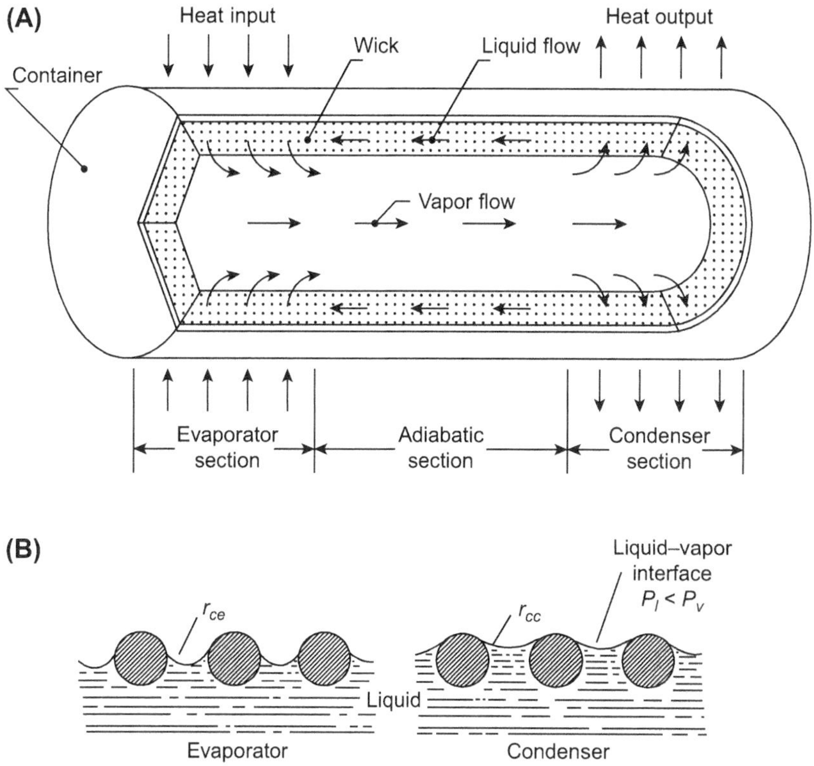

FIG. 1.5 (A) Components and principle of operation of a conventional heat pipe; (B) radii of curvature of the liquid-vapor interface in the condenser and evaporator [12].

Depending on its applications in particular in Nuclear Reactor Industry where these reactors are a source of electric power generation, heat pipes are used as a cooling part of secondary loop of inherent shut down system, which typically you see them as liquid heat pipe (i.e., Mercury or Sodium as cooling environment within heat pipe) there might be some consideration for a section of heat pipe that is known as Adiabatic zone where heat pipe is cast into structure of cooling assembly. A typical example of such approach could be seen in early study of companies such as Westinghouse Electric on their core design of Liquid Metal Fast Breeder Reactor (LMFBR), the Mercury Heat Pipe was considered as part fully inherent shutdown system of this particular reactor design (where author was involved with such design and Westinghouse was awarded few patents). This sort of approach was giving better safety factor for any incidental melt down of reactor and was providing better tool to release excessive heat and help to reduce the core temperature below critical point without any operator within the loop. Modern design and new generation of LMFBR reactor such as French built Phoenix are utilizing such heat pipes.

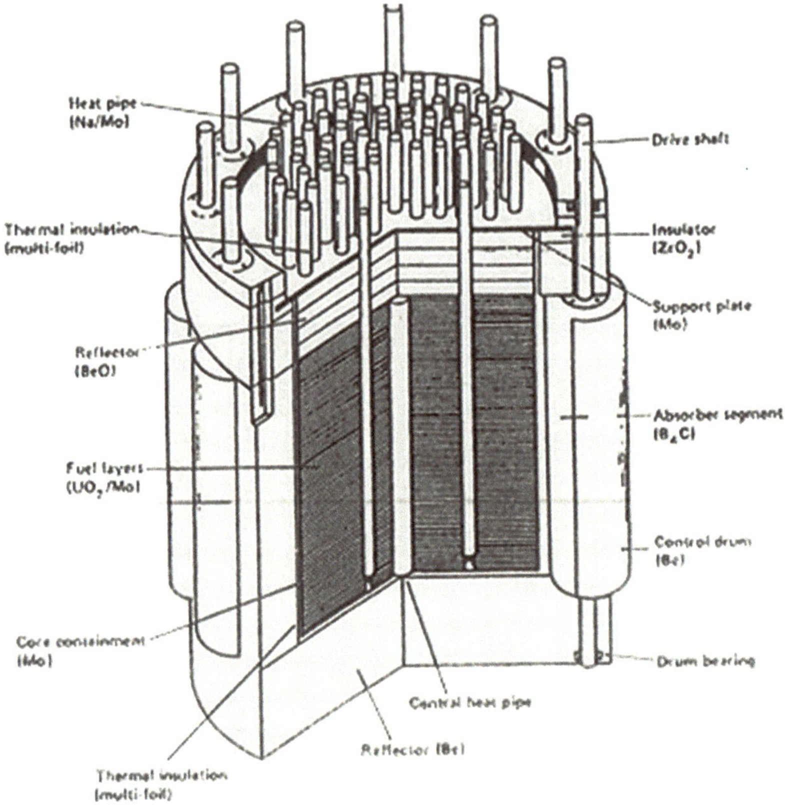

FIG. 1.6 Usage of sodium/molybdenum heat pipes in the thermal control of nuclear power reactor [11].

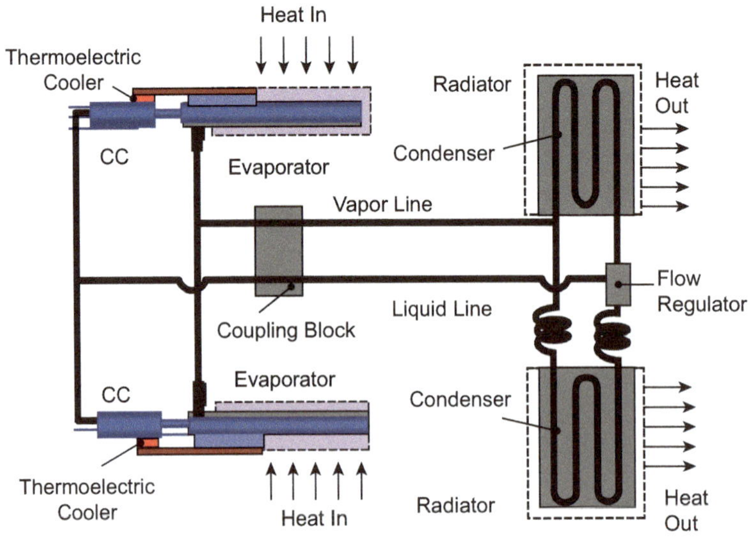

FIG. 1.7 Loop heat pipe (LHP) containing two parallel evaporators and two parallel condensers.

A typical conventional heat pipe with its adiabatic section is depicted in Fig. 1.5 [12]. These types of heat pipes with their adiabatic zone are designed for use in thermal control of nuclear reactor cores that are used in form of convection, conduction and radiation heat transfer device and they are shown in Fig. 1.6. In case of rapid reduction of core temperature additional radiation surface area in form fin are built on top of evaporation section of heat pipe or variable heat pipes are utilized which are described in later part of this section.

This type of Loop Heat Pipe have been utilized as part of new NASA series of experiments for space worthy value of heat pipe studies as illustrated in Fig. 1.7.

Loop Heat Pipe operation involves complex physical processes such as:

- Fluid dynamics, heat transfer, and thermodynamics
- Gravitational, inertial, viscous and capillary forces

The first orbital test demonstrating heat pipe operation under zero-gravity conditions took place in 1967. The launch vehicle for the ATS-A satellite carried a heat pipe with thermocouples to determine its temperature uniformity and performance under varying heat loads throughout different portions of its earth orbit. This successful demonstration was followed 1 year later by the launching of GEOS-2, using heat pipes designed by Johns Hopkins. GEOS-2 was the first satellite designed to use heat, pipes as an integral part of its overall therti1 control system. See Fig. 1.8.

The large-scale satellite and the International Space Station have made great progress in the past several decades. One of the problems urgent to be solved is the heat dissipation. There exists a large amount of heat that should be transferred and radiated into the outer space. Single-phase liquid loops were

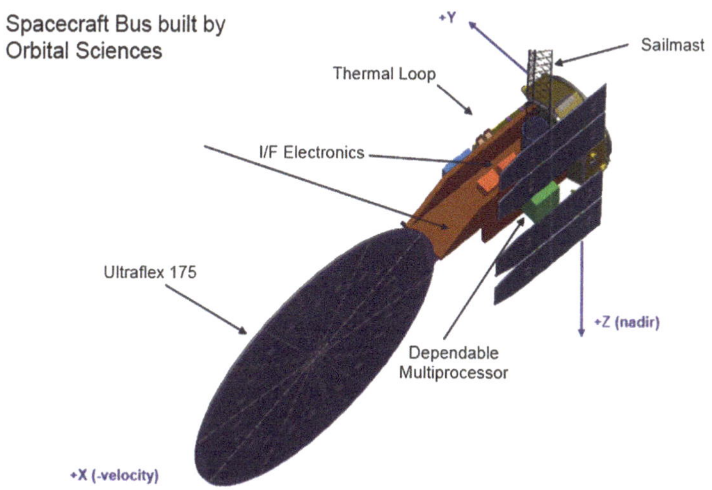

FIG. 1.8 Part of NASA's New Millennium Space Program Satellite.

the major method used in large-scale spacecrafts for heat transfer and dissipation in the past decade. Since 1980s researches worldwide have focused their efforts on two-phase liquid loop technology to be used in the spacecraft thermal control systems on the International Space Station, telecommunication and technological satellites. Space nuclear systems require large area radiators to reject the unconverted heat to space. A conceptual design of a waste heat radiator has been developed for a thermoelectric space nuclear power system [13]. The basic shape of the heat pipe radiator was a frustum of a right circular cone. The design included stringer heat pipes to carry reject heat from the thermoelectric modules to the radiator skin that was composed of small-diameter, thin-walled cross heat pipes. The stringer heat pipes were armored to resist puncture by a meteoroid. The cross heat pipes were designed to provide the necessary unpunctured radiating area at the mission end with a minimum initial system mass. Several design cases were developed in which the individual stringer survival probabilities were varied and the radiator system mass was calculated. Results are presented for system mass as a function of individual stringer survival probability for six candidate container materials, three candidate heat pipe fluids, two radiator operating temperatures, two meteoroid shield types, and two radiating surface cases. Results are also presented for radiator reject heat as a function of system mass, area, and length for three system sizes. We will discuss this in later chapters.

Heat pipe operation on Earth is dominated by the force of gravity, which makes it difficult to predict the performance of a heat pipe in space and requires thermal system engineers to adopt conservative designs and ground-test programs to reduce the risks of systems failures after launch. A heat pipe is a very efficient heat transfer device commonly used for cooling electronic components and sensors. Recently, the results from a Heat Pipe Performance experiment have led to the development and validation of an improved analytical heat pipe model. The accuracy of this computer model, known as GAP, now allows engineers to be less conservative in their designs, which leads to fewer heat pipes per spacecraft, thereby achieving significant cost and weight savings.

Loop Heat Pipe (LHP) and Capillary Pumped Loop (CPL) with "natural" circulation of two-phase flow are used on satellites to ensure the thermal transfer from core module equipment to a radiator. LHP and CPL are considered as reliable thermal management devices that are able of operating at any orientation in a gravitational field and heat can be transported over long distances. The main components of LHP and CPL are evaporator that is responsible for the generation of capillary forces that drive the working fluid via a porous structure and condenser [14].

Electric blankets are also well-known but often provide uneven heating and subject the user to low-level electromagnetic radiation. Nonetheless, whether used for heating or cooling, such suits and blankets generally disadvantageously require external connections and/or require external power sources. Numerous devices have been developed for regional, therapeutic heat transfer. Faghri's invention [15] meets the need for lightweight, comfortable suits and blankets for body temperature regulation by using heat pipes to redistribute body heat and to provide supplemental heat from external sources. A temperature regulation system for the human body will result, taking the form of garments, blankets and pads. This invention further provides for an improved pad incorporating heat pipes for use in regional, therapeutic heat transfer. The heat pipes are positioned to provide heat transfer between one or more separate portions of the body. A garment for use in cold environments, such as a body suit, pants or jacket, may, thus, include heat pipes which extend from the torso of the body, which is typically warmer, to an extreme level. For example, in a garment such as a body suit, this fourth embodiment provides a means to overcome problems with damage to heat-sensitive organs when whole body hyperthermia is induced for medical treatment. Heat may be applied to major portions of the body to induce hyperthermia with one heat exchanger having means for heating, while portions of the body facing heat sensitive organs may be cooled with another heat exchanger having means for cooling. Fig. 1.9. Temperature control is of particular interest where the present invention is used for deliberate inducement of hyperthermia for medical treatment, or to provide controlled heating or cooling for hypothermia or hyperthermia patients.

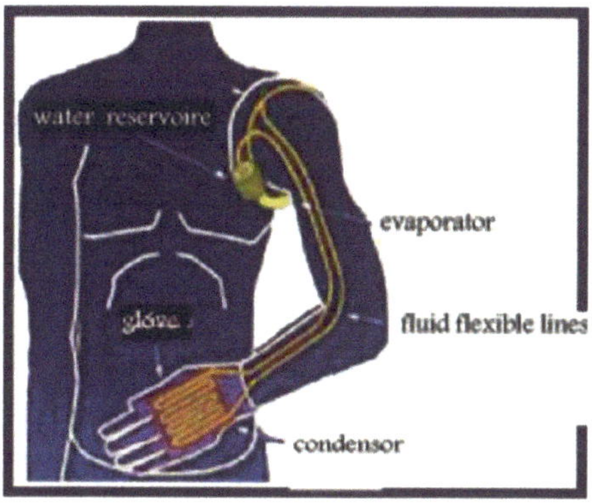

FIG. 1.9 A method for temperature regulation in hand.

Note that:

Heat transfer mechanisms can be grouped into 3 broad categories:

Conduction: Regions with greater molecular kinetic energy will pass their thermal energy to regions with less molecular energy through direct molecular collisions, a process known as conduction. In metals, a significant portion of the transported thermal energy is also carried by conduction-band electrons.

Convection: When heat conducts into a static fluid it leads to a local volumetric expansion. As a result of gravity-induced pressure gradients, the expanded fluid parcel becomes buoyant and displaces, thereby transporting heat by fluid motion (i.e., convection) in addition to conduction. Such heat-induced fluid motion in initially static fluids is known as **free convection**.

For cases where the fluid is already in motion, heat conducted into the fluid will be transported away chiefly by fluid convection. These cases, known as **forced convection**, require a pressure gradient to drive the fluid motion, as opposed to a gravity gradient to induce motion through buoyancy.

Radiation: All materials radiate thermal energy in amounts determined by their temperature, where the energy is carried by photons of light in the infrared and visible portions of the electromagnetic spectrum. When temperatures are uniform, the radiative flux between objects is in equilibrium and no net thermal energy is exchanged. The balance is upset when temperatures are not uniform, and thermal energy is transported from surfaces of higher to surfaces of lower temperature.

Generally speaking typically, there are two classes of Heat Pipes, "Variable Conductance" Heat Pipe (VCHP) or "Conventional" Heat Pipe also known as Constant Conductance Heat Pipe (CCHP) or Fixed Conductance Heat Pipe (FCHP). A typical Conventional Heat Pipe is illustrated in Fig. 1.9, while a Variable Conductance Heat Pipe is depicted in Fig. 1.10. The distinctive feature of these types of heat pipes from conventional ones are their abilities and functionality to operate in a specific desired temperature range along certain portions of the pipe, in spite of variations in the source and sink conditions. When such conditions are desired based on the application of heat pipe, it is important to actively or passively control the heat pipe so that the desired temperature range can be maintained.

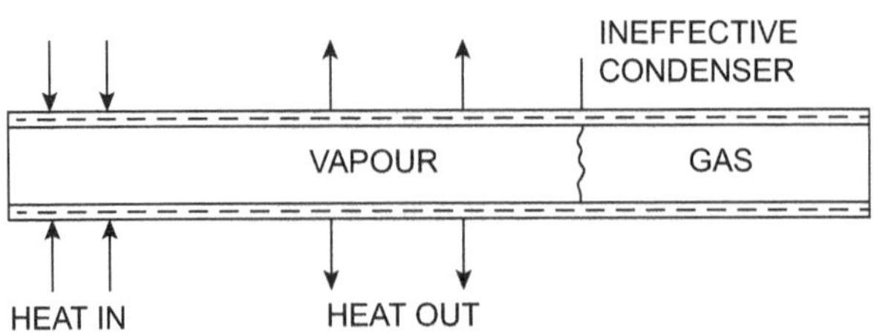

FIG. 1.10 Equilibrium state of gas-loaded heat pipe [12].

Fig. 1.10 is the simple form variable heat pipe structure with gas buffered section on the condenser part of the pipe. Later on, there was an add on reservoir was structured to give a downstream form of condenser (Fig. 1.11) which was allowing the heat pipe to have adequate effective length on condenser side of the pipe to operate at maximum of its capability and provide more accurate control of the vapor temperature [12].

In early days of designing the Cold-Reservoir Variable Conductance Heat Pipe (VCHP) there was problem associated with diffusion of vapor into the reservoir, followed by condensation. We will discuss more about variable heat pipe in later on chapter and some approaches taken by designer of such heat pipes to improve their structure and applications in industry. Computer codes were generated to have better feeling for these type of heat pipes and of the most know computer code "GASPIPE" was developed by Marcus [16] and associates at TRW were they studied and developed VCHP for NASA in early days of 1970s.

A new type of variable conductance heat pipe, the Liquid Controlled Heat Pipe (LCHP), has been developed. While the gas controlled heat pipe is able to stabilize the temperature of the heating zone, the LCHP limits the temperature of the cooling zone to a certain adjustable value. The physical principle is to regulate the heat transfer capability by regulating the amount of liquid inside the heat pipe. The liquid is partly stored in a reservoir with a variable volume, as for example, a bellows. The temperature of the cooling zone, corresponding to the vapor pressure inside the heat pipe, can be adjusted by the outer pressure (gas or spring) on the bellows. The LCHP is applicable where heat is needed at a constant temperature or where the vapor pressure inside a heat pipe has to be limited.

In Fig. 1.12A, a gradual increase in working temperature is accompanied by a rapid increase in saturated vapor pressure of the working fluid. In contrast there is only a slight increase in the temperature on the non-condensable gas and, since the relationship between pressure and cubic capacity is constant, the boundary surface is pushed out by the working fluid so that the "condenser section that is effective for radiation with admixture of working fluid" migrates toward the gas reservoir. In Fig. 1.12B, the region occupied by the working fluid is reduced, and the internal heat transfer rate is still small. If the temperature is further increased, the condenser section that is effective for

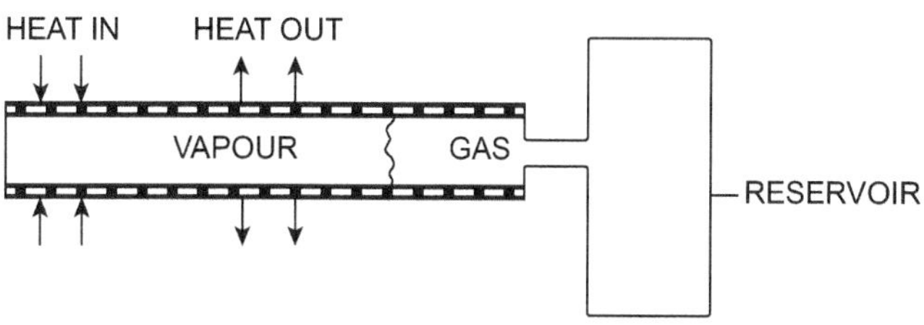

FIG. 1.11 Cold reservoir variable conductance heat pipe [12].

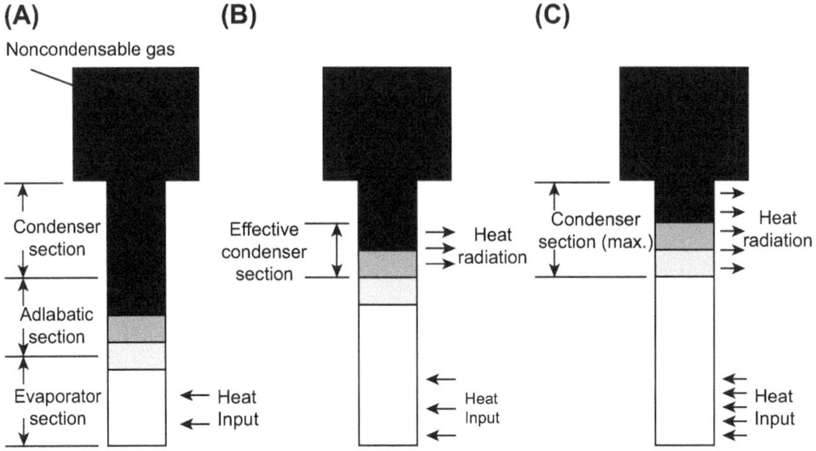

FIG. 1.12 Heat transfer during VCHP operation. (A) QL: No heat radiation, (B) Boundary surface migration, (C) QH: Maximum heat transfer.

radiation becomes larger, and the heat radiation rate increases. At the point of maximum heat radiation, the non-condensable gas is totally contained within the gas reservoir. In Fig. 1.12C, if the heat radiation rate is greater than the maximum heat input to the evaporator section by the heat source, sufficient heat radiation can be obtained, and the temperature of the evaporator section will not increase further [15].

Constant Conductance Heat Pipe (CCHP) also known as Fixed Conductance Heat Pipe (FCHP).

Constant conductance heat pipes (CCHPs) transport heat from a heat source to a heat sink with a very small temperature difference. Axial groove capillary wick structures are utilized because of the relative ease of manufacturing (aluminum extrusions) and their demonstrated heritage in spacecraft and instrument thermal control applications. CCHPs can transport heat in either direction and are typically used to transfer heat from specific thermal loads to a radiator panel or as part of an integrated heat pipe radiator panel. Common working fluids include: ammonia, propylene, ethane, and water.

Fixed Conductance Heat Pipes (FCHPs) filled with working fluid at low or moderate temperatures develop a volume of excess liquid when operated at high temperatures. The excess liquid forms as either a puddle or a slug at the coldest end of the condenser and creates a temperature differential between the evaporator and the condenser end cap. Simple algebraic expressions are presented for predicting the thermal performance of an FCHP operating with a liquid slug formed by the combined influence of liquid density temperature dependence and meniscus depression. Both differential and two-node models are developed to account for condensation modeled either as a constant flux processor based on an isothermal vapor with a constant internal film coefficient. Numerical examples are included to illustrate the behavior of two axially grooved pipes operating over a range of heat loads with both real and ideal fluids. Prediction of evaporator temperature and liquid slug length is observed to have a weak dependence on the choice of model and mode of condensation and a strong dependence on real fluid effects.

Fig. 1.13 shows a structural comparison of a conventional heat-pipe and the Variable Conductance Heat Pipe (VCHP). In the conventional type, a small volume of working fluid is sealed into an evacuated metal container. The working fluid repeatedly vaporizes and condenses as a result of the small temperature difference (or temperature gradient) between the evaporator and condenser sections, and heat is transferred by the latent heat of the working fluid. The heat-pipe comprises an evaporator section, an adiabatic section and a condenser section, and a wick or mesh is provided in the container to facilitate the circulation of the working fluid. The maximum heat transfer rate, which is the measure of heat-pipe performance, is determined by the wick, the volume of working fluid, and so on, while the working temperature is determined passively by the external heat source, the temperature of the heat radiator (condenser) section, and so on [15].

"Fig. 1.14 compares the heat-radiation characteristics of conventional and variable-conductance heat-pipes. Unlike the conventional heat-pipe, in which the heat radiation rate has a constant gradient relative to temperature, the radiation rate in the VCHP increases rapidly from a given temperature (the radiation onset temperature) until the radiation limit is reached. When the radiation limit is exceeded, it traces a constant gradient, like a conventional

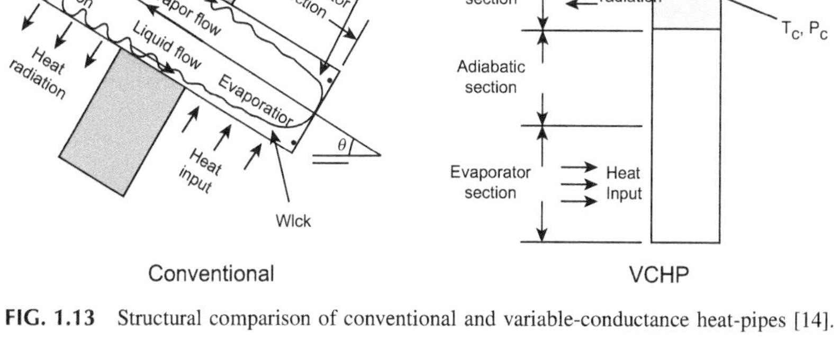

FIG. 1.13 Structural comparison of conventional and variable-conductance heat-pipes [14].

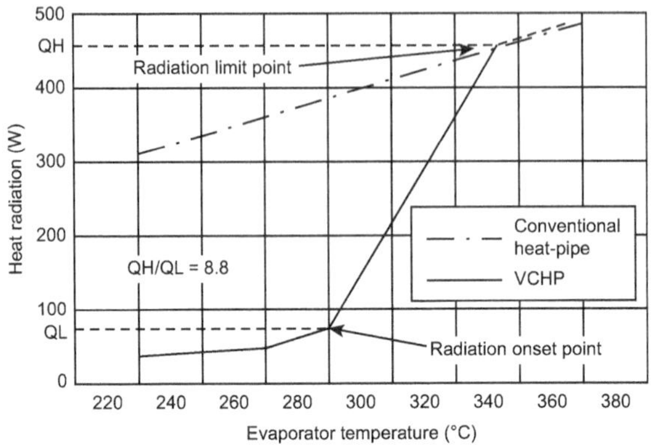

FIG. 1.14 Heat-radiation characteristics of conventional and variable-conductance heat-pipes [14].

heat-pipe. The term 'variable conductance' has its origin in this characteristic, and the slope of the line between the radiation onset point and the radiation limit point (hereinafter referred to as the radiation gradient) is an important characteristic of the VCHP" [15].

1.5 Principles of operation

The three basic components of a heat pipe are:

1. The container
2. The working fluid
3. The wick or capillary structure

1.5.1 Container

The function of the container is to isolate the working fluid from the outside environment. It has to therefore be leak-proof, maintain the pressure differential across its walls, and enable transfer of heat to take place from and into the working fluid.

Selection of the container material depends on many factors. These are as follows:

- Compatibility (both with working fluid and external environment)
- Strength to weight ratio
- Thermal conductivity
- Ease of fabrication, including welding, machineability and ductility
- Porosity
- Wettability

Most of the above are self-explanatory. A high strength to weight ratio is more important in spacecraft applications. The material should be non-porous to prevent the diffusion of vapor. A high thermal conductivity ensures minimum temperature drop between the heat source and the wick.

1.5.2 Working fluid

A first consideration in the identification of a suitable working fluid is the operating vapor temperature range. Within the approximate temperature band, several possible working fluids may exist, and a variety of characteristics must be examined in order to determine the most acceptable of these fluids for the application considered. The prime requirements are:

- Compatibility with wick and wall materials
- Good thermal stability
- Wettability of wick and wall materials
- Vapor pressure not too high or low over the operating temperature range
- High latent heat
- High thermal conductivity
- Low liquid and vapor viscosities
- High surface tension
- Acceptable freezing or pour point

The selection of the working fluid must also be based on thermodynamic considerations which are concerned with the various limitations to heat flow occurring within the heat pipe like viscous, sonic, capillary, entrainment and nucleate boiling levels.

In heat pipe design, a high value of surface tension is desirable in order to enable the heat pipe to operate against gravity and to generate a high capillary driving force. In addition to high surface tension, it is necessary for the working fluid to wet the wick and the container material i.e., contact angle should be zero or very small. The vapor pressure over the operating temperature range must be sufficiently great to avoid high vapor velocities, which tend to setup large temperature gradient and cause flow instabilities.

A high latent heat of vaporization is desirable in order to transfer large amounts of heat with minimum fluid flow, and hence to maintain low pressure drops within the heat pipe. The thermal conductivity of the working fluid should preferably be high in order to minimize the radial temperature gradient and to reduce the possibility of nucleate boiling at the wick or wall surface. The resistance to fluid flow will be minimized by choosing fluids with low values of vapor and liquid viscosities. Tabulated below are a few mediums with their useful ranges of temperature. See Fig. 1.15.

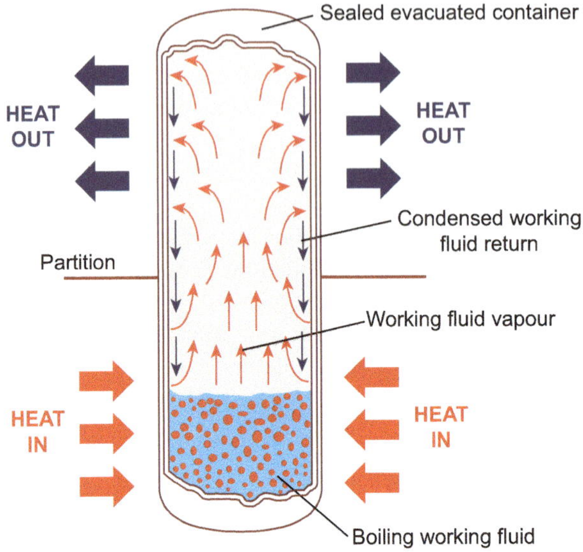

FIG. 1.15 Schematic representation of the heat pipe.

1.5.3 Wick or capillary structure

It is a porous structure made of materials like steel, aluminum, nickel or copper in various ranges of pore sizes. They are fabricated using metal foams, and more particularly felts, the latter being more frequently used. By varying the pressure on the felt during assembly, various pore sizes can be produced. By incorporating removable metal mandrels, an arterial structure can also be molded in the felt.

Fibrous materials, like ceramics, have also been used widely. They generally have smaller pores. The main disadvantage of ceramic fibers is that, they have little stiffness and usually require a continues support by a metal mesh. Thus, while the fiber itself may be chemically compatible with the working fluids, the supporting materials may cause problems. More recently, interest has turned to carbon fibers as a wick material. Carbon fiber filaments have many fine longitudinal grooves on their surface, have high capillary pressures and are chemically stable. A number of heat pipes that have been successfully constructed using carbon fiber wicks seem to show a greater heat transport capability.

The prime purpose of the wick is to generate capillary pressure to transport the working fluid from the condenser to the evaporator. It must also be able to distribute the liquid around the evaporator section to any area where heat is likely to be received by the heat pipe. Often these two functions require wicks of different forms. The selection of the wick for a heat pipe depends on many factors, several of which are closely linked to the properties of the working fluid.

The maximum capillary head generated by a wick increases with decrease in pore size. The wick permeability increases with increasing pore size. Another feature of the wick, which must be optimized, is its thickness. The heat transport capability of the heat pipe is raised by increasing the wick thickness. The overall thermal resistance at the evaporator also depends on the conductivity of the working fluid in the wick. Other necessary properties of the wick are compatibility with the working fluid and wettability.

The most common types of wicks that are used are as described in the following sub-sections.

1.5.4 Sintered powder

This process will provide high power handling, low temperature gradients and high capillary forces for anti-gravity applications. The photograph shows a complex sintered wick with several vapor channels and small arteries to increase the liquid flow rate. Very tight bends in the heat pipe can be achieved with this type of structure.

1.5.5 Grooved tube

The small capillary driving force generated by the axial grooves is adequate for low power heat pipes when operated horizontally, or with gravity assistance. The tube can be readily bent. When used in conjunction with screen mesh the performance can be considerably enhanced.

1.5.6 Screen mesh

This type of wick is used in the majority of the products and provides readily variable characteristics in terms of power transport and orientation sensitivity, according to the number of layers and mesh counts used. See Figs. 1.16 and 1.17.

Heat transfer in a heat pipe is limited by: the rate at which liquid can flow through the wick; "choking" (the inability to increase vapor flow with increasing pressure differential, also called "sonic limit"); entrainment of liquid in the vapor stream, such that liquid flow to the evaporator is reduced; the rate at which evaporation can take place without excessive temperature differentials in the evaporator section.

Isothermalizer heat pipes will transport heat in either direction, and for a given configuration, the heat flow will depend entirely on the temperature difference between the heat source and heat sink. The isothermalizer heat pipe is therefore a basically passive device with a fixed conductance, provided that none of its limiting conditions are exceeded (boiling, sonic, entrainment, capillary limits). The isothermalization function of a heat pipe can be modified to produce active devices in two ways: diode heat pipes, where the pipe

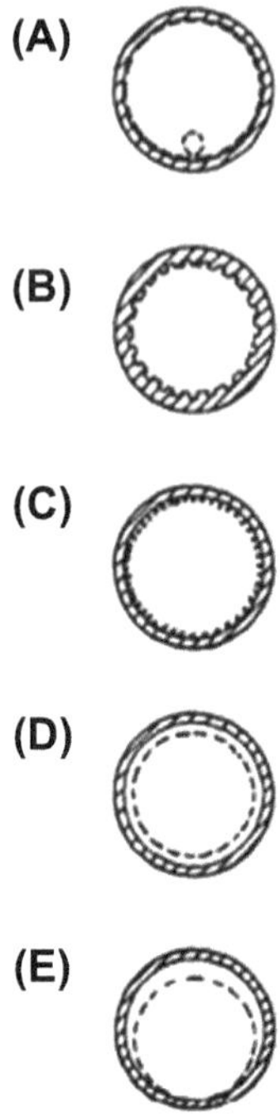

FIG. 1.16 Cross section of various wick structures [17]. (A) Artery. (B) Channels. (C) Screen. (D) Concentric annulus. (E) Crescent annulus.

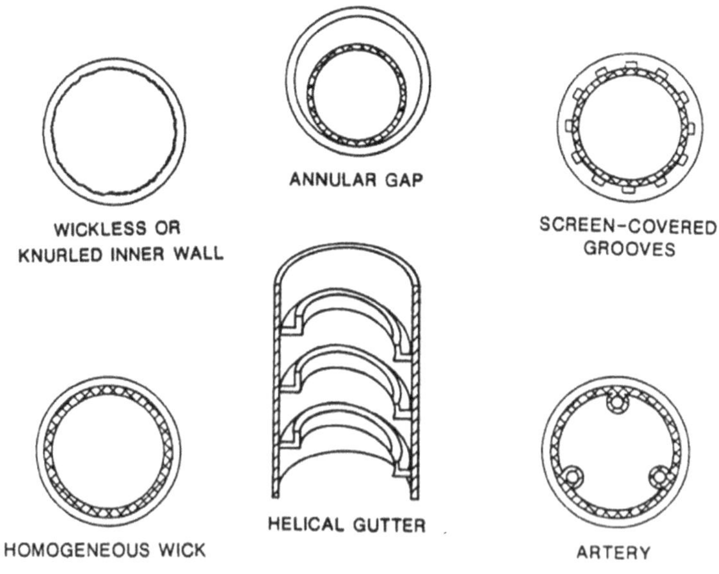

FIG. 1.17 Heat pipe liquid return geometries [18].

operates as an isothermalizer in the forward mode and shuts off in the reverse mode; or variable conductance heat pipes, where the conductance in the forward mode can be actively controlled and again shuts down in the reverse mode.

Fig. 1.18 demonstrates several common wicking structures presently in use, along with more advanced concepts under development [18].

1.5.7 How the heat pipe is working

Inside the container is a liquid under its own pressure, that enters the pores of the capillary material, wetting all internal surfaces. Applying heat at any point along the surface of the heat pipe causes the liquid at that point to boil and

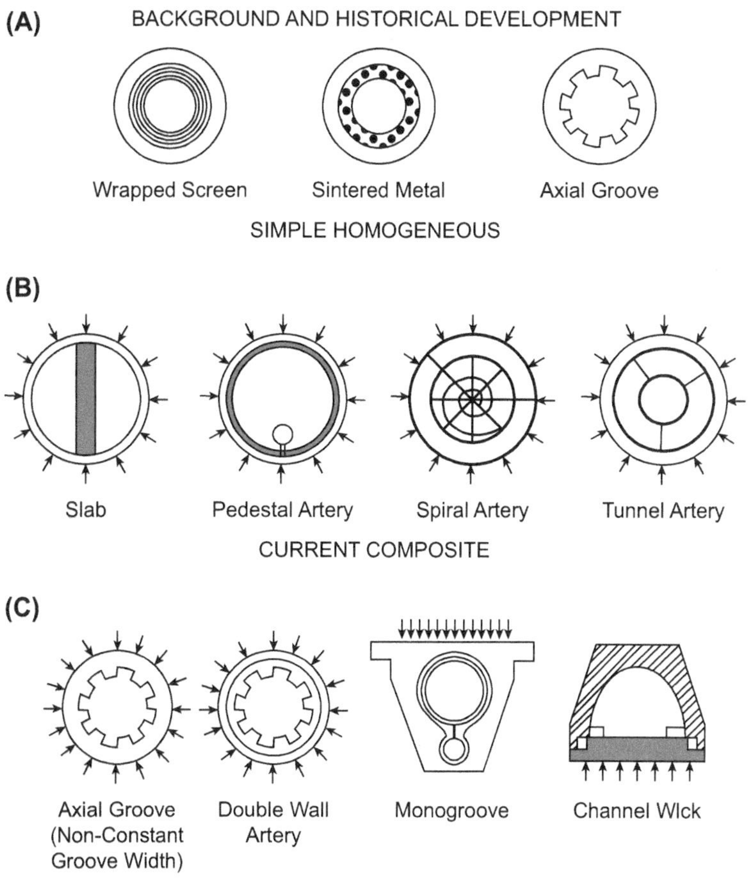

FIG. 1.18 Typical heat pipe wicking configurations and structures [19]. (A) Simple homogeneous. (B) Current composite. (C) Advanced designs.

enter a vapor state. When that happens, the liquid picks up the latent heat of vaporization. The gas, which then has a higher pressure, moves inside the sealed container to a colder location where it condenses. Thus, the gas gives up the latent heat of vaporization and moves heat from the input to the output end of the heat pipe. See Fig. 1.19.

Heat pipes have an effective thermal conductivity many thousands of times that of copper. Its "Axial Power Rating (APC)" specifies the heat transfer or transport capacity of a heat pipe. It is the energy moving axially along the pipe. The larger the heat pipe diameter, greater is the APR. Similarly, longer the heat pipe lesser is the APR. Heat pipes can be built in almost any size and shape. A simple Heat Pipe Assemblies and Design Guidelines can be found on so many manufactures of heat pipe. Fig. 1.20 is the best that author has found.

Web site of Aavid Engineering, which was founded in 1964 as subsidiary of Aavid Thermal Technologies, Inc., is recommending certain and simple criteria as role of thump for any Heat Pipe Assemblies Design as follow;

1.5.8 Heat pipe assemblies design guidelines

The following approach is sort of quick and back envelope type analysis of choosing your optimum design before you model your heat pipe for best optimum point of your design that falls within constrain and limit of operating range of heat pipe such as Sonic, Entrainment, Wicking and Boiling limits.

1.5.9 Orientation with respect to gravity

For the best performance, the application should have gravity working with the system; that is, the evaporator section (heated) should be lower, with respect to gravity, than the condenser (cooling) section. In other orientations where gravity is not aiding the condensed liquid return, the overall performance will

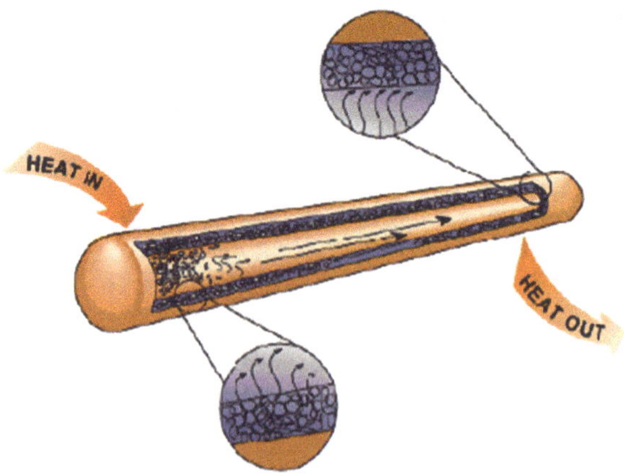

FIG. 1.19 Principal of conserved energy and heat transfer in heat pipe.

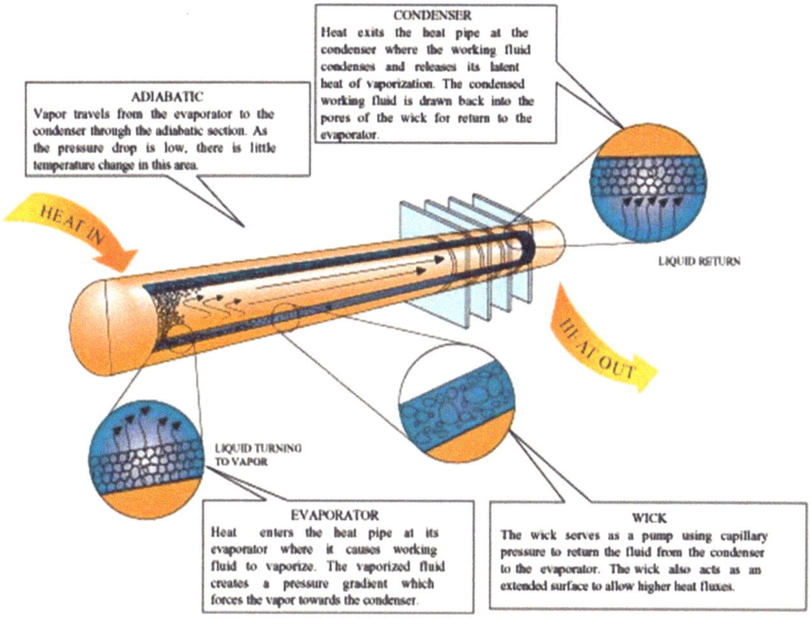

FIG. 1.20 Heat pipe operation [20].

be degraded. Performance degradation depends on a number of factors including wick structure, length and working fluid of the heat pipe along with heat flux of the application. Careful design can minimize the performance loss and allow an accurate prediction performance.

1.5.10 Temperature limits

Most pipes use water and methanol/alcohol as the working fluids. Depending on the wick structure, pipes will operate in environments with as low as −40 °C. Upper temperature limits depend on the fluid, but 60−80 °C is the average limit.

1.5.11 Heat removal

Heat can be removed from the condenser using air cooling in combination with extrusion, bonded-fin heat sinks, or flat-fin stock. Enclosing the condenser in a cooling jacket allows liquid cooling.

1.5.12 Reliability

Heat pipes have no moving parts and have demonstrated life of over 20 years. The largest contributor to heat pipe reliability comes from control of the manufacturing process. The seal of the pipe, purity of the materials used in the wick structure and cleanliness of the internal chamber have measurable effect on the long term performance of a heat pipe. Any leakage will eventually render the pipe inoperable. Contamination of the internal chamber and wick structure will contribute to the formation of Non-Condensable Gas (NCG) that will degrade performance over time. Well-developed processes and rigorous testing are required to ensure reliable heat pipes.

1.5.13 Forming or shaping

Heat pipes are easily bent or flattened to accommodate the needs of the heat sink design. Forming heat pipes may affect the power handling capability as the bends and flattening will cause a change in fluid movement inside the pipe. Therefore, design rules that take heat pipe configurations into consideration and the effect on thermal performance ensure the desired solution performance.

1.5.14 Effects of length and pipe diameter

The vapor pressure differential between the condenser end and the evaporator end controls the rate at which the vapor travels from one end to the other. Diameter and length of the heat pipe also affect the speed at which the vapor moves and must be considered when designing with heat pipes. The larger the diameter, the more cross sectional area available to allow vapor to move from the evaporator to the condenser. This allows for greater power carrying capacity. Conversely, length when in opposition to gravity has a negative effect on heat transport as the rate at which the working fluid returns from the condenser end to the evaporator end is controlled by the capillary limit of the wick, which is an inverse function of the length of the pipe. Therefore, shorter heat pipes carry more power than longer pipes when used in application not assisted by gravity.

1.5.15 Wick structures

Heat pipe inner walls can be lined with a variety of wick structures. The four most common wicks are:

a. Groove
b. Wire mesh
c. Sintered powder metal
d. Fiber/spring

The wick structure provides a path for liquid to travel from condenser to the evaporator using capillary action. Wick structures have performance advantages and disadvantages depending on the desired characteristics of the heat sink design. Some structures have low capillary limits making them unsuitable for applications where they must work without gravity assist. The plots as it is in Figs. 1.21A and B are demonstration standard operating range of simple heat pipe.

1.6 Heat pipe operating ranges

No matter what type of application you may consider the heat pipe for, there certain limitation that one has to look upon it for the heat pipe to operate properly and be able to perform the requirements that are imposed on the heat pipe and its application in that environment. These limitations are briefly described below as well as demonstrated in Fig. 1.22.

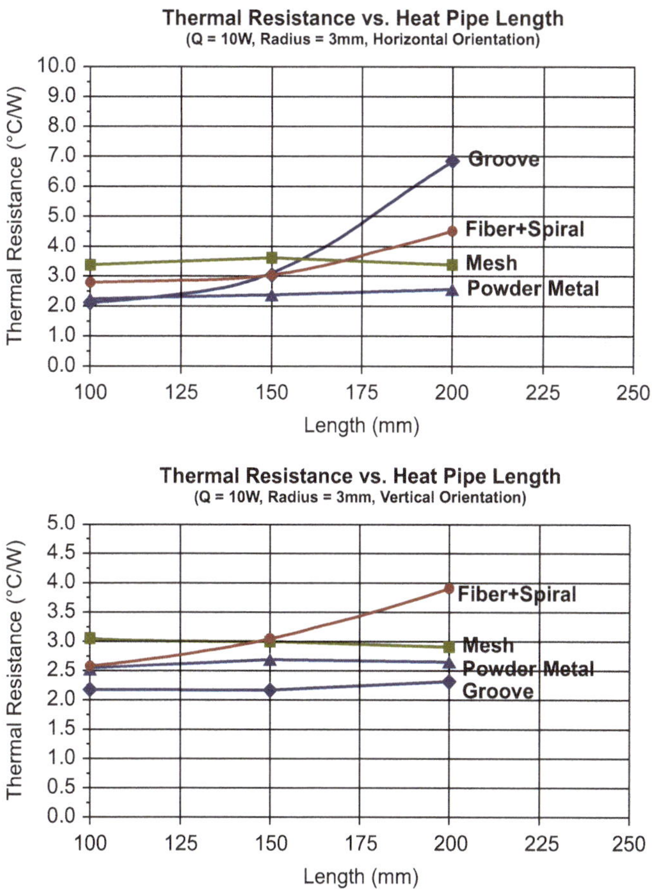

FIG. 1.21 Standard operating range of simple heat pipe.

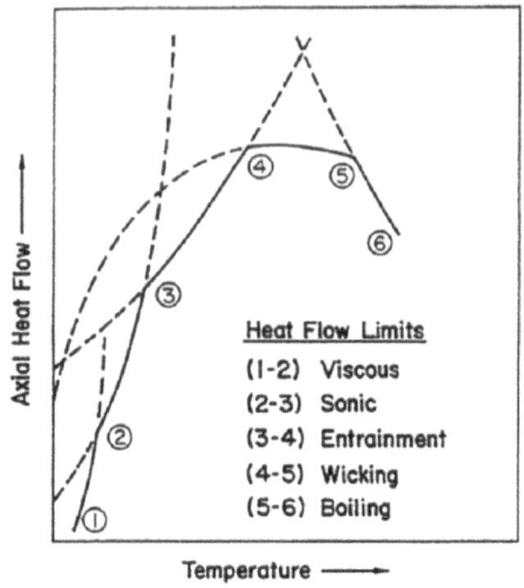

FIG. 1.22 Typical limit of heat pipe operating limits.

1. ***Viscous limit:*** In long pipes and at low temperatures, the vapor pressure is low and the effect of viscous friction on the vapor flow may dominate over the inertial forces. In this situation the circulation of the working fluid is limited, which consequently, limits the heat transfer through the pipe.
2. ***Sonic limit:*** At low vapor pressures, the velocity of the vapor at the exit of the evaporator may reach the speed of sound. Then the evaporator cannot respond to further decrease in the condenser pressure. That is, the vapor flow is chocked, which limits the vapor flow rate.
3. ***Capillary limit:*** A capillary structure is able to provide circulation of a given fluid up to a certain limit. This limit depends on the permeability of the wick structure and the properties of the working fluid.
4. ***Entrainment limit:*** The vapor flow exerts a shear force on the liquid in the wick which flows opposite the direction of the vapor flow. If the shear force exceeds the resistive surface tension of the liquid, the vapor flow entrains small liquid droplets (Kelvin-Helmholtz instabilities). The entrainment of liquid increases the fluid circulation but not the heat transfer through the pipe. If the capillary force cannot accommodate the increased flow, dry out of the wick in evaporator may occur.
5. ***Boiling limit:*** At high temperatures, nucleate boiling may take place which produces vapor bubbles in the liquid layer. The bubbles may block the wick pores and decrease the vapor flow. Furthermore, the presence of the bubbles decreases the conduction of heat through liquid layer which limits the heat transfer from the heat pipe shell to the liquid which is by conduction only.

Later chapters will show how to calculate each of these limits and its mathematical modeling. These analyses will allow the designer to plot the above curves and best design and fabrication of a heat pipe is where it operates within these envelope limits which is the area below all the curves. This area is known as best optimum design of heat pipe for the particular application of designer.

1.7 Constraints

There are many parameters which affect the performance of a heat pipe like compatibility of materials, operating temperature range, heat transport limitation, thermal resistance, operating orientation, dimension and geometric constraints, etc. For example, in miniature heat pipes, the maximum heat transport capacity was found to be primarily governed by the capillary pressure [21].

All heat pipes have three physical elements in common. These include an outer container, a small amount of working fluid, and a capillary wick structure. In addition to these basic components, heat pipes may also include gas reservoirs (variable conductance/diode heat pipes) and liquid or gas traps (diodes). Functionally, the heat pipe consists of three sections: evaporator, condenser section, and adiabatic regions. The evaporator section is mounted to the heat-producing components, while the condenser is thermally coupled to a heat sink or radiator. The adiabatic section allows heat to be transferred from the evaporator to the condenser with very small heat losses and temperature drops. Fig. 1.23 depicts the basic heat pipe.

Heat pipes can operate in the fixed conductance, variable conductance, or diode mode. The fixed conductance heat pipe can transfer heat in either direction and operates over broad temperature ranges but has no inherent temperature control capability. Constant conduction heat pipes allow isothermalization of shelves, radiators, and structures; spread heat from high

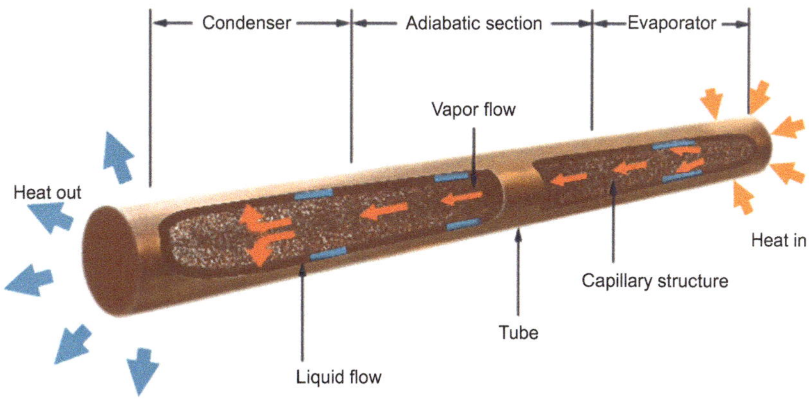

FIG. 1.23 Depicts the basic heat pipe.

heat dissipating components; and conduct heat away from heat-producing devices embedded within instruments and satellites. In the Variable Conductance Heat Pipe (VCHP), a small quantity of Non-Condensable Gas (NCG) is loaded into the heat pipe. The VCHP can be used to control the temperature of equipment within very narrow limits; control is possible to less than 1K by using careful design techniques. This is accomplished by controlling the location of the NCG/vapor interface within the condenser end of the heat pipe, thereby varying the active length of the condenser and causing a modulation in the condenser heat rejection capability. Temperature control of the attached device is achieved by an active feedback system consisting of a temperature sensor at the heat source and a controller for a heater at the NCG reservoir.

The heater causes the gas in the reservoir to expand, thus moving the gas/vapor interface. Diode heat pipes permit heat to flow in one direction and inhibit heat flow in the opposite direction.

Specific benefits of heat pipes are:

(1) Heat pipes have enormously more heat transfer capability than other methods on a weight and size basis,
(2) Heat pipes permit configuration flexibility in contact areas with heat sources and heat sinks,
(3) Heat can be transported over considerable distances with insignificant temperature drop,
(4) Capillary pumping in the wick is generated by the heat transfer process and requires no other power or moving parts to pump the condensate, and,
(5) Heat pipes operate satisfactorily in a zero gravity environment.

The choice of working fluid is dictated by several considerations, including operating temperature, latent heat of vaporization, liquid viscosity, toxicity, chemical compatibility with container material, wicking system design, and performance requirements. Figs. 1.24 and 1.25 and Table 1.1 depict some of

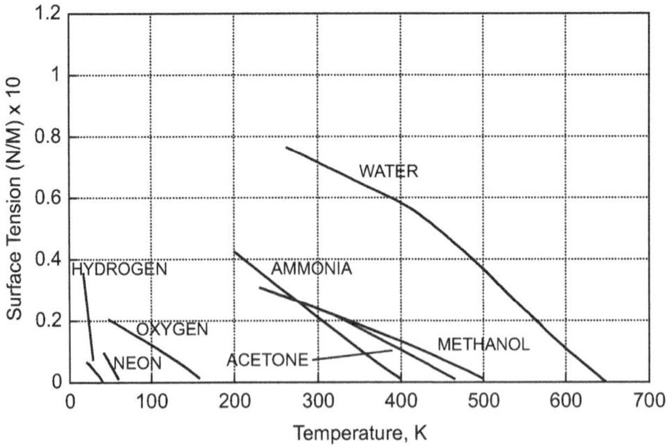

FIG. 1.24 Surface tension for typical heat pipe fluid.

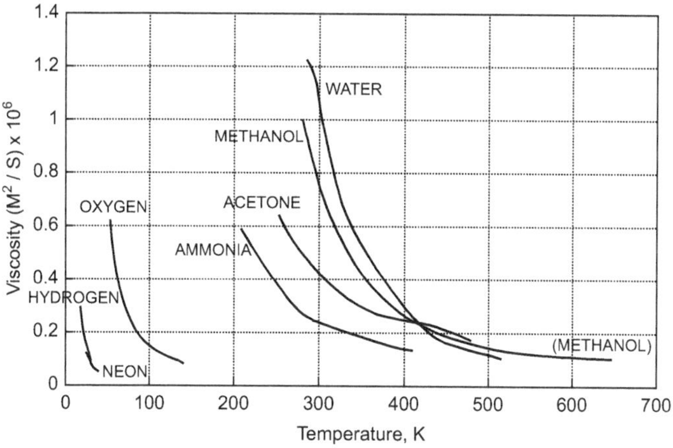

FIG. 1.25 Viscosity for typical heat pipe fluid.

the above characteristics for several fluids. The highest performance from a heat pipe is obtained by utilizing a working fluid that has a high surface tension (s), a high latent heat (l), and a low liquid viscosity (n_l). These fluid properties are contained in the parameter N_l the Liquid Transport Factor. Fig. 1.26 is a plot of N_l for five typical heat pipe working fluids. These data are used as selection criteria for heat pipe working fluids. Once an application is defined, the heat pipe designer reviews the requirements and selects the best working fluid. Below the freezing point of water and above about 200K, ammonia is an excellent working fluid. Regardless of the fluid chosen, minimum purity must be at least 99.999%. A careful analysis of the purity of the ammonia should be obtained from an independent laboratory prior to use.

Image Description - 0698table2

A table is captioned "Table 1.2. Material Compatibility for Heat Pipe/Fluid Combinations." A column of six fluid names is on the left of the table.

The fluid names are separated by horizontal lines that extend to the right.

In descending order, the fluid names are "WATER", "AMMONIA", "METHANOL", "ACETONE", "SODIUM", and "POTASSIUM."

Five columns are on the right of the table. Each column is headed by a name of a metal. The names of the metals are separated by vertical lines that extend downward. From left to right the names of the metals are "ALUMINUM", "STAINLESS STEEL", "COPPER", "NICKEL, and "TITANIUM."

The extended horizontal lines and the extended vertical lines intersect to form boxes. There are five boxes in a horizontal line adjacent to each fluid name. Each box contains one to two symbol or is left blank.

TABLE 1.1 Comparison of latent heat to specific heat for typical heat pipe fluids.

Fluid	Boiling Point 176K	Latent Heat kJ/kg h_{fg}	Specific Heat kJ/kg- K c_p	Ratio K h_{fg}/c_p
Fluid properties				
Helium	4	23	4.60	5
Hydrogen	20	446	9.79	46
Neon	27	87	1.84	47
Oxygen	90	213	1.90	112
Nitrogen	77	198	2.04	97
Argon	87	162	1.14	142
Propane	231	425	2.20	193
Ethane	184	488	2.51	194
Methane	111	509	3.45	147
Toluene	384	363	1.72	211
Acetone	329	518	2.15	241
Heptane	372	318	2.24	142
Ammonia	240	1180	4.80	246
Mercury	630	295	0.14	2107
Water	373	2260	4.18	541
Benzene	353	390	1.73	225
Cesium	943	49	0.24	204
Potassium	1032	1920	0.81	2370
Sodium	1152	3600	1.38	2608
Lithium	1615	19330	4.27	4526
Silver	2450	2350	0.28	8393

The symbols are identified as follows:

C = COMPATIBLE
I = INCOMPATIBLE
** = SENSITIVITY TO CLEANING*

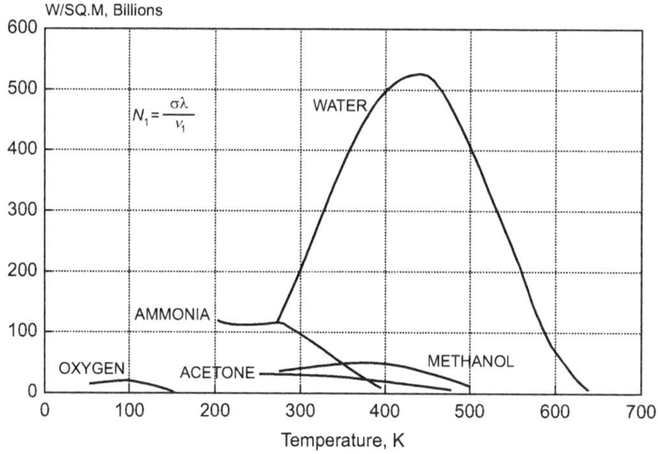

FIG. 1.26 Comparison of liquid transport factor for typical heat pipe working fluids.

TABLE 1.2 Material composite for heat pipe/fluid combinations.

	Aluminum	Stainless steel	Copper	Nickel	Titanium
Water	I	C*	C	C	
Ammonia	C	C		C	
Methanol	I	C	C	C	
Acetone	C	C	C		
Sodium		C		C	I
Potassium				C	I

C - Compatible, I - Sensitivity to Cleaning, * - Incompatible.

The fluid names and the contents of the horizontal row of boxes adjacent to them are as follows:

"WATER" Aluminum - I, Stainless Steel - C, Copper - C, Nickel - C, Titanium - blank*

"AMMONIA" Aluminum - C, Stainless Steel - C, Copper - blank, Nickel - C, Titanium - blank

"METHANOL" Aluminum - I, Stainless Steel - C, Copper - C, Nickel - C, Titanium - blank

"ACETONE" Aluminum - C, Stainless Steel - C, Copper - C, Nickel - blank, Titanium - blank

"SODIUM" Aluminum - blank, Stainless Steel - C, Copper - blank, Nickel - C, Titanium - I
"POTASSIUM" Aluminum - blank, Stainless Steel - blank, Copper - blank, Nickel - C, Titanium

The outer container usually consists of a metal tube to provide mechanical support and pressure containment. The chosen design and processing of the container are extremely important in selecting the metal, because they can affect the useful life of the heat pipe. In addition, a compatibility must exist between the pipe material and the working fluid. For heat pipes, working fluid/container compatibility issues encompass any chemical reactions or diffusion processes occurring between the fluid and wall/wick materials that can lead to gas formation and/or corrosion. Table 1.2 lists the compatibilities of several metals and working fluids. Along with the metal/fluid compatibility, other considerations in the metal selection are ease of working the material, extrusion capability of the material, and its weldability. Proper container cleaning and heat pipe processing procedures are of extreme importance, since residual contamination within the heat pipe may also lead to gas generation. Steps must also be taken to ensure the purity of the fluid charge; trace amounts of water in ammonia can lead to a reaction with the aluminum container and the formation of hydrogen gas. Chi [12] and B & K Engineering [20] list standard cleaning and filling methods for a variety of working fluid/wall material combinations. Special consideration must be given to the processing of heat pipes to be used at temperatures below 250K. As the temperature drops, the vapor pressure of the fluid falls off. This allows any non-condensable gas created by contamination to expand, thus creating an even larger problem.

The heat pipe wick structure provides a porous medium for the formation of liquid menisci (which cause the capillary pumping action) and a vehicle for returning the working fluid from the condenser to the evaporator. To accomplish these wick functions effectively, the designer must provide pores, cavities, or channels of the right size, shape, quantity and location. An optimization technique is used in wick design to find the desired combination of ultimate heat transfer capacity, pumping capability, and temperature drop. The designer must also consider ease of wick fabrication, compatibility with the working fluid, wetting angle and permeability of the selected wick material. Fig. 1.26 depicts a cross-sectional view of an axial groove wick; this design probably is the most commonly used for space application. Fig. 1.27.

In addition, X-ray certification of all welds at the end caps and fill tube is required to ensure good weld penetration and the absence of voids. The heat pipe container must be pressure tested to at least twice its Maximum Expected Operating Pressures (MEOPs) prior to filling [22]. Other qualification procedures include performance tests at adverse tilt angles to demonstrate proper wick function, and gas pocket tests performed with the heat pipe in the reflux mode. Heat pipes should be handled with care, especially those that

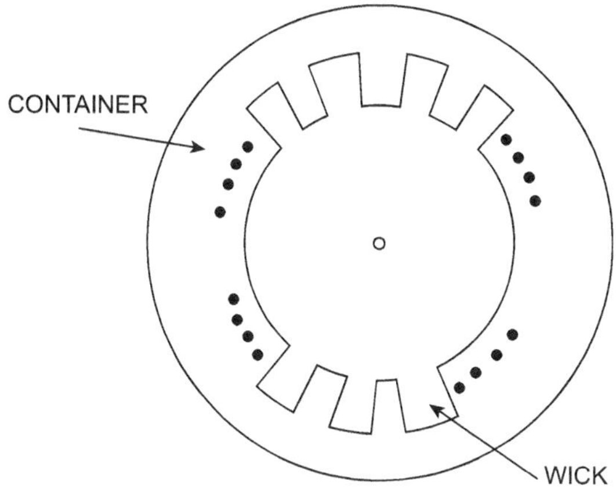

FIG. 1.27 Axially grooved wick.

contain ammonia or other high vapor pressure fluids. They should be treated as any other pressure vessel, and appropriate safety precautions must be exercised. Exposure to ammonia vapor can cause severe irritation to eyes and other mucous membranes. Exposure to ammonia liquid can cause severe burns to the skin. Whenever possible, heat pipes should be stored in a cold, dry environment. This will inhibit any internal chemical reactions which produce non-condensable gas.

1.8 Lessons learned

Heat transfer by means other than the heat pipe can have the following impacts:

1. A price paid with respect to weight and size of the heat transfer equipment,
2. Significant heat lost in transfer over considerable distances,
3. Electromotive devices, such as liquid pumps, required to move the heat, and
4. Possible problems presented by operation in zero gravity.

Non-adherence to the implementation methods presented above could result in the following possible impacts: improper cleaning and processing of the aluminum container could result in contaminants reacting with the ammonia to form NCG, which will interfere with the flow of vapor and reduce the heat transfer effectiveness. Contaminants reacting with ammonia normally produce hydrogen, and the gas collects in the condenser region. As more and more of the condenser is blocked, the surface area available for heat rejection decreases, reducing the heat transfer effectiveness; ultimately, the heat pipe may cease to function. Failure to certify welds at the end caps and the fill tube could result in improper or defective welds permitting leaks or catastrophic failure of the pressure vessel. For long-term space missions, working fluids in the appropriate temperature range, such as methanol and water, exhibit an incompatibility with aluminum, and should not be used.

1.9 Applications

There are many applications for heat pipes, which are well proven and may now be regarded as routine. In conventional use, heat pipes are integrated into a total thermal subsystem to transport heat from the heat source to remote areas. The heat pipes' ability to act as a primary heat conductive path allows engineers to solve thermal problems in applications with space constraints or other limitations. Thus, you can use heat pipes to carry heat away from the heat-sensitive components to the finned array or a heat sink located in an area where more space for heat dissipation is allowed — leaving room for electronics layout flexibility.

A high-capacity power electronics cooler is an example of a thermal solution where no sufficient space is available to mount directly a finned heat sink to the heat source.

In addition to acting as a heat conductive path and aiding in remote heat transfer, heat pipes can improve thermal solution efficiency. You can accomplish this by embedding heat pipes into the heat sink base or passing the heat pipes through the fins. In most cases, embedding heat pipes into the conventional thermal solution results in size or weight reduction.

The most appropriate application for heat pipe integration into the heat sink base is when the base is large compared with the heat source. In such applications, the heat source location produces the highest temperature. The smaller the heat source, the more spreading has to occur over the heat sink base, resulting in a greater temperature rise in the center of the base. Integrating heat pipes into the base of the heat sink decreases the temperature gradient across the base, yielding a more efficient solution.

You can also improve heat sink fin efficiency with heat pipe integration. Fin efficiency is related to the rate at which the fin can dissipate heat energy. The maximum rate at which the fin can dissipate energy is the rate that would exist if the fin were at base temperature. Therefore, the efficiency of the fin can be improved by passing a heat pipe through the fin. Compared with the traditional finned heat sink, the use of a heat pipe configuration with implementation of fin as part of condenser reduces footprint area of the power heat sink and improves heat dissipation capability.

While external factors such as shock, vibration, force impact, thermal shock, and corrosive environment can affect heat pipe life, its integration into a thermal solution also delivers many benefits. If manufactured and designed properly, heat pipes are highly reliable and have no moving parts. In addition, heat pipes are economical, having little effect on the overall cost of the total thermal.

The heat pipe itself is not a heating or cooling device. Heat pipe assemblies are used for moving heat away from the input area (cooling - the most common application) or for moving heat into the output area (heating). Heat pipe assemblies typically have three parts:

1. The heat input assembly;
2. The heat transport assembly (the heat pipe);
3. The heat output assembly.

Heat pipe assemblies provide thermal management solutions in all mediums: liquid, solid, and gas. Compared to heat pipes, the traditional methods of cooling (extruded metal heat sinks, fans, water, air conditioning, etc.), all have inherent limitations in size, weight and efficiency. More and more the limiting factor in systems of all kinds is the inability to dissipate heat. The desire for more power in smaller packages with less weight often cannot be achieved because of excess heat.

Use of heat pipes in high-power (>150 W) cooling applications has been limited to custom applications requiring either low thermal resistance or having a severely restricted enclosure area. The cost of these larger diameter heat pipes was high due to a limited number of manufacturers and handmade assembly times.

A new and valuable addition to the heat transfer community, a heat transport device known as a Loop Heat Pipe (LHP), is discussed in this work. This body of research is very important as the LHP is becoming increasingly prevalent in heat transfer applications. US commercial use of the Loop Heat Pipe will begin on the next generation of communications satellites being developed and built by Hughes Space and Communications Company. These satellites take advantage of the passive nature of the LHP, requiring no external means of pumping, along with its ability to transport large quantities of heat over significant distances. This device comes to the heat transfer community at an ideal time, as the aerospace industry is demanding higher and higher power payloads and this increasing power must be handled by the most efficient means possible. The LHP is also being investigated for uses in ground-based applications such as solar collectors and computer cooling. This dissertation focuses on experimentation conducted with a space-based satellite application in mind; however, results are applicable to other implementations as well. The LHP is a descendant of the conventional heat pipe. The LHP utilizes the advantages of the conventional heat pipe while overcoming some of the conventional heat pipe's inherent disadvantages. This dissertation serves as a complete body of work on this new device; from background and literature review on the development and history of the LHP, to important computer simulation and experimental work, both ground-based and space-based, performed on the LHP in an effort to gain a thorough understanding of the workings of the Loop Heat Pipe and to investigate novel new applications for the LHP such as the ability to control the temperature of an entire spacecraft payload with a minute fraction of the heater power once required. The LHP introduces important new opportunities to the heat transfer community and the research presented here furthers the knowledge and understanding of this breakthrough device.

The application of heat pipes can be as diverse as their structure and shapes. We can claim these unique heat transfer devices are used in many field in industry and they have played very important roles from simple heat exchanger to Electronic, Space Application, Nuclear Reactor, Oil Lines and even for constructing ice pontoons through marshes and the foundations of drilling towers, as well as roads in permafrost regions. Reference 16 has variety flavor of heat pipes applications in industry and the companies and manufactures who are involved with their unique design and application of such devices.

For example, within USA there are applications and development in progress in a drill for ultra deep drilling of a bore in the form of a miniature fast-neutron reactor cooled by means of heat pipes.

Other application of heat pipe can be seen in centrifugal heat pipe shapes that are used for cooling asynchronous motors with short-circuited cast rotors. Such motors are used mainly in mechanical engineering. With the use of centrifugal heat pipes in a rotor, it has become possible to control the motor speed electrically, eliminating the need for complex transmissions and gear trains [23]. Investigations are currently in progress in the USSR to explore the possible use of heat pipes to cool transformers, both air-filled and oil-filled, miniature and high-power, and for cooling of electrical busbars.

The West German firm Brown Boveri Corporation has developed a system of electronic devices with heat pipes.

A. Thyristor systems of power greater than 1 kW; the thermal resistance R of the heat pipe is 0.035K/W and the cooling air velocity is $v = 6$ m/s.
B. A device for a portable current rectifier system (700 W; thermal resistance 0.055∼ cooling air velocity $v = 6$ m/s).

Heat pipes have proved adaptable to the incorporation of electronic equipment, thereby increasing the cooling effect by factors of 10.

Products of the British SRDE laboratory (Signal Research and Development Establishment) include the following: heat pipes in the form of planar electrical insulators, a heat pipe of very small diameter, and various combinations of heat pipes and thermally insulating modules.

Very interesting possibilities have opened up for producing static batteries and thermal energy converters based on heat pipes, thermal diodes, vapor chambers, etc., and materials which vary their aggregate state (fused salts, metals, sulfur with halogens, etc.); operating temperature 500–800∼ the material of the heat pipes is stainless steel, and the heat-transfer agent is sodium. The thermal power stored can be up to 10–100 kW/h. High-temperature heat pipes using alkali metals can be employed successfully as electrodes in plasma generators.

In the energy industry there is a trend to build electric stations using solar energy and hot springs. At present, an electric station of power of at least 100 kW is under construction in the southern USA; it takes the form of a battery of high-temperature heat pipes, heated by the sun, and working into water-vapor generators or thermoelectric converters. Such batteries of heat pipes, linked to heat-storage units, will make it possible to develop electrical energy around the clock. There are plans to use heat pipes as electric cables and distribution lines.

On October 4, 1974, a sounding rocket was launched into space (the Black Brant Sounding Rocket), which carried heat pipes made by the NASA/Goddard Space Flight Center; ESRO; GFW; Hughes Aircraft Company, NASA/AMES;

A. ESRO constructed two aluminum heat pipes of length 885 mm and diameter 5 mm. The wick was a single layer of stainless steel mesh with an artery diameter of 0.5 mm. One pipe was filled with ammonia and the other with acetone. The acetone heat pipe transmitted 8.4 W of power, and the ammonia pipe transmitted 21 W. The heat sink was an aluminum block;

B. GFW (Geselfsehaft fiir Weltraumforschung) of the West German Ministry of Technology constructed a flat aluminum heat pipe in the form of a disk of diameter 150 mm and a titanium heat pipe of length 600 mm, charged with methanol, with its end face joined to the disk by an aluminum tube. The flat heat pipe was filled with acetone, and the other end was joined to a heat-storage device (a canister with a molten substance - “Eicosane” -with a fusion temperature of 35 °C. This system transmitted 26 W of power;

C. The Hughes Aircraft Company constructed two flexible heat pipes made of stainless steel (6.4 mm in diameter; 270 g in length). The working liquid is methanol, and the wick is a metal mesh;

D. NASA/Ames constructed two stainless steel heat pipes of length 910 mm and diameter 12.7 ram. The liquid is methanol, and the inert gas is nitrogen. The wick is a screw thread on the body, and the artery is a wafer of metallic felt. This kind of artery is insensitive to the presence of non-condensable gas;

E. NASA constructed a cryogenic heat pipe made of aluminum with longitudinal channels of length 910 mm and diameter 16 mm, charged with methanol.

Thus, in the international experiment on October 4, 1974, the organizations NASA/GSFC (Grumman and TRW), NASA/Ames (Hughes), Hughes (Hughes), ESRO (the IKE Institute in Stuttgart), and GFW (Dornier) took part in the testing of heat pipes in space. Of these, Grumman constructed five different groups of heat pipes, and TRW constructed three.

In addition to the sounding rockets, NASA has used a number of satellites for testing heat pipes, to evaluate the effect of long-term weightless conditions on heat-pipe parameters (the spacecraft Skylab, OAO-III, ATS-6, CTS, etc.).

The French National Center for Space Research, CNS, independently of the American and European Space Center (AESC), developed and operated a program of space experiments with heat pipes, constructed by the Aerospatiale and SABCA companies. In November 1974, the French sounding rocket ERIDAN 214 was launched, carrying a radiator of heat pipes. The aim of the experiment was to verify the operational capability of heat pipes under weightless conditions; to verify that the heat pipes would be ready to operate at the start of a rocket flight; and to select various heat-pipe structures for spacecraft equipment.

Three types of heat pipe were investigated.

(1) A curved heat pipe made by SABCA, of length 560 mm and diameter 3.2 mm, made of stairs steel, the filter being a stainless steel mesh, with ammonia as the heat-transfer agent. The transmitted power was 4 W. The pipe was flexible.

(2) A heat pipe made by the CENG organization (the atomic center in Grenoble) of length 270 mm and diameter 5 ram, made of copper, with a wick made of sintered bronze powder. The heat-transfer agent was water. The transmitted power was 20 W.

(3) A SABCA heat pipe, similar to No. 1, but straight. The transmitted power was 5 W. The heat sink was a box with a variable-phase fusible substance $T_f = 28.5$ °C (n-octadecane). The energy source was an electric battery with $U = 27$ V. The total weight of the experimental equipment was 2.3 kg.

These investigations point very clearly to positive gains at present, and we can confidently assert that heat pipes will find wide applications in space in the near future. For example, the USA plans to use heat pipes for thermal control and thermal protection of the reusable shuttle and also for the Spacelab space laboratory.

For these, the heat-sensitive equipment will be located in boxes or canisters within which the temperature will be held constant by means of heat pipes located in the walls of the enclosure.

Ref. [23] Provide a vast variety application of heat pipes in present industry and future trend of it.

1.10 Summary

Heat pipes

General

A heat pipe is a passive energy recovery heat exchanger that has the appearance of a common plate-finned water coil except the tubes are not interconnected. Additionally, it is divided into two sections by a sealed partition. Hot air passes through one side (evaporator) and is cooled while cooler air passes through the other side (condenser). While heat pipes are sensible heat transfer exchangers, if the air conditions are such that condensation forms on the fins there can be some latent heat transfer and improved efficiency. Fig. 1.28.

Heat pipes are tubes that have a capillary wick inside running the length of the tube, are evacuated and then filled with a refrigerant as the working fluid and are permanently sealed. The working fluid is selected to meet the desired temperature conditions and is usually a Class I refrigerant. Fins are similar to conventional coils - corrugated plate, plain plate, spiral design. Tube and fin spacing are selected for appropriate pressure drop at design face velocity. HVAC systems typically use copper heat pipes with aluminum fins; other materials are available.

Advantages

- Passive heat exchange with no moving parts,
- Relatively space efficient,
- The cooling or heating equipment size can be reduced in some cases,
- The moisture removal capacity of existing cooling equipment can be improved,
- No cross-contamination between air streams.

Disadvantages

The use of the heat pipe.

- Adds to the first cost and to the fan power to overcome its resistance,
- Requires that the two air streams be adjacent to each other,
- Requires that the air streams must be relatively clean and may require filtration.

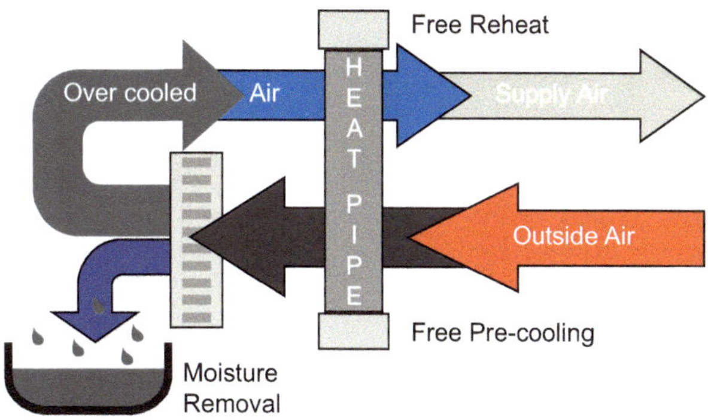

FIG. 1.28 Heat pipe application concept.

Applications

Heat pipe heat exchanger enhancement can improve system latent capacity. For example, a 1 °F dry bulb drop in air entering a cooling coil can increase the latent capacity by about 3%. The heat pipe's transfer of heat directly from the entering air to the low-temperature air leaving the cooling coil saves both cooling and reheating energy. It can also be used to precool or preheat incoming outdoor air with exhaust air from the conditioned spaces.

Best applications

- Where lower relative humidity is an advantage for comfort or process reasons, the use of a heat pipe can help. A heat pipe used between the warm air entering the cooling coil and the cool air leaving the coil transfers sensible heat to the cold exiting air, thereby reducing or even eliminating the reheat needs. Also the heat pipe pre-cools the air before it reaches the cooling coil, increasing the latent capacity and possibly lowering the system cooling energy use.
- Projects that require a large percentage of outdoor air and have the exhaust air duct in closeness, proximity to the intake can increase system efficiency by transferring heat in the exhaust to either pre-cool or preheat the incoming air.

Possible applications

- Use of a dry heat pipe coupled with a heat pump in humid climate areas.
- Heat pipe heat exchanger enhancement used with a single-path or dual-path system in a supermarket application.
- Existing buildings where codes require it or they have "sick building" syndrome and the amount of outdoor air intake must be increased,
- New buildings where the required amount of ventilation air causes excess loads or where the desired equipment does not have sufficient latent capacity.

Applications to avoid

- Where the intake or exhaust air ducts must be rerouted extensively, the benefits are likely not to offset the higher fan energy and first cost.
- Use of heat pipe sprays without careful water treatment. Corrosion, scale and fouling of the heat pipe where a wetted condition can occur needs to be addressed carefully.

Technology types (resource)

Hot air is the heat source, flows over the evaporator side, is cooled, and evaporates the working fluid. Cooler air is the heat sink, flows over the condenser side, is heated, and condenses the working fluid. Vapor pressure difference drives the evaporated vapor to the condenser end and the condensed liquid is wicked back to the evaporator by capillary action. Performance is affected by the orientation from horizontal. Operating the heat pipe on a slope with the hot (evaporator) end below horizontal improves the liquid flow back to the evaporator. Heat pipes can be applied in parallel or series.

Efficiency

Heat pipes are typically applied with air face velocities in the 450–550 feet per minute range, with 4–8 rows deep and 14 fins per inch and have an effectiveness of 45%–65%. For example, if entering air at 77 °F is cooled by the heat pipe evaporator to 70 °F and the air off the cooling coil is reheated from 55 °F to 65 °F by the condenser section, the effectiveness is 45 % [=(65 − 55)/(77 − 55) = 45%]. As the number of rows increases, effectiveness increases but at a declining rate. For example, doubling the rows of a 48% effective heat pipe increases the effectiveness to 65%.

Tilt control can be used to:

- Change operation for seasonal changeover,
- Modulate capacity to prevent overheating or overcooling of supply air,
- Decrease effectiveness to prevent frost formation at low outdoor air temperatures.

Tilt control (6 degrees maximum) involves pivoting the exchanger about its base at the center with a temperature-actuated tilt controller at one end. Face and bypass dampers can also be used.

Manufacturers

Heat Pipes

1. American Heat Pipes, Inc.
   6914 E. Fowler Ave.
   Suite E
   Tampa, FL 33617
   1-800-727-6511
2. Dectron Inc.
   4300 Blvd. Poirier
   Montreal, PQ H4R 2C5
   Canada (514) 334-9609 mail@dectron.com
3. Des Champs Laboratories Inc.
   P.O.Box 220
   Douglas Way
   Natural Bridges Station, VA 24579 (703) 291-1111
4. EcoTech Consultants, Inc.
   3466 Holcombe Bridge Road
   Suite 1000
   Norcross, GA 30092 (404) 723-6564
5. Heat Pipe Technology Inc.
   P.O. Box 999
   Alachua, FL 32615-0999
   1-800-393-2041
6. Munters Dry Cool
   16900 Jordan Rd.
   Selma, TX 78154-1272
   1-800-229-8557 moreinfo-dc@americas.munters.com

7. Nautica Dehumidifiers, Inc.
   9 East Carver St.
   Huntington, NY 11743 (516) 351-8249 dehumidify@aol.com.
8. Octagon Air Systems
   1724 Koppers Road.
   Conley, GA 30288 (404) 609-8881
9. Power-Save International
   P.O. Box 880
   Cottage Grove, OR 97424
   1-800-432-5560
10. Seasons 4 Inc.
    4500 Industrial Access Road
    Douglasville, GA 30134 (770) 489-0716
11. Temprite Industries
    1555 Hawthorne Lane
    West Chicago, IL 60185
    1-800-552-9300
12. Venmar CES
    2525 Wentz Ave.
    Saskatoon, SK S7K 2K9
    Canada
    1-800-667-3717 customerservice@venmarvent.com

References

[1] R.S. Gaugler, Heat Transfer Device, U.S. Patent 2, 350, 348, June 6, 1944.

[2] L. Trefethen, On the Surface Tension Pumping of Liquids or a Possible Role of the Candlewick in Space Exploration, G. E. Tech. Info., Feb. 1962. Ser. No. 615 D114.

[3] T. Wyatt, Wyatt (Johns Hopkins/Applied Physics Lab.), Satellite Temperature Stabilization System. Early Development of Spacecraft Heat Pipes for Temperature Stabilization, U.S. Patent No. 3,152,774, October 13, 1964. Application was files June 11, 1963.

[4] G.M. Grove, T.P. Cotter, G.F. Erikson, Structures of very high thermal conductivity, J. Appl. Phys. 35 (1964) 1990.

[5] J.E. Deverall, J.E. Kemme, High Thermal Conductance Devices Utilizing the Boiling of Lithium and Silver, Los Alamos Scientific Laboratory report LA-3211, October 1964.

[6] G.M. Grover, T.P. Cotter, G.F. Erickson, Structures of very high thermal conductance, J. Appl. Phys. 35 (6) (1964) 1990–1991.

[7] G.M. Grover, J. Bohdansky, C.A. Busse, The Use of a New Heat Removal System in Space Thermionic Power Supplies, European Atomic Energy Community - EURATOM report EUR 2229.e, 1965.

[8] T.P. Cotter, J. Deverall, G.F. Erickson, G.M. Grover, E.S. Keddy, J.E. Kemme, E.W. Salmi, Status report on theory and experiments on heat pipes at Los Alamos, in: Proceedings of the International Conference on Thermionic Power Generation, London, September 1965, 1965.

[9] W.A. Ranken, J.E. Kemme, Survey of Los Alamos and EURATOM Heat Pipe Investigations, in: Proc. IEEE Thermionic Conversion Specialist Conf., San Diego, California, October 1965; Los Alamos Scientific Laboratory, report LA-DC-7555, 1965.

[10] J.E. Kernme, Heat pipe capability experiments, in: Proc. of Joint AEC Sandia Laboratories Report SC-M-66-623, 1, October 1966. Expanded Version of This Paper, Los Alamos Scientific Laboratory Report LA-3585-MS (August 1966), Also as LA-DC-7938. Revised Version of LA-3583-MS, Proc. EEE Thermionic Conversion Specialist Conference, Houston, Texas, November 1966.

[11] P.D. Dunn, D.A. Reay, Heat Pipes, third ed., Pergamon, New York, 1982.

[12] S.W. Chi, Heat Pipe Theory and Practice, McGraw-Hill, New York, 1976.

[13] G.A. Bennett, Conceptual Design of a Heat Pipe Radiator, LA-6939-MS Technical Report, Los Alamos Scientific Lab, N. Mex., USA, September 1, 1977.

[14] Y.F. Gerasimov, Y.F. Maidanik, G.T. Schegolev, Low-temperature heat pipes with separated channels for vapor and liquid, Eng. Phys. J. 28 (6) (1975) 957–960 (in Russian).

[15] K. Watanabe, A. Kimura, K. Kawabata, T. Yanagida, M. Yamauchi, Development of a Variable-Conductance Heat-Pipe for a Sodium-Sulfur (NAS) Battery, Furukawa Review, No. 20, 2001.

[16] B.D. Marcus, Theory and Design of Variable Conductance Heat Pipes: Control Techniques, Research Report No. 2, July 1971. NASA 13111-6027-R0-00.

[17] J.E. Kemme, Heat Pipe Design Considerations, Los Alamos Scientific Laboratory report LA-4221-MS, August 1, 1969.

[18] K.A. Woloshun, M.A. Merrigan, D. Elaine, Best. HTPIPE: a steady-state heat pipe analysis program, a user's manual, 1988.

[19] G.P. Peterson, An Introduction to Heat Pipes: Modeling, Testing, and Applications, John Wiley & Sons, Inc., 1994, pp. 175–210.

[20] D. Scott, PE. Garner, Thermacore Inc.

[21] A. Faghri, US Patent 5269369 - Temperature Regulation System for the Human Body Using Heat Pipes.

[22] P.J. Brennan, E.J. Kroliczek, Heat Pipe Design Handbook, B & K Engineering, Inc., Towson, Maryland, 1979.

[23] MIL-STD-1522A (USAF), Military Standard General Requirements for Safe Design and Operation of Pressurized Missile and Space Systems, May 1984.

Chapter 2

Application of heat pipe in industry

Chapter outline

2.1 Introduction 48
2.2 Overview industrial application of heat pipes 51
2.2.1 Cooling of electronic components 53
2.2.2 Spacecraft 53
2.2.3 Energy conservation 54
2.2.4 Heat pipe driven heat exchanger (HPHX) 54
2.2.5 Preservation of permafrost 55
2.2.6 Snow melting and deicing 55
2.2.7 Heat pipe inserts for thermometer calibration 56
2.2.8 High-temperature heat pipe furnace 56
2.2.9 Miscellaneous heat pipe applications 57
2.3 Energy-dependent boundary equations 58
2.4 Heat pipe in space 61
2.4.1 Radioisotope systems 62
2.4.1.1 Ulysses 62
2.4.1.2 Galileo 63
2.4.1.3 Cassini-huygens 64
2.4.1.4 New Horizons 64
2.4.2 Fission systems: heat 68
2.4.3 Fission systems: propulsion 69
2.4.4 Nuclear thermionic technology development 69
2.4.4.1 Conductively coupled, multicell thermionic fuel element 71
2.4.4.2 Cylindrical inverted multicell 72
2.4.5 Potential space nuclear thermionic missions 74
2.4.6 Heat pipe power system 77
2.4.7 Space reactor power systems 78
2.4.7.1 Heat pipe operated mars exploration reactor (HOMER) 78
2.4.7.2 Heat pipe reactor HOMER-15 and Homer-25 designs 80
2.4.7.3 Heat pipe and fuel pins configuration 82
2.4.8 Stirling engine system 85
2.4.9 Heat pipe design 86
2.4.10 Nuclear reactor power system 87
2.4.11 Material choices 92
2.4.12 Safety considerations 92
2.4.13 Reactor control 92
2.4.14 Neutron shielding 93

Functionality, Advancements and Industrial Applications of Heat Pipes.
https://doi.org/10.1016/B978-0-12-819819-3.00002-X

2.4.15 Reactor sitting 94
2.4.16 Nuclear energy propulsion of aircraft (NEPA) 95
2.4.17 Project prometheus 2003 97
2.4.18 Mars one mission 99
2.4.19 Kilopower reactor using stirling technology (KRUSTY) experiment 101

2.5 Space shuttle orbiter heat pipe applications 104

2.6 Heat pipe in electronics 107

2.6.1 Electronic and electrical equipment cooling 109

2.7 Heat pipe in defense and avionics 113

2.7.1 On the ground application 114
2.7.2 In the sea application 115
2.7.3 In the air application 116
2.7.4 In the space application 116

2.8 Heat pipe as heat exchanger 118

2.9 Heat pipe in residential building 121

2.10 Heat pipe applications in thermal energy storage systems 123

2.10.1 Energy storage methods 124
2.10.1.1 Electrical storage 124
2.10.1.2 Thermal energy storage 125
2.10.2 Latent heat thermal energy storage 126
2.10.2.1 Thermochemical 127
2.10.3 Latent heat thermal storage materials 127
2.10.3.1 Thermal properties 127
2.10.3.2 Physical properties 127
2.10.3.3 Chemical properties 128
2.10.4 Phase Change Material (PCM) classification 128
2.10.4.1 PCMs for different thermal storage applications 129
2.10.5 Latent heat thermal energy storage systems assisted by heat pipes 134

2.11 Passive thermal technical discipline lead (TDL) for luna lander 153

2.12 Heat pipe driving home energy system 159

2.13 Heat pipe driving heat exchangers and heat pumps 165

2.14 Gas turbine engines and the automotive industry 166

2.15 Heat pipes driving production tools 169

2.16 Medicine and human body temperature control via heat pipe 169

2.17 Heat pipes driving ovens and furnaces 171

2.18 Heat pipes driving Permafrost Stabilization 172

2.19 Heat pipes driving transportation systems and deicing 173

References 174

2.1 Introduction

Grover and his colleagues [1,2] were working on cooling systems for nuclear power cells for spacecraft, where extreme thermal conditions are found. Heat pipes have since been used extensively in spacecraft as a means for managing internal temperature conditions.

Heat pipes have been applied in many ways since their introduction in 1964 (Vasiliev; Mochizuki et al.) [3,4]. Depending on their intended use, heat pipes can operate over a temperature range from 4.0 to 3000 K. In all cases, their applications can be divided into three main categories: separation of heat source and sink, temperature equalization, and temperature control. Due to their extremely high thermal conductivity, heat pipes can efficiently transport heat from a concentrated source to a remotely mounted sink. This property can enable dense packing of electronics, for example, without undue regard for heat sink space requirements.

Another benefit of the high thermal conductivity is the ability to provide an accurate method of temperature equalization. For example, a heat pipe mounted between two opposing faces of an orbiting platform will enable both faces to maintain constant with equal temperatures, thus minimizing thermal stresses. The temperature control is a result of the capability of heat pipes to transport large quantities of heat very rapidly. This feature enables a source of varying flux to be kept at a constant temperature as long as the heat flux extremes are within the operating range of the heat pipe.

Heat pipes are extensively used in many modern computer systems, where increased power requirements and subsequent increases in heat emission have resulted in greater demands on cooling system. Heat pipes are typically used to move heat away from components such as Central Processing Units (CPUs)xe "Central Processing Units (CPUs)" and Graphics Processing Units (GPUs)xe "Graphics Processing Units (GPUs)" to heat sinks where thermal energy may be dissipated into the environment.

In solar thermal water heating applications, an evacuated tube collector can deliver up to 40% more efficiency compared to more traditional "flat-plate" solar water heaters. Evacuated tube collectors eliminate the need for antifreeze additives to be added as the vacuum helps prevent heat loss. These types of solar thermal water heaters are frost protected down to more than −3°C and are being used in Antarctica to heat water. Also due to the enormous increase in the global energy demand and possible depletion of conventional energy resources such as fossil fuel, renewable energy resources turned out to be promising options to supply clean and low-cost energy, and heat pipe shows very appropriate and promising apparatus as part of heating and cooling system of such demand.

Heat pipes are used to dissipate heat on the Trans-Alaska Pipeline System. Without them residual ground heat remaining in the oil as well as that produced by friction and turbulence in the moving oil would conduct down the pipe's support legs. This would likely melt the permafrost on which the supports are anchored. This would cause the pipeline to sink and possibly sustain damage. To prevent this each vertical support member has been mounted with four vertical heat pipes (see Fig. 2.1).

Heat pipes are also being widely used in solar thermal water heating applications in combination with evacuated tube solar collector arrays. In these applications, distilled water is commonly used as the heat transfer fluid inside

FIG. 2.1 Alaska pipeline support legs cooled by heat pipes to keep permafrost frozen.

a sealed length of copper tubing that is located within an evacuated glass tube and oriented toward the sun.

A company such as TrueLeaf offers fined heat pipe assembly for the greenhouse application using DuoFin shape structure for the condenser side of heat pipe in order to provide incredibly even heating across the width of your greenhouse growing areas. Its unique tapered fin design delivers a combination of convective and radiant heating quietly and efficiently. Having only two fins means that more loops are needed to meet your heat load than with sister product, StarFin. But it also means you never have to clean debris from the fins. Also, the large inside diameter means less pumping energy is required to provide greater evenness than any competitive system (See Fig. 2.2).

Below is the illustration of DuoFin assemblies and connectivities produced by TrueLeaf Corporation (See Figs. 2.3 and 2.4).

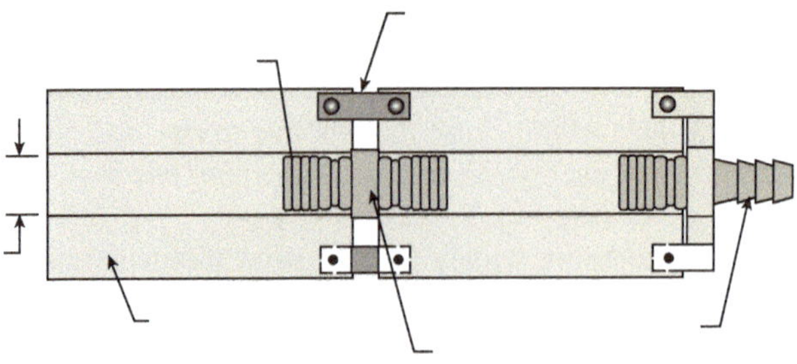

FIG. 2.2 A green house application of finned heat pipe structure.

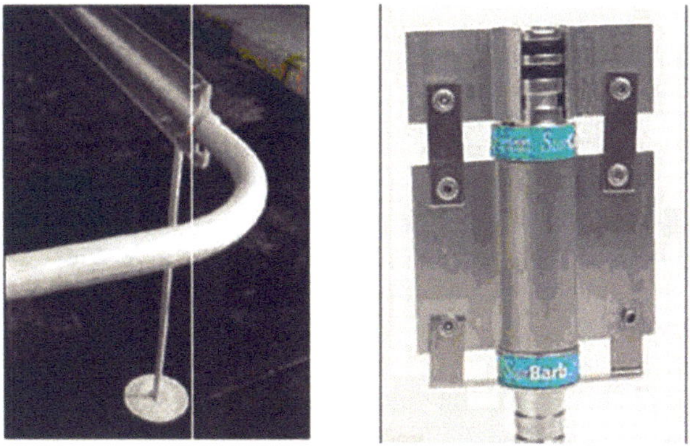

FIG. 2.3 Finned heat pipe.

FIG. 2.4 General use of finned heat pipe.

Below is the illustration of DuoFin assemblies and connectivity's produced by TrueLeaf Corporation.

2.2 Overview industrial application of heat pipes

In general, the applications come within a number of broad groups, each of which describes a property of the heat pipe. These groups are [5]:

- Separation of heat source and sink
- Temperature flattening

- Heat flux transformation
- Temperature control
- Thermal diodes and switches

The high effective thermal conductivity of a heat pipe enables heat to be transferred at high efficiency over considerable distances. In many applications where component cooling is required, it may be inconvenient or undesirable thermally to dissipate the heat via a heat sink or radiator located immediately adjacent to the component. For example, heat dissipation from a high-power device within a module containing other temperature-sensitive components would be affected by using the heat pipe to connect the component to a remote heat sink located outside the module. Thermal insulation could minimize heat losses from intermediate sections of the heat pipe [5].

The second property listed above, temperature flattening, is closely related to source–sink separation. As a heat pipe, by its nature, tends toward operation at a uniform temperature, it may be used to reduce thermal gradients between unevenly heated areas of the body. The body may be the outer skin of a satellite, part of which is facing the sun, the cooler section being in shadow. Alternatively, an array of electronic components mounted on a single pipe would tend to be subjected to feedback from the heat pipe, creating temperature equalization (Fig. 2.5).

The third property listed above, heat flux transformation, has attractions in reactor technology. In some manufacturer such as Thermionics Corporation,

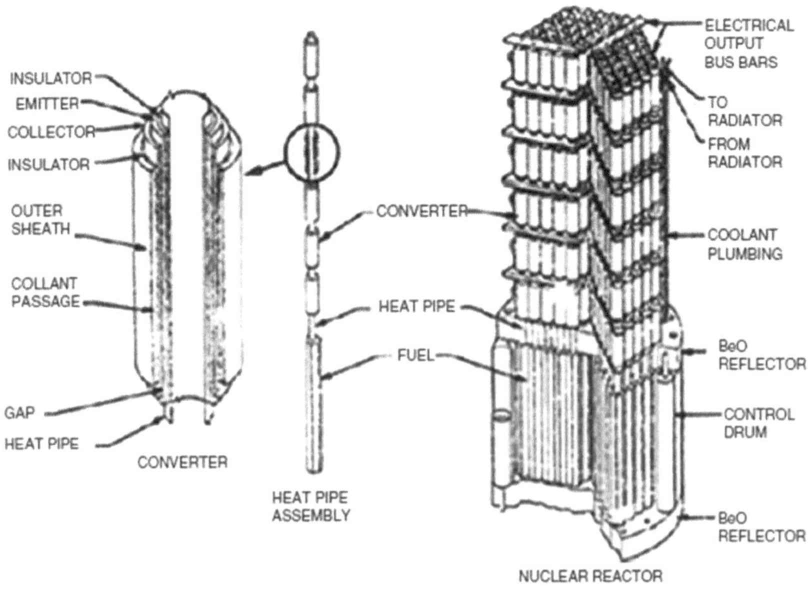

FIG. 2.5 Out of core of thermionic reactor concept [6].

for example, the transformation of a comparatively low heat flux, as generated by radioactive isotopes, into sufficiently high heat fluxes capable of being utilized effectively in thermionic generators has been attempted.

The fourth area of application, temperature control, is best carried out using the variable conductance heat pipe. This can be used to control accurately the temperature of devices mounted on the heat pipe evaporator section, while the variable conductance heat pipe found its first major applications in many more mundane applications, ranging from temperature control in electronic equipment to ovens and furnaces [5].

As with any other device, the heat pipe must fulfill a number of criteria before it becomes fully acceptable in applications in industry. For example, in the die-casting and injection molding, the heat pipe has to:

- Be reliable and safe
- Satisfy a required performance
- Be cost-effective
- Be easy to install and remove

Obviously, each application must be studied in its own right, and the criteria vary considerably. A feature of the molding processes, for example, is the presence of high-frequency accelerations and decelerations. In these processes, therefore, heat pipes should be capable of operating when subjected to this motion, and this necessitates development work in close association with the potential users [5].

2.2.1 Cooling of electronic components

At present the largest application of heat pipes in terms of quantity used is the cooling of electronic components such as transistors, other semiconductor devices, and integrated circuit packages [1].

There are two possible ways of using heat pipes:

1. Mount the component directly onto the heat pipe, and
2. Mount the component onto a plate into which heat pipes are inserted.

2.2.2 Spacecraft

Heat pipes, certainly at vapor temperatures up to 200°C, have probably gained more from developments associated with spacecraft applications than from any other area. The variable conductance heat pipe is a prime example of this "technological fallout." In the literature can be found details about the following types of application:

- Spacecraft temperature equalization
- Component cooling, temperature control, and radiator design

- Space nuclear power sources:
  - Moderator cooling.
  - Removal of the heat from the reactor at emitter temperature. (Each fuel rod would consist of a heat pipe with externally attached fuel).
  - Elimination of troublesome thermal gradients along the emitter and collector.

2.2.3 Energy conservation

The heat pipe, because of its effectiveness in heat transfer, is a prime candidate for applications involving the conservation of energy and has been used to advantage in heat recovery systems and energy conversion devices.

Energy conservation is becoming increasingly important as the cost of fuel rises and the reserves diminish, and the heat pipe is proving a particularly effective tool in a large number of applications associated with conservation.

2.2.4 Heat pipe driven heat exchanger (HPHX)

There are a large number of techniques for recovering heat from exhaust air or gas streams or from hot water streams. Details and explanations about heat pipe heat exchangers can be found in this material.

Heat Pipes (HPs)xe “Heat Pipes (HPs)” and Thermosyphons (TSs)xe “Thermosyphons (TSs)” are widely recognized as being excellent passive thermal transport devices that can have effective thermal conductivities orders of magnitude higher than similarly-dimensioned solid materials. The integration of heat pipes into heat exchangers (HXs) and heat sinks (HPHXs and HPHSs, respectively) have been shown to have strong potential for energy savings, especially in response to the significant reduction in the manufacturing costs of heat pipes in recent years. This review documents HPHXs applications, general design procedures, and analysis tools based on the thermal network approach. The thermal network approach is a robust engineering tool that is easy to implement and program, is user friendly, straightforward, computationally efficient, and serves as a baseline methodology to produce results of reasonable accuracy.

In summary, Heat pipes (HPs)xe “Heat pipes (HPs)” and Thermosyphons (TSs)xe “Thermosyphons (TSs)” are passive devices which operate by utilizing the latent heat of an internal working fluid to transfer large amounts of heat, nearly isothermally, with a minimal driving temperature difference through a small cross sectional area and is divided into three segments: evaporator, adiabatic and condenser sections denoted based on their external thermal boundary conditions. A HP/TS functions when heat is applied to the evaporator section, which causes vaporization of the working fluid. The vapor flows through the adiabatic section to the lower temperature condenser section, within which condensation of the HP/TS working fluid occurs.

Features of heat pipe heat exchangers that are attractive in industrial heat recovery applications are:

- No moving parts and no external power requirements, implying high reliability.
- Cross-contamination is totally eliminated because of a solid wall between the hot and cold fluid streams.
- Easy to clean.
- A wide variety of sizes are available, and the unit is in general compact and
- suitable for all.
- The heat pipe heat exchanger is fully reversible—i.e., heat can be transferred in
- either direction.
- Collection of condensates in the exhaust gases can be arranged, and the flexibility
- accruing to the use of a number of different fins spacing can permit easy cleaning if required.

The application of heat pipe heat exchangers falls into three main categories:

1. Recovery of waste heat from processes for reuse in the same process or in another, e.g., preheating of combustion air. This area of application is the most diverse and can involve a wide range of temperatures and duties.
2. Recovery of waste heat from a process to preheat air for space heating.
3. Heat recovery in air-conditioning systems, normally involving comparatively low temperatures and duties.

More details on the subject of Heat Pipe Heat Exchangers (HPHXs) can be found in Chapter Five of this book including their types and classifications and how each one operates along with their applications in industry.

2.2.5 Preservation of permafrost

One of the largest contracts for heat pipes was placed with McDonnell Douglas Corporation by Alaska Pipeline Service Company for nearly 100,000 heat pipes for the Trans-Alaska pipeline.

The function of these units is to prevent thawing of the permafrost around the pipe supports for elevated sections of the pipeline. Diameters of the heat pipes used are 5 and 7.5 cm, and lengths vary between 8 and 18 m. The system developed by McDonnell Douglas uses ammonia as the working fluid, heat from the ground being transmitted upward to a radiator located above ground level.

2.2.6 Snow melting and deicing

An area of application, and one which works in Japan, has been particularly intense. This has been the use of heat pipes to melt snow and prevent icing.

The operating principle of the heat pipe snow melting (or deicing) system is based upon the use of heat stored in the ground as the heat input to the evaporators of the heat pipes.

2.2.7 Heat pipe inserts for thermometer calibration

Heat pipe inserts have been developed at IKE, Stuttgart, for a variety of duties, including thermocouple calibration. The heat pipes are normally operated inside a conventional tubular furnace. The built-in enclosures provide isothermal conditions, a necessary prerequisite for temperature sensor calibration. The isothermal working spaces can also be used for temperature-sensitive processes, such as fixed-point cell heating, crystal growing, and annealing.

2.2.8 High-temperature heat pipe furnace

Under contract from the European Space Agency, IKE developed a high-temperature heat pipe surface, for material processing in a microgravity environment in the temperature range of 900–1500°C.

Recently, also need for high-temperature thermal management is on rise, and for efficient high-temperature (heat source of 300–2000°C) heat transfer and dissipation, thermal spreading, high-heat flux cooling, and other high-temperature applications such as Advanced High-Temperature Reactor (AHTR)xe "Advanced High-Temperature Reactor (AHTR)" of Generation Four (GEN-IV), high-temperature heat pipes are in demand for the thermal solution of choice as part of a fully inherent system for heat transfer and safety factors.

This demand has been established with the background of heat pipe applications around the 1960s time frame in nuclear reactor thermal management such as Liquid Metal Fast Breeder Reactor (LMFBR)xe "Liquid Metal Fast Breeder Reactor (LMFBR)" research studies in the USA with project such as Clinch River project by Westinghouse Electric Corporation at their Advanced Reactor Division by this author around 1970s and finally full production of this reactor in France under Phoenix-II plant.

Today, companies such as Thermacore® with their high-temperature heat pipe technology are offering the use and application of such heat pipes from the ocean floor to lunar surfaces, and they are satisfying the demand requirements with these applications. Aerospace and chemical processing such as annealing, furnace isothermally status, semiconductor material crystal growth, oil shale extraction, and wide range of high-tech electronics are also in need of such high-temperature heat pipes, where the heat dissipation and heat uniformity applications are a must circumstance.

Some of the advantages that are offered by this family of heat pipes are listed below:

- High-power heat transfer capabilities (>25 kW).
- High-heat flux cooling capability (>100 kW = cm^2).
- High-precision temperature control and rapid temperature recovery.
- Isothermality (i.e., equal or constant temperatures) at high temperature.
- Uniform material crystal growth.
- Thermal-to-electric energy conversion energy savings.

Today's cutting-edge technologies need high-temperature heat pipes to deliver the performance their applications require.

2.2.9 Miscellaneous heat pipe applications

To assist the reader in lateral thinking, a number of other applications of heat pipes are listed below:

- Heat pipe roll-bond panels for warming bathroom floors (Japan)
- Heat pipe-cooled dipstick for cooling motorbike engine oil (Japan)
- Passive cooling of remote weather station equipment (Canada)
- Cooling of drills (Russia)
- Thermal control of thermoelectric generators (USA)
- Cooling of gas turbine blades (Czech Republic)
- Thermal control of electric storage heaters (Byelorussia, UK)
- Cooling of semiautomatic welding equipment (Russia)
- Deicing fish farms and ornamental ponds (Romania)
- Heating heavy oil in large tanks (Romania)
- Cooling of soldering iron bit (UK)
- Cooling of bearings for emergency feedwater pumps (UK)
- Cooling of targets in particle accelerators (UK)
- Isothermalization of bioreactors (China)
- Cooling of snubber pins in the synthetic fiber industry (UK)
- Thermal control in electric battery dehumidifiers (USA)
- Car passenger compartment heating
- Domestic warm air heaters (USA)

As far as other applications of heat pipe are concerned, enormous field and usage can be named which are beyond the scope of this book, and readers should do their own investigations and research for further information; however, it is also worth to mention that for efficient data center cooling as part of inherent heat transfer system, computer server microprocessors, CPUs, and other concentrated heat loads can be coupled directly to cooling water (chiller) circuits while keeping water outside the cabinet, and Thermacore's Therma-Bus® technology is offering such solution. Therma-Bus gives you the benefits of water as a cooling medium—with much better heat transfer properties than air—without concerns about introducing water to an electronic environment. Improved thermal efficiency (reduced ΔT) also cools major data center loads without active refrigeration.

2.3 Energy-dependent boundary equations

Heat pipe-cooled reactors offer operation redundancy and simplicity in the reactor startup from a frozen state. The redundancy is based on the facts that the reactor could continue to operate with one or more failed heat pipes and has no single-point failure in the removal of its thermal power.

Heat pipe has a long history of more than 40 years up to now and will have more and more application due to demand by market and technology that is available to researcher and designer of this unique thermal device.

A heat pipe-cooled nuclear reactor has been designed to provide 3.2 MW of power to an out-of-core thermionic conversion system. The reactor is a fast reactor to design to operate at a nominal heat pipe temperature of 1675 °K. Each reactor fuel element consists of a hexagonal molybdenum block which is bonded along its axis to one end of molybdenum, lithium vapor, heat pipe. The block is perforated with an array of longitudinal holes which are loaded with UO_2 pellets. The heat pipe transfers heat directly to a string of six thermionic converters which are bonded along the other end of the heat pipe. An assembly of 90 such fuel elements forms a hexagonal core. The core is surrounded by a thermal radiation shield, a mal neutron absorber, and a BeO reflector containing boron-loaded control [6]. This study describes the conceptual design of a space nuclear reactor which produces 3.2 MW of power to a 500-kWe out-of-core thermionic conversion system described elsewhere. The reactor is a fast reactor and heat pipe cooled such that each fuel element of the core is directly coupled via a heat pipe to a sting of thermionic converters as shown in Fig. 4.6.

A Heat Pipe-Operated Mars Exploration Reactor (HOMER)xe "Heat Pipe-Operated Mars Exploration Reactor (HOMER)" providing between 50 and 250 kWe has been proposed for life support, operations, in-situ propellant production, scientific experiments, high-intensity lamps for plant growth, and other activities on Mars mission. It is crucial, since a solar array providing the same power on Mars would require a surface area of several football fields. In addition, environmental side effects such as day and night seasonal variations, geographical sunlight from the sun, and dust storm and other solar phenomena would not affect a fission reactor system. Fig. 2.6 shows the core design of such a nuclear power structure which is producing 125 kW of power. The rotating drums around the circumference achieve power-level control. These consist of a neutron absorbing side and a neutron scattering and reflecting side, allowing power control without the need for terrestrial-used control rods. Moving parts are also eliminated by the use of heat pipes transforming heat for rejection by radiation to space without the uses of any mechanical moving parts such as pumps [7].

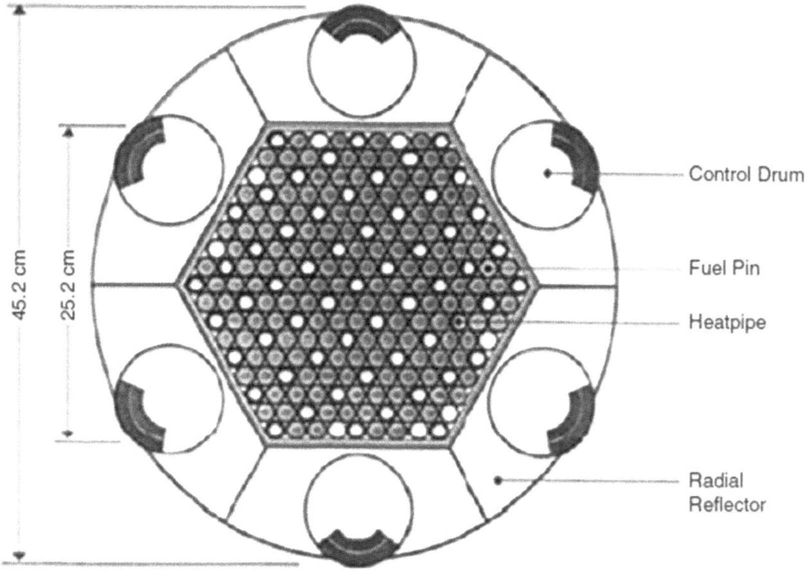

FIG. 2.6 Cross section of heat pipe space reactor of 125 KWth power, showing the peripheral control drums [7].

As we stated, the Heat pipe Power System (HPS)xe "Heat pipe Power System (HPS)" is a near-term, low-cost, space fission power system that has been under development at Los Alamos National Laboratory (LANL)xe "Los Alamos National Laboratory (LANL)" for 5 years. The goal of the HPS project is to devise an attractive space fission system that can be developed quickly and affordably. The primary ways of doing this are by using existing technology and by designing the system for inexpensive testing. If the system can be designed to allow highly prototypic testing with electrical heating, then an exhaustive test program can be carried out inexpensively and quickly and thorough testing of the actual flight unit can be performed—which is a major benefit to reliability. Over the past 4 years, LANL has conducted three HPS proof-of-concept technology demonstrations—each has been highly successful. The Heat Pipe-Operated Mars Exploration Reactor (HOMER) is a derivative of the HPS designed especially for producing electricity on the surface of Mars. The key attributes of the HOMER are described in this paper, as well as a 20-kWe point design, system scalability, and the current technology status.

Major power conversion technique in space where the source is a nuclear heat source is depicted in Fig. 2.7 below.

Fig. 2.8 also illustrates chronology of space nuclear power development.

The HOMER is a robust, low technical risk fission system that can provide electric power on the surface of Mars.

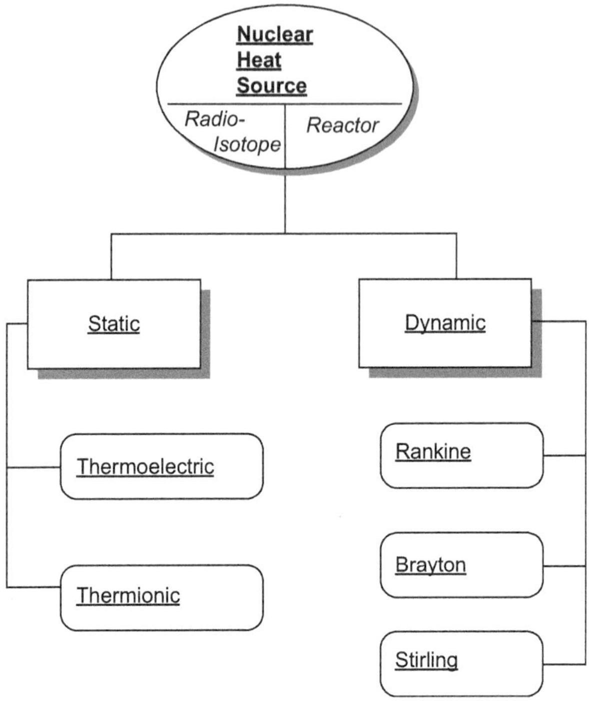

FIG. 2.7 Major power conversion techniques in space.

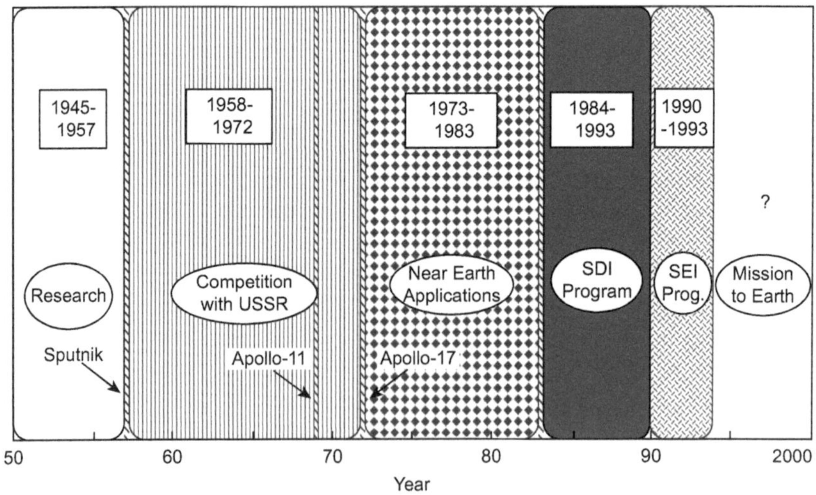

FIG. 2.8 Chronology of space nuclear power development.

This system is based on the Heat-Pipe Power System (HPS)xe "Heat-Pipe Power System (HPS)" concept that has been under development for several years at LANL. The HOMER is designed to be a near-term, low-cost system by using existing technology, designing the system for prototypic electrically heated testing, and keeping the system simple. Hardware testing currently is being conducted that, if successful, will help verify that the HOMER is a viable and attractive option for Mars surface power [12].

2.4 Heat pipe in space

"In 1996, three Los Alamos heat pipes, prototypes of liquid-metal heat pipes to be used in advanced spacecraft, were flown and tested aboard the space shuttle Endeavor. They operated at temperatures exceeding 900°F and performed flawlessly."

"Heat pipes work well in a zero-gravity environment." Commercially developed heat pipes operating near room temperature are now routinely used on geostationary communications satellites [8].

High-capacity heat pipe radiator panels have been proposed as the primary means of heat rejection for Space Station Freedom. In this system, the heat pipe would interface with the thermal bus condensers. Changes in system heat load can produce large temperature and heat load variations in individual heat pipes. Heat pipes could be required to start from an initially cold state, with heat loads temporarily exceeding their low-temperature transport capacity. The present research was motivated by the need for accurate prediction of such transient operating conditions. In this work, the cold startup of a 6.7-m long high-capacity heat pipe is investigated experimentally and analytically. A transient thermohydraulic model of the heat pipe was developed which allows simulation of partially primed operation. The results of cold startup tests using both constant temperature and constant heat flux evaporator boundary conditions are shown to be in good agreement with predicted transient response [9].

After a gap of several years, there is a revival of interest in the use of nuclear fission power for space missions [10].

While Russia has used over 30 fission reactors in space, the USA has flown only one—the 10A System for Nuclear Auxiliary Power(SNAP)xe "System for Nuclear Auxiliary Power(SNAP)" in 1965.

Early on, from 1959 to 1973, there was a US nuclear rocket program—Nuclear Engine for Rocket Vehicle Application (NERVA)xe "Nuclear Engine for Rocket Vehicle Application (NERVA)"—which was focused on nuclear power replacing chemical rockets for the latter stages of launches. NERVA used graphite-core reactors heating hydrogen and expelling it through a nozzle. Some 20 engines were tested in Nevada and yielded thrust up to more than half that of the space shuttle launchers. Since then, "nuclear rockets" have been about space propulsion, not launches. The successor to NERVA is today's Nuclear Thermal Rocket (NTR)xe "Nuclear Thermal Rocket (NTR)" [11].

Another early idea was the US project Orionxe "US project Orion", which would launch a substantial spacecraft from the Earth using a series of small nuclear explosions to propel it. The project commenced in 1958 and was aborted in 1963 when the atmospheric test ban treaty made it illegal, but radioactive fallout could have been a major problem. The Orion idea is still alive as other means of generating the propulsive pulses are considered.

2.4.1 Radioisotope systems

So far, Radioisotope Thermoelectric Generators (RTGs)xe "Radioisotope Thermoelectric Generators (RTGs)" have been the main power source for US space work over more than 40 years, since 1961. The high decay heat of plutonium-238 (0.56 W/g) enables its use as an electricity source in the RTGs of spacecraft, satellites, navigation beacons, etc. Heat from the oxide fuel is converted to electricity through static thermoelectric elements (solid-state thermocouples), with no moving parts. RTGs are safe, reliable, and maintenance-free and can provide heat or electricity for decades under very harsh conditions, particularly where solar power is not feasible.

So far 45 RTGs have powered 25 US space vehicles including Apollo, Pioneer, Viking, Voyager, Galileo, Ulysses, and New Horizons space missions as well as many civil and military satellites. The Cassini spacecraft carries three RTGs providing 870 W of power enroot to Saturn. Voyager spacecraft which has sent back pictures of distant planets has already operated for over 20 years and is expected to send back signals powered by their RTGs for another 15–25 years. The Viking and Rover landers on Mars depended on RTG power sources, as will the Mars rovers have launched in 2009.

The latest RTG is a 290-W system known as the General Purpose Heat Source – Radioisotope Thermoelectric Generator (GPHS-RTG)xe "General Purpose Heat Source – Radioisotope Thermoelectric Generator (GPHS-RTG)" and it is depicted in Fig. 2.9.

The thermal power source for this system is the General-Purpose Heat Source (GPHS)xe "General-Purpose Heat Source (GPHS)". Each GPHS contains four iridium-clad Pu-238 fuel pellets, stands 5 cm tall and 10 cm square, and weighs 1.44 kg. Eighteen GPHS units power one GPHS RTG. The Multi-Mission RTG (MMRTG)xe "Multi-Mission RTG (MMRTG)" will use eight GPHS units producing 2 kW which can be used to generate 100 W of electricity and is a focus of current research.

GPHS-RTG or General Purpose Heat Source — Radioisotope Thermoelectric Generator, is a specific design of the Radioisotope Thermoelectric Generator (RTG) used on US space missions. The GPHS-RTG was used on [34]:

2.4.1.1 Ulysses

Ulysses is a decommissioned robotic space probe whose primary mission was to orbit the Sun and study it at all latitudes. It was launched in 1990, made

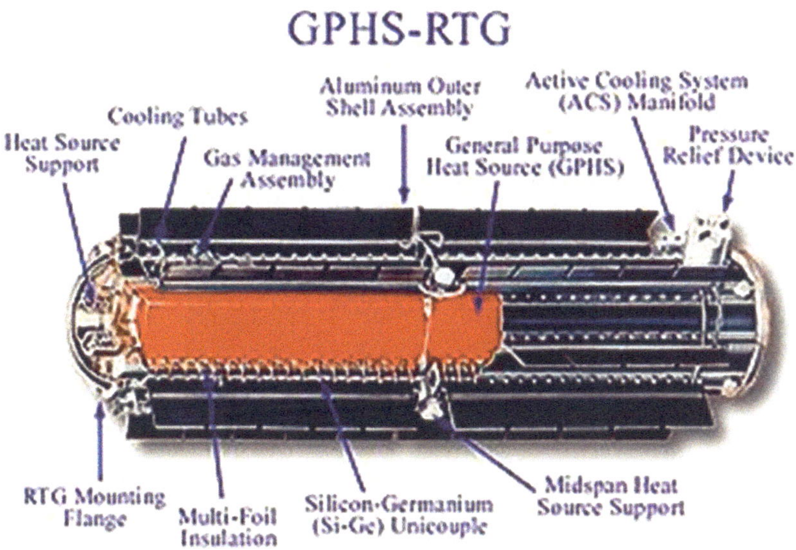

FIG. 2.9 Diagram of an RTG used on the cassini probe.

three "fast latitude scans" of the Sun in 1994/1995, 2000/2001, and 2007/2008. In addition, the probe studied several comets. Ulysses was a joint venture of NASA and the European Space Agency (ESA) with participation from Canada's National Research Council [35]. The last day for mission operations on Ulysses was June 30, 2009 [36,37].

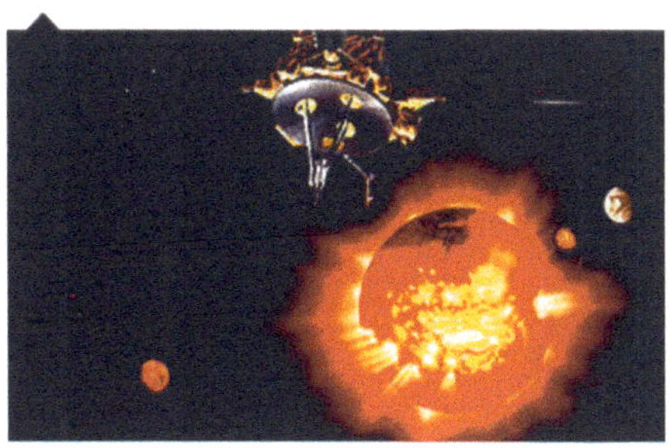

2.4.1.2 Galileo

The Galileo Probe was an atmospheric-entry probe carried by the main Galileo spacecraft to Jupiter, where it directly entered a hot spot and returned data from the planet [38]. The 339-kg (747 lb.) probe was built by Hughes Aircraft Company [39] at its El Segundo, California plant, measured about 1.3 m (4.3 ft) across. Inside the probe's heat shield, the scientific instruments were protected from extreme heat and pressure during its high-speed journey into the Jovian atmosphere, entering at 47.8 km (29.7 mi) per second. It entered Jupiter on December 7, 1995, 22:04 UTC and stopped functioning 23:01 UTC, 57 min and 36 s later.

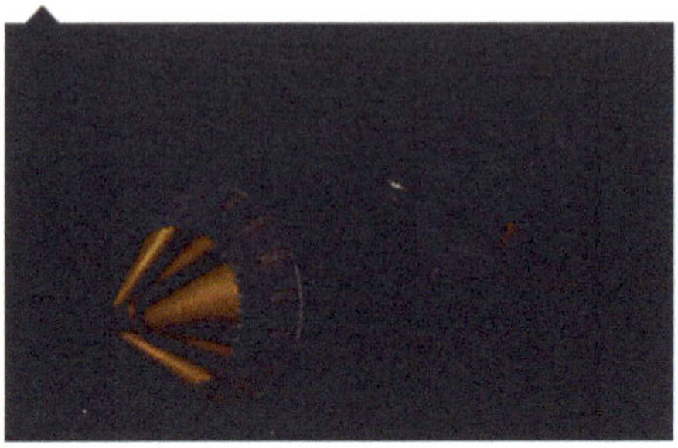

2.4.1.3 Cassini-huygens

The Cassini—Huygens mission, commonly called Cassini, was a collaboration between NASA, the European Space Agency (ESA)xe "European Space Agency (ESA)", and the Italian Space Agency (ASI)xe "Italian Space Agency (ASI)" to send a probe to study the planet Saturn and its system, including its rings and natural satellites. The Flagship-class robotic spacecraft comprised both NASA's Cassini probe, and ESA's Huygens lander which landed on Saturn's largest moon, Titan [40]. Cassini was the fourth space probe to visit Saturn and the first to enter its orbit. The craft were named after astronomers Giovanni Cassini and Christiaan Huygens.

2.4.1.4 New Horizons

New Horizons is an interplanetary space probe that was launched as a part of NASA's New Frontiers program [41]. Engineered by the Johns Hopkins University Applied Physics Laboratory (APL)xe "University Applied Physics Laboratory (APL)" and the Southwest Research Institute (SwRI), with a team led by S. Alan Stern, the spacecraft was launched in 2006 with the primary mission to perform a flyby study of the Pluto system in 2015, and a secondary mission to fly by and study one or more other Kuiper Belt Objects (KBOs)xe "Kuiper Belt Objects (KBOs)" in the decade to follow. It is the fifth space probe to achieve the escape velocity needed to leave the Solar System.

The GPHS-RTG has an overall diameter of 0.422 m and a length of 1.14 m [42]. Each GPHS-RTG has a mass of about 57 kg and generates about 300 Watts of electrical power at the start of mission (5.2 We/kg), using about 7.8 kg of Pu-238 which produces about 4,400 Watts of thermal power [42]. The plutonium oxide fuel is in 18 GPHSs. Note that the GPHS are cuboid although they contain cylindrical plutonium based pellets.

The GPHS-RTG units used on spacecraft were not created by NASA. They were designed and built by General Electric Space Division (later part of Martin-Marietta, subsequently part of Lockheed Martin), in King of Prussia, Pennsylvania. The generators were filled with plutonium by Department of Energy laboratories in Miamisburg, Ohio and Idaho Falls, Idaho.

After the Ulysses and Galileo RTGs were fueled, their launches were postponed by four and three years respectively. As a result, the missions were slightly adapted to utilize the lower power that would be available [43]. The decay heat reduces by about 0.8% per year, so the thermoelectric converter 'ages' or degrades to some extent. See Fig. 2.10.

The thermoelectric elements convert the heat energy from the isotope into electricity. The GPHS-RTG use SiGe thermoelectric elements ('Uncouples') which are no longer in production [44]. Missions after 2010 requiring RTGs, such as the Mars Science Laboratory, will use the Multi-Mission Radioisotope Thermoelectric Generators instead.

- Ulysses, mission completed 2007 in heliocentric orbit (orbiting the Sun)
- Galileo, mission completed 2003 entry into planet Jupiter
- Cassini, mission completed 2017 entry into planet Saturn
- New Horizons, mission ongoing departing Solar System (escape trajectory)

The Stirling Radioisotope Generator (SRG) is based on a 55-W electric converter powered by one GPHS unit. The hot end of the Stirling converter reaches 650°C, and heated helium drives a free piston reciprocating in a linear

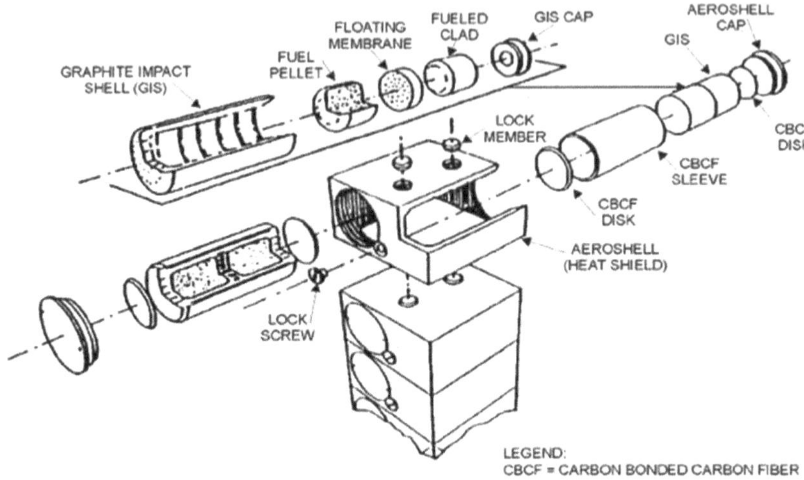

FIG. 2.10 Diagram of general purpose heat source module.

alternator, heat being rejected at the cold end of the engine. The AC is then converted to 55-W DC. This Stirling engine produces about four times more electric power from the plutonium fuel than an RTG. Thus, each SRG will utilize two Stirling converter units with about 500 W of thermal power supplied by two GPHS units and will deliver 100–120 W of electric power. The SRG has been extensively tested but has not yet flown.

Russia has also developed RTGs using Po-210; two are still in orbit on 1965 Cosmos navigation satellites. But it concentrated on fission reactors for space power systems. As well as RTGs, Radioactive Heater Units (RHUs)xe "Radioactive Heater Units (RHUs)" are used on satellites and spacecraft to keep instruments warm enough to function efficiently. Their output is only about 1 W and they mostly use Pu-238—typically about 2.7 g of it. Dimensions are about 3 cm long and 2.5-cm diameter, weighing 40 g. Some 240 have been used so far by the USA, and two are in shutdown Russian lunar rovers on the moon. There will be eight on each of the US Mars rovers launched in 2003.

Both RTGs and RHUs are designed to survive major launch and reentry accidents intact, as is the SRG.

Note that in recent Space Power Reactors (SPRs)xe "Space Power Reactors (SPRs)" activities, the Radioisotope Thermoelectric Generators (RTGs)xe "Thermoelectric Generators (RTGs)" convert thermal power from the alpha decay of Pu $_{238}$ to electrical power by way of solid state thermoelectric elements.

RTGs have also been used on the surface of Mars on the two Viking landers. The Mars Exploration Rovers relied on radioactive heater units for internal thermal control keeping the electronics and charged batteries from freezing during the Martian nights.

Electrical power on the surface of Mars was generated by solar panels in spite of atmospheric dust conditions that limit the amount of solar radiation that reaches the surface of the panels. For power requirements in space, all USA missions have relied almost solely on fuel cells. RTGs, and solar cells for energy. See Fig. 2.11, where Spirit Rover used on surface of Mars during assembly and testing by NASA.

The single exception is the SNAP-10A 45 kWth fission reactor, that was launched in 1965. Russia has utilized fission reactors on more than 30 satellite surveillance mission though. These power sources offer distinct advantages for extended missions on the moon or Mars. RTGs become prohibitively massive at high electrical powers.

The Cassini spacecraft mission to Saturn and its moons like Titan, carries three RTGs and 32.8 kg of Pu $_{238}$ fuel that provide a total electrical power of 0.870 kWe. High-efficiency thin film silicon solar cell arrays can produce 0.676 kWe/Kg and triple-junction InGaAs solar cell arrays can produce 0.360 kW/kg at geosynchronous orbit [46].

FIG. 2.11 Spirit rover. *Courtesy of NASA.*

Fuel cells on the Space Shuttle produced electricity at 0.130 kWe/kg at a continuous output of 7 kWe. For an estimated power budget of 100 kWe for surface missions, electricity produced exclusively by these technologies becomes impractical. However, a combination of these technologies and nuclear fission driven reactors with Stirling Cycle engines may provide a more practical solution to electrical power needs and thermal control for surface exploration.

2.4.2 Fission systems: heat

Over 100 kWe, fission systems have a distinct cost advantage over RTGs. The US SNAP-10A launched in 1965 was a 45-kWt thermal nuclear fission reactor which produced 650W using a thermoelectric converter and operated for 43 days but was shut down due to a satellite (not reactor) malfunction. It remains in orbit.

The last US space reactor initiative was a joint NASA-DOE-Defense Department program developing the SP-100 reactor—a 2-MWt fast reactor unit and thermoelectric system delivering up to 100 kWe as a multiuse power supply for orbiting missions or as a lunar/Martian surface power station. This was terminated in the early 1990s after absorbing nearly $1 billion. It used uranium nitride fuel and was lithium cooled.

There was also a Timberwind pebble bed reactor concept under the Defense Dept Multi-Megawatt (MMW) space power program during the late 1980s, in collaboration with DOE. This had power requirements well beyond any civil space program.

Between 1967 and 1988, the former Soviet Union launched 31 low-powered fission reactors in Radar Ocean Reconnaissance Satellites (RORSATs)xe "Radar Ocean Reconnaissance Satellites (RORSATs)" on Cosmos missions. They utilized thermoelectric converters to produce electricity, as with the RTGs. Romashka reactors were their initial nuclear power source, a fast spectrum graphite reactor with 90% enriched uranium carbide fuel operating at high temperature. Then the Bouk fast reactor produced 3 kW for up to 4 months. Later reactors, such as the one on Cosmos-954 which reentered over Canada in 1978, had U–Mo fuel rods and a layout similar to the US heat pipe reactors described below.

These were followed by the Topaz reactors with thermionic conversion systems, generating about 5 kWe of electricity for onboard uses. This was a US idea developed during the 1960s in Russia. In Topaz-2 each fuel pin (96% enriched UO_2) sheathed in an emitter is surrounded by a collector, and these form the 37 fuel elements which penetrate the cylindrical ZrH moderator. This in turn is surrounded by a beryllium neutron reflector with 12 rotating control drums in it. NaK coolant surrounds each fuel element.

Topaz-1 was flown in 1987 on Cosmos 1818 and 1867. It was capable of delivering power for 3–5 years for ocean surveillance. Later Topaz was aiming for 40 kWe via an international project undertaken largely in the USA from 1990. Two Topaz-2 reactors (without fuel) were sold to the USA in 1992. Budget restrictions in 1993 forced cancellation of a Nuclear Electric Propulsion Spaceflight Test Program associated with this.

2.4.3 Fission systems: propulsion

For spacecraft propulsion, once launched, some experience has been gained with Nuclear Thermal Rocket (NTR)xe "Nuclear Thermal Rocket (NTR)" propulsion systems which are said to be well developed and proven. Nuclear fission heats a hydrogen propellant which is stored as liquid in cooled tanks. The hot gas (about 2500°C) is expelled through a nozzle to give thrust (which may be augmented by injection of liquid oxygen into the supersonic hydrogen exhaust). This is more efficient than chemical reactions. Bimodal versions will run electrical systems onboard a spacecraft, including powerful radars, as well as providing propulsion. Compared with nuclear electric plasma systems, these have much more thrust for shorter periods and can be used for launches and landings.

However, attention is now turning to nuclear electric systems, where nuclear reactors are a heat source for electric ion drives expelling plasma out of a nozzle to propel spacecraft already in space. Superconducting magnetic cells ionize hydrogen or xenon, heat it to extremely high temperatures (million°C), accelerate it, and expel it at very high velocity (e.g., 30 km/s) to provide thrust. Research for one version, the Variable Specific Impulse Magnetoplasma Rocket (VASIMR)xe "Variable Specific Impulse Magnetoplasma Rocket (VASIMR)", draws on that for magnetically confined fusion power (tokamak) for electricity generation, but here the plasma is deliberately leaked to give thrust. The system works most efficiently at low thrust (which can be sustained), with small plasma flow, but high thrust operation is possible. It is very efficient, with 99% conversion of electric to kinetic energy.

2.4.4 Nuclear thermionic technology development

A 1998 report published by the National Research Council's Committee on Advanced Space Technology (NRC 1998) stated as follows:

> *Advanced space nuclear power systems will probably be required to support deep space missions, lunar and planetary bases, extended human exploration missions, and high-thrust, high-efficiency propulsion systems. A major investment will eventually be needed to develop advanced space nuclear power sources .... Unless NASA supports R&T in areas such as innovative conversion methodologies or innovative packaging and integration, future space nuclear power systems will probably be more expensive and less efficient.*

For some space propulsion missions that require high power, or where nuclear power is a critical requirement, the potential performance advantages of a nuclear thermionic system are compelling. The demonstrated state of the art of thermionic systems in terms of lifetime and device-level power output, coupled with their low mass and compactness, make this technology attractive and suggest that it could satisfy future space power requirements in the low to mid tens of kilowatts to megawatts.

In some cases, fully developed thermionic technology may be mission enabling. However, the committee acknowledges that the technical risks in developing a functional thermionic system are high. The technical uncertainty in developing an operational system that could achieve the desired performance is especially high for power systems that use thermionic converters powered by nuclear reactors.

There is no capability in the United States to test nuclear thermionic fuel materials for fuel swelling issues because those fast-flux test facilities were deactivated. A possible alternative to reestablishing test facilities in this country is to coordinate with Russia in future thermionic materials testing.

The National Research Council's Committee recommends orienting the near-term thermionics research and development program toward a solar thermionics conversion technology aimed at competing with other energy conversion technologies available today, such as solar photovoltaics. Basic research and long-term planning, however, should be oriented to establish a technology base that could be used by a future space mission requiring nuclear power. This chapter details the current state of nuclear thermionic research and the path that should be followed to establish a long-term nuclear thermionic capability.

The National Research Council's Committee as part of their recommendation, also states that the thermionics research and development program should be directed in the near term toward the development of a solar thermionic system. Long-term thermionic program goals, however, should be directed toward establishing and maintaining an option for a nuclear thermionic system, a position stated in Recommendation 4 above in this chapter. Balancing two sets of requirements to meet these short-and long-term goals will not be easy. It will not, for example, be possible for the sponsoring agency to design a solar thermionic system that simultaneously addresses issues such as the radiation damage to materials and mechanical stress caused by nuclear fuel swelling. However, for a viable nuclear thermionic system to be built, those are exactly the issues that must be addressed.

Despite the difficulty, the committee believes that the sponsoring agency should work to develop a technology base that can advance systems that will meet both sets of requirements. However, the area where the two technologies overlap is difficult to define, so the sponsoring agency needs to carefully decide which specific technologies or systems will be developed.

The committee feels that the Cylindrical Inverted Multicell (CIM)xe "Cylindrical Inverted Multicell (CIM)" thermionic converter, proposed by General Atomics in conjunction with the High-Power, Advanced, Low-Mass (HPALM)xe "High-Power, Advanced, Low-Mass (HPALM)" system (Described in https://www.nap.edu/read/10254/chapter/6Chapter 4) [47], and the Conductively Coupled Multi-Cell (CC/MC)xe "Conductively Coupled Multi-Cell (CC/MC)" thermionic fuel element, also proposed by General Atomics, may be specific examples of technologies that can be used for solar thermionic applications and adapted to nuclear thermionic applications in the future. The following discussions of the CC/MC and CIM are not meant to indicate that these technologies are the perfect (or the only) technologies that can be developed for use in both nuclear and solar conversion systems. Rather, these technologies should be considered indicative of the type of devices that offer promise of being compatible with both heat source systems.

2.4.4.1 Conductively coupled, multicell thermionic fuel element

A traditional multicell Thermionic Fuel Element (TFE)xe "Thermionic Fuel Element (TFE)" consists of thermionic converters connected in series. Each element is loaded with nuclear fuel and both ends of the element are sealed. The nuclear fueled "flashlight" TFE, which was the baseline for most in-core reactor concepts in the United States as well as the former Soviet Union, is a stack of these individual thermionic converters connected in series to form a thermionic generator. Each fueled thermionic converter gives the impression of a standard D-cell sized battery, hence the term "flashlight configuration." General Atomics has performed most of the work on flashlight TFEs in the United States.

By the early 1990s, program planners and researchers realized that the ability to conduct nonnuclear ground testing of flight units prior to launch would be useful for acceptance or flight qualification testing. Unfortunately, conventional sealed thermionic cells used in the thermionic flashlight generator are difficult to heat-test electrically because of the mechanical design of the TFE.

Moreover, nuclear heating is not practical for flight system verification on the ground. When units are tested for flight qualification on the ground, radioactive fission products are produced in the nuclear core. Having these products in the nuclear core during a rocket launch creates additional complications for launch safety assurance. A predecessor of the only US nuclear reactor to fly in space, SNAP-10A, was ground tested. The second, untested unit was then flown on the experimental space mission.

Radiative coupling is one way around the dilemma of not being able to test a TFE individually prior to combining it with the nuclear heat source. With a TFE designed for radiative heating, a vacuum gap electrically isolates the nuclear fuel from the thermionic converters. An electrical heat source is then used to mimic the radiative heat properties of a nuclear heat source. In this way, each individual TFE and the entire energy conversion subsystem can be tested before loading the reactor with nuclear fuel. The reactor can then be fueled relatively late in the checkout process before launch.

The major disadvantages of using radiative electrical heat testing are that the radiative heating introduces an additional temperature gradient between the heat source and thermionic emitter due to the vacuum gap between the two. This situation requires that the fuel maintain a temperature roughly 200 K higher than the emitter surface to compensate for the gap.

An alternative approach for a design that keeps the thermionic converter separate from the nuclear fuel is conduction coupling. There, the heat from the nuclear fuel is transferred conductively using a ceramic insulator that electrically isolates the fuel from the emitter. As is the case for radiatively coupled converters, the conduction coupled converter can be heated electrically by placing a heating element inside the hollow center of the cylindrical thermionic converter to replicate the heating properties of nuclear fuel. This is the method that General Atomics proposes to use with its CC/MC concept.

Although conceptually simple, using a heat-conducting medium between the nuclear fuel heat source and the thermionic emitter is a challenge because of the combination of high temperature, voltage gradient, and nuclear radiation, which cause electrolytic dissociation of most ceramic insulators.

Russian research offers a possible solution. It has shown that Scandia (Sc_2O_3) insulators offer extraordinarily high temperature capability. An emitter trailer may be fabricated with a fuel cladding layer, followed by a Scandia insulator layer and then the thermionic emitter. However, the lifetime for a device such as a CC/MC has not been conclusively demonstrated, especially under the combined influence of temperature, voltage gradient, and irradiation (Streckert et al. 2000a, b) [48,49].

Since the CC/MC was designed to be heated by sources other than nuclear fuel, it could be used with some other method of transporting heat into the CC/MC cylinder, such as with a heat pipe assembly. However, the addition of a heat pipe system to a solar concentrator, for example, may negate some of the potential weight advantages of the solar thermionic system. So, although such a system is possible, the system tradeoffs to make such a system viable need to be examined before any final conclusions can be made.

Finding: The conductively coupled multicell (CC/MC) thermionic fuel element, as proposed by General Atomics, needs to be evaluated for nuclear in-core use by resolving radiation-induced fuel swelling and insulator degradation issues before the concept can be declared viable.

2.4.4.2 Cylindrical inverted multicell

The Cylindrical Inverted Multicell (CIM) thermionic converter is essentially an inverted version of the CC/MC device and was proposed specifically for use with the HPALM concept (See Fig. 2.12). So, while one challenge with the

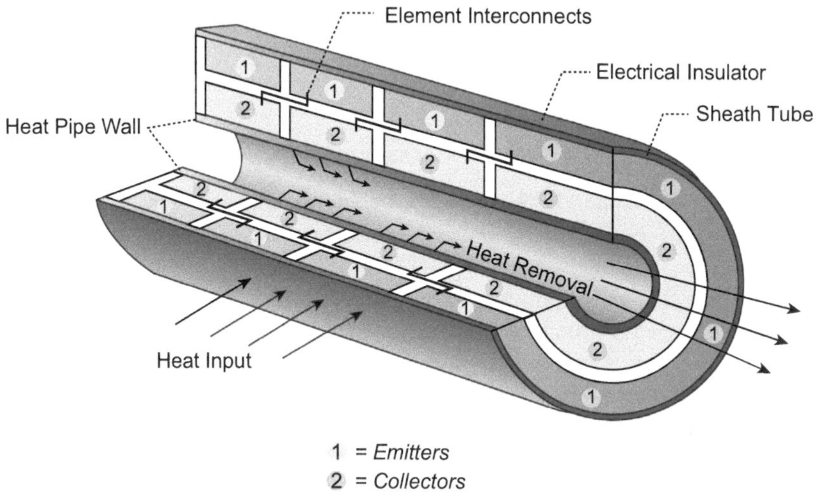

FIG. 2.12 Cylindrical inverted multicell cross-section. *L. Begg, General Atomics, Presentation to the Committee on Thermionic Research and Technology, August 2000.*

CC/MC device may be to find a suitable solar candidate mission, a challenge with the CIM device is to find a suitable nuclear candidate application.

In the CIM thermionic converter, the heat is applied externally, where the emitter is located, and the waste heat is removed from the hollow interior of the cylinder. This conceptual device is intended for use with the HPALM system and is, therefore, inherently capable of being tested using an electrical heater. The CIM converter would be immersed in the heat receiver of the HPALM system, so a comparable nuclear system could be the space thermionic advanced reactor compact (Star-C) concept as was proposed by General Atomics. The STAR-C concept is a nuclear reactor in which the heat from the uranium carbide nuclear fuel is captured by a graphite block. The CIM thermionic converter could be placed into this graphite block so that the CIM is heated radially inward. One potential advantage of this design is that, since the nuclear fuel is not in contact with the CIM, fuel swelling issues may not be as large a concern for the thermionic converter portion of the device. However, the CIM would still be inside the nuclear core, and so any development would have to contend with radiation damage to materials just as would the CC/MC development. This configuration could introduce the added complexity of carbon diffusion into the thermionic device.

Finding: Both the inverted thermionic element and the planar thermionic converter that could be used the HPALM concept are compelling, but the system has not been built or tested.

Finding: The cylindrical inverted multicell (CIM) thermionic fuel element, as proposed by General Atomics, needs to be proven for solar conversion applications. The same technology, if chosen for use with a nuclear in-core system, needs to be evaluated for nuclear in-core use by resolving radiation-induced insulator degradation issues before the concept can be considered viable.

2.4.5 Potential space nuclear thermionic missions

Nuclear heat conversion is an alternative to solar heat conversion for providing power to a spacecraft. Nuclear power applications can be divided roughly into those with low and high power requirements. Low power requirements can be satisfied by radioisotope power systems that generate a few kilowatts of power at most. High power requirements, near or exceeding 100 kW, may require the use of nuclear reactors, depending on the specific mission.

In general, thermionic converters would not be used with radioisotope heat sources because other conversion devices are better suited to operate at the lower temperatures typical of radioisotope heat sources.

The decision to use a nuclear power source on a spacecraft will generally be driven by compelling mission requirements. To date, the United States has only flown one reactor in space. There have been other missions that use nonreactor-based radioisotope power. All of the spacecraft that have flown beyond the orbit of Mars have been powered by radioisotope power sources because there is simply not enough sunlight for photovoltaic arrays. Fig. 2.13 illustrates that the solar energy available decreases rapidly as an interplanetary spacecraft flies to the outer planets of the solar system.

Currently, there are no planned or approved space missions that use a nuclear reactor. Smaller spacecraft, such as NASA's planned Europa orbiter

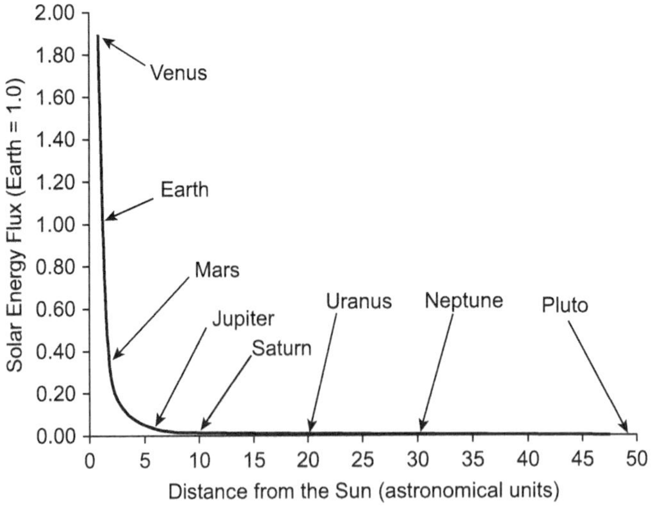

FIG. 2.13 Fig. 5.2 solar energy flux as a function of distance from the sun.

and the Pluto/Kuiper Express, meet the criteria for needing a nuclear power source. However, both spacecraft require a fairly small amount of power and could therefore make use of the smaller radioisotope power systems. In addition, the Pluto/Kuiper Express program at NASA has been canceled at the time of this report's publication. A number of potential future missions have been postulated that would use nuclear reactors based on the following conditions:

- *High power.* Nuclear power is the only practical source of continuous, high power levels in space (more than 100 kW), especially where solar energy is not adequately available. This is due to an economy of scale effect: little size or mass is added as power levels increase.
- *Self-sufficiency.* Nuclear power sources make the spacecraft more independent of potentially unreliable external solar or chemical heat sources. For example, a nuclear power source can be used for missions to the outer planets where there is not sufficient solar energy for a solar-powered system. Or, for missions on the Martian surface, a nuclear power system would not be affected as much by dust storms that reduce the available sunlight.
- *Survivability.* Nuclear power sources are generally less vulnerable to external radiation (e.g., the radiation belts around Earth and Jupiter) and to other potentially hostile environments, such as meteoroids, Martian dust storms, space weapons, and extreme temperatures such as those experienced on the lunar surface.

Examples of missions that are considered for nuclear reactor use include human and cargo missions to Mars, human lunar or planetary bases in harsh conditions, electric propulsion missions to the outer planets, and missions to the outer planets with high power science instruments or high information data rates.

Some of these missions combine a reactor power system with a high power electric propulsion system for enhanced deep space travel. The dual mode systems are described briefly in https://www.nap.edu/read/10254/chapter/6Chapter 4 [47] for a solar thermionic power system, and https://www.nap.edu/read/10254/chapter/13Appendix D [47] discusses the potentially significant benefits of combining a lightweight power system with emerging electric propulsion technology.

Some future military missions may have a need for nuclear reactor systems. These include high power radar systems and space-based electric weapons. However, current studies of these types of missions indicate that they can be accomplished with nonnuclear power systems. Most space-based radar concepts being studied use a combination of low orbits and low duty cycles to reduce the continuous power level required by the vehicle to between 4 and 30 kW. This requirement can currently be met by state-of-the-art solar power systems.

None of the approved NASA far-term missions seem to require power that cannot be provided by solar arrays or advanced radioisotope power sources. The exception is the establishment of a lunar base and a human mission to Mars, mentioned above, which have not been approved but are being considered.

Rationalizing the development of nuclear power supplies for spacecraft is difficult, especially in the near term, because of the absence of current missions and the effects of other factors associated with nuclear development, such as cost, development risk, and potential nuclear safety issues. These risks (or perceived risks) have halted the development of space-based nuclear reactors. However, most studies that explore the concept of space bases and their power requirements assume the future availability of nuclear power.

The advanced conversion technologies that are currently being pursued by various government agencies, as discussed in other sections of this report, are aimed primarily at solar energy and isotope heat sources with heat to electric conversion via:

- Free piston Stirling engine,
- Brayton engine,
- Alkali Metal Thermal to Electric Converter (AMTEC)
- Thermophotovoltaics, and
- Advanced thermoelectric generators.

Should a permanent human mission to the Moon or Mars be authorized and a nuclear heat source for the power be selected, thermionic conversion might well be able to compete with these conversion technologies. While NASA and DOE are sponsoring development work on the other conversion systems, the Defense Threat Reduction Agency (DTRA) program is the only US program working on thermionic converters in relation to space nuclear power. The committee believes that the research on thermionic converters should continue so that there will be a technology base on which future nuclear power reactor system development programs can draw.

Finding: The current thermionic research and development program sponsored by the DTRA is the only thermionics work being conducted today in the United States related to space nuclear power. However, the program does not include efforts to address nuclear issues related to incorporating the technology into a reactor system, namely radiation-induced fuel swelling or radiation damage to converter materials for use with nuclear in-core systems.

While the DTRA program is important because it is the only funded effort in this area, there are limits to what the program can realistically accomplish given the relatively meager funding. Also, if the sponsoring agency follows this committee's recommendation to establish a US thermionic research program that focuses on near-term solar thermionic applications, the program will be limited in its ability to achieve a nuclear thermionic capability.

Previous thermionic technology programs have identified as a problem the fuel swelling that occurs over time in an in-core thermionic fuel element. Not only is DTRA not examining fuel swelling effects, but there is also no materials or device testing being conducted in nuclear environments, presumably due to cost and the lack of fast-flux test facility availability in the United States. The absence of nuclear in-core testing could invalidate any nuclear thermionic design that is developed by the current program.

2.4.6 Heat pipe power system

Heat Pipe Power System (HPS)xe "Heat Pipe Power System (HPS)" reactors are compact fast reactors producing up to 100 kWe for about 10 years to power a spacecraft or planetary surface vehicle. They have been developed since 1994 at the Los Alamos National Laboratory as a robust and low technical risk system with an emphasis on high reliability and safety. They employ heat pipes to transfer energy from the reactor core to make electricity using Stirling or Brayton cycle converters. Energy from fission is conducted from the fuel pins to the heat pipes filled with sodium vapor which carry it to the heat exchangers and thence in hot gas to the power conversion systems to make electricity. The gas is 72% helium and 28% xenon.

The reactor itself contains a number of heat pipe modules with the fuel. Each module has its central heat pipe with rhenium-clad fuel sleeves arranged around it. They have the same diameter and contain 97% enriched uranium nitride fuel, all within the cladding of the module. The modules form a compact hexagonal core. Control is by six stainless steel-clad beryllium drums each 11- or 13-cm diameter with boron carbide forming a 120° arc on each. The drums fit within the six sections of the beryllium radial neutron reflector surrounding the core and rotate to effect control, moving the boron carbide in or out. Shielding is dependent on the mission or application, but lithium hydride in stainless steel cans is the main neutron shielding.

The SAFE-400 (Safe Affordable Fission Engine) space fission reactor is a 400-kWt HPS producing 100 kWe to power a space vehicle using two Brayton power systems—gas turbines driven directly by the hot gas from the reactor. Heat exchanger outlet temperature is 880°C. The reactor has 127 identical heat pipe modules made of molybdenum or niobium with 1% zirconium. Each has three fuel pins 1-cm diameter, nesting together into a compact hexagonal core 25 cm across. The fuel pins are 70 cm long (fueled length 56 cm); the total heat pipe length is 145 cm, extending 75 cm above the core, where they are coupled with the heat exchangers. The core with reflector has a 51-cm diameter. The mass of the core is about 512 kg and each heat exchanger is 72 kg. SAFE has also been tested with an electric ion drive.

A smaller version of this kind of reactor is the HOMER-15—the Heat Pipe-Operated Mars Exploration Reactor. It is a 15-kW thermal unit similar to the larger SAFE model and stands 2.4 m tall including its heat exchanger and

3-kWe Stirling engine (see above). It operates at only 600°C and is therefore able to use stainless steel for fuel pins and heat pipes, which are 1.6-cm diameter. It has 19 sodium heat pipe modules with 102 fuel pins bonded to them, 4 or 6 per pipe, and holding a total of 72 kg of fuel. The heat pipes are 106 cm long and the fuel height is 36 cm. The core is hexagonal (18 cm across) with six BeO pins in the corners. The total mass of reactor system is 214 kg, and the diameter is 41 cm.

2.4.7 Space reactor power systems

In the 1980s the French ERATO program considered three 20-kWe turbo-electric power systems for space. All used a Brayton cycle converter with a helium–xenon mix as working fluid. The first system was a sodium-cooled UO_2-fuelled fast reactor operating at 670°C, the second a high-temperature gas-cooled reactor (thermal or epithermal neutron spectrum) working at 840°C, and the third a lithium-cooled UN-fueled fast reactor working at 1150°C (Table 2.1).

Fission reactors are expected to play a critical role in upcoming human planetary mission. The primary source of electrical power for the Apollo spacecraftxe "Apollo spacecraft" were fuel cells, but nuclear power was utilized during these missions to the moon to operate surface science experiments.

2.4.7.1 Heat pipe operated mars exploration reactor (HOMER)

A Heat Pipe Operated Mars Exploration Reactor (HOMER) providing between 50 and 25 kWe has been proposed for life support, operations, in-situ propellant production, scientific experiments, high-intensity lamps for plant growth and other activities on a Mars mission. This is crucial, since a solar array providing the same power on Mars would require a surface area of several football fields. In addition, day and night, geographical sunlight issues, seasonal variations and dust storm environments would not affect a fission reactor system. As we stated before, Fig. 2.6 shows the core assembly of such a design producing 125 kWth of power. The rotating drums around the circumference achieve power level control. These consist of a neutron absorbing side and a neutron scattering and reflecting side, allowing power control without the need for terrestrial used control rods. Moving parts are also eliminated by the use of heat pipes transferring heat for rejection by radiation to space without the use of pumps and moving parts.

The core contains stainless steel clad Uranium Dioxide (UO_2) fuel. The fuel pins are structurally and thermally bounded to a sodium heat pipe. Heat is conducted from the fuel pins to the heat pipes which carry the heat to the power conversion system.

The core design is compatible with different types of power conversion cycles as: Thermoelectric, Thermionic, Brayton, Stirling, Rankine or Alkaline Metal Thermal to Electric Converter (AMTEC)xe "Alkaline Metal Thermal to

TABLE 2.1 Space reactor power systems [12].

	SNAP-10 US	SP-100 US	Romashka Russia	Bouk Russia	Topaz-1 Russia	Topaz-2 Russia-US	SAFE-400 US
dates	1965	1992	1967	1977	1987	1992	2007
kWt	45.5	2000	40	<100	150	135	400
kWe	0.65	100	0.8	<5	5–10	6	100
converter	t'electric	t'electric	t'electric	t'electric	t'ionic	t'ionic	t'electric
fuel	U-ZrH_x	UN	UC_2	U–Mo	UO_2	UO_2	UN
reactor mass, kg	435	5422	455	<390	320	1061	512
neutron spectrum	thermal	fast	fast	fast	thermal	thermal/ epithermal	fast
control	Be	Be	Be	Be	Be	Be	Be
coolant	NaK	Li	none	NaK	NaK	NaK	Na
core temp. °C, max	585	1377	1900	?	1600	1900?	1020

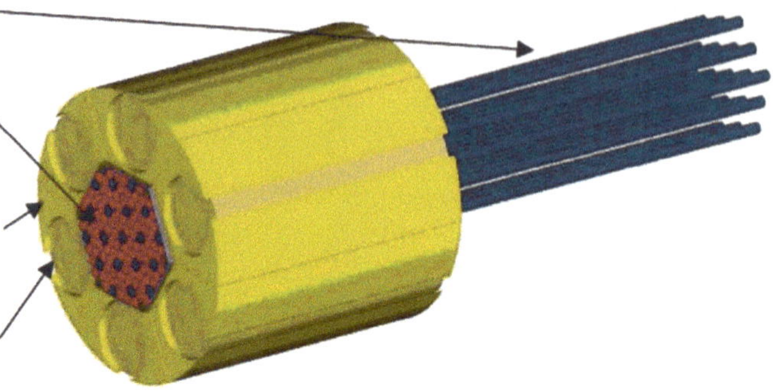

FIG. 2.14 Heat pipe reactor HOMER-15 Configuration.

Electric Converter (AMTEC)" using high pressure Sodium (Na) Vapor for the Heat Pipe (HP)xe "Heat Pipe (HP)".

It will take at least a decade of research and development, with an expense of at least $50 Billion to prepare for a Mars mission. NASA has been trying a strategy of "faster, yet cheaper and better) in its exploration of Mars, leading to about a 2 out of 3 as a success rate. With manned space missions, a higher degree of reliability will be needed.

2.4.7.2 Heat pipe reactor HOMER-15 and Homer-25 designs

The Heat pipe Operated Mars Exploration Reactor-15 (HOMER-15)xe "Heat pipe Operated Mars Exploration Reactor-15 (HOMER-15)" is a nuclear fission reactor concept for future Lunar and Martian surface missions. See Fig. 2.14.

The reactor core contains Uranium Nitride (U_2N_3) fuel pellets contained in stainless steel fuel pins that produce a total of 15 kWth of thermal power.

Sodium-filled heat pipes transfer the thermal energy to a Stirling engine that produces 3 kWe of electrical power for an overall thermal conversion efficiency of:

$$\eta_{th} = \frac{P_e}{P_{th}} = \frac{3}{15} = \frac{1}{5} = \%20 \tag{2.1}$$

A 25kWe HOMER-25 version uses Uranium Dioxide (UO_2) fuel and transfers heat to six Stirling engines through Potassium (K) filled Heat Pipe (HP)xe "Heat Pipe (HP)".

Martian Surface

The study of surface characteristics (or surface properties and processes) is a broad category of Mars science that examines the nature of the materials making up the Martian surface. The study evolved from telescopic and remote-sensing techniques developed by astronomers to study planetary surfaces. However, it has increasingly become a subdiscipline of geology as automated spacecraft bring ever-improving resolution and instrument capabilities. By using characteristics such as color, albedo, and thermal inertia and analytical tools such as reflectance spectroscopy and radar, scientists are able to study the chemistry and physical makeup (e.g., grain sizes, surface roughness, and rock abundances) of the Martian surface. The resulting data help scientists understand the planet's mineral composition and the nature of geological processes operating on the surface. Mars' surface layer represents a tiny fraction of the total volume of the planet yet plays a significant role in the planet's geologic history.

Contingency Sample Collection on Lander

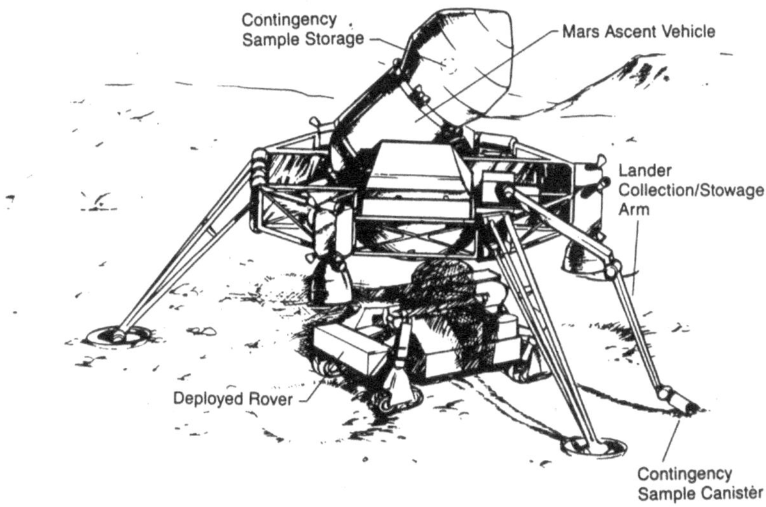

Understanding physical surface properties is also very important in determining safe landing sites for spacecraft.

Note that: For the Lunar base mission, the United States of America has targeted placing a base on the moon by 2020 as a precursor to launching a manned mission to Mars. Humans have been to the moon already in the 1970s as part of the Apollo Moon missions.

The HOMER-15 reactor is a modular reactor design with an arrangement of fuel pins, heat pipes, and neutron reflectors. The mass of the reactor is shown in Table 2.2. There are a total of 19 heat pipes and 102 fuel pins in the core design, including 13 six-pin modules and 6 four-pin modules. Six-pin modules are located near the center and four-pin modules are located near the outside of the reactor core.

Four-pin modules, arranged around a heat pipe as shown in Fig. 2.13, experience higher thermal stress because of their asymmetrical arrangement. They are located near the exterior where the temperature is lowest.

TABLE 2.2 Mass summary for HOMER-15.

Component	Mass [kg]
Fuel pins	95.0
BeO pines	1.8
Heat pipes	9.8
Tricusps	7.9
Slats	2.3
Baffle	4.4
Radial reflector	41.5
Control drams	47.0
Support	4.7

Fuel pin modules are arranged in a hexagonal core shape with a Beryllium Oxide (BeO) neutron reflector pin located at each corner. The core measures 18.1 cm from one edge of the hexagon to the opposite edge.

2.4.7.3 Heat pipe and fuel pins configuration

Heat pipes are bonded to a stainless steel tri-cups that run along the length of a fuel pin as shown in Fig. 2.15.

As a safety feature, the hollow tri-cusps are filled with Boron Carbide (B_4C) as a thermal neutron absorber. This allows the reactor to remain subcritical in the event of an accident in which the core is flooded with water such as from a leak in the reactor shielding.

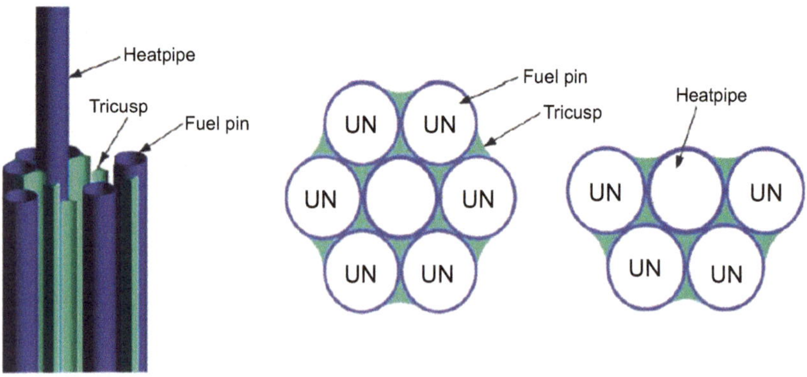

FIG. 2.15 Cross section of four pin and six heat pipe modules.

Fuel pins are constructed with SS-316 with an outside diameter of 1.59 cm and thickness of 0.635 mm. Each fuel pin contains a 36 cm stack of fuel pellets contained within a stainless steel sleeve.

A 4 cm stack of Beryllium Oxide (BeO) pellets surrounded the fuel pellets at the end of each pin. Including an end cap on each fuel pin, the total of each fuel pin is approximately 44 cm. The fuel pellets are made of 97 wt percent enriched Uranium Nitride (UN). Uranium Nitride is a high density, high thermal conductivity fuel. This results in a smaller core size with lower total mass and lower operating fuel temperature. The stability of Uranium Nitride (UN) fuel is lower than Uranium Dioxide (UO_2). The fabricability of UN is more difficult than UO_2 and the most recently produced UN fuel in the United States of America was during the late 1980's for the 100 kWe Space Reactor Prototype (SP-100) (Fig. 2.16) radiation experiment. In the event of a cladding failure of a UN pin, there would be little effect on the thermal or neutronic characteristic of the core. In the Martian atmosphere, which is primarily Carbon Dioxide (CO_2), this would result in surface carbonization, but UC and UN have similar densities and thermal conductivities.

Note that: SP-100 (**S**pace reactor **P**rototype was a US research program for nuclear fission reactors useable as small fission power systems for spacecraft. It was started in 1983 by NASA, the US Department of Energy and other agencies. A reactor was developed with heat pipes transporting the heat to thermionic converters. It was cooled with Lithium. The project never advanced to flight hardware and was terminated in 1994.

The HOMER-25 is a large scale version of the HOMER-15 reactor. The mass of HOMER-25 subsystems are listed in Table 2.3 and key reactor parameters are listed in Table 2.4.

The reactor core consists of 156 fuel pins and 61 heat pipes, arranged as shown in Fig. 2.6. The fuel pins and heat pipes are not bonded together in individual modules as in the HOMER-15 reactor. Instead, they are arranged in a hexagonal monolith lattices as shown in Fig. 2.15.

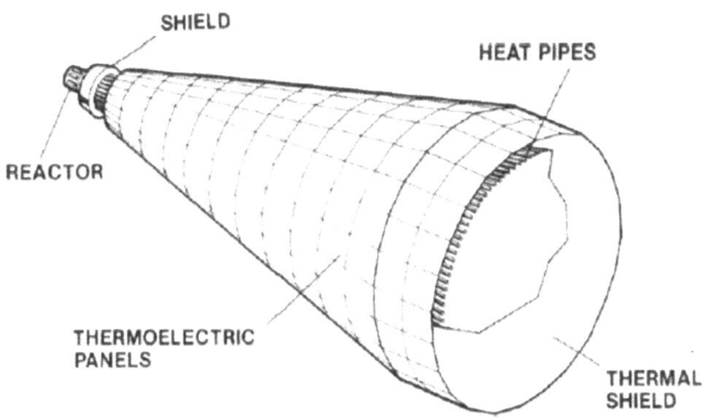

FIG. 2.16 Nuclear power system SP-100.

TABLE 2.3 Technical characteristic for the HOMER-25 design.

Parameter	Value
Peak fuel temperature [K]	931.7
Average fuel temperature [K]	914.9
Peak fuel burnup	0.27 %
Average fast flux (>100 keV) [n/(cm^2/s)]	7.04×10^{12}
Average moderated flux (<100 keV) [n/(cm^2/s)]	1.37×10^{12}
Average total flux in fuel [n/(cm^2/s)]	8.41×10^{12}
Peak fast fluence (>100 keV) [n/cm^2]	1.64×10^{21}
Average heat pipe temperature [K]	880
Boiler saturation temperatine [K]	860
Stirling temperature, hot end [K]	847.8
Stirling temperature, cold end [K]	414.4
Net electrical power [kWe]	25
Stirling output power [kWe]	26.3
Reactor thermal power [kWth]	94.5
Rejected power through radiator [kWth]	67.3
Radiator temperature [K]	400
Required radiator area [m^2]	75.8

TABLE 2.4 Mass summary for HOMER-25.

Component	Mass [kg]	
Reactor (fuel, heat pipes in-core, monolith, reflector)	461.9	
Internals and controls	63.0	
Heat pipes above core and boiler	130.9	
Shield	569.2	
Reactor module subtotal		1225.0
Stirling engines	372.5	
Power management and distribution	75.0	
Cabling	37.5	
Power conversion system subtotal		485.0
Radiator	178.5	
Secondary heat transfer	60.9	
Secondary heat subtotal		239.4
Integrating superstructure		183.7
Total mass		2,133.1

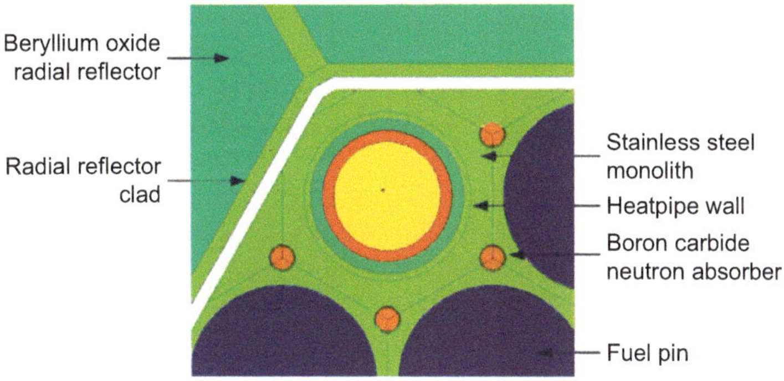

FIG. 2.17 Fuel pins and heat pipes configuration in the HOMER-25 design.

Heat pipes are located at the four corners of the hexagonal core instead of Beryllium Oxide (BeO) reflector pin. Heat pipes in the HOMER-25 reactor use Potassium (K) instead of Sodium (Na) as a working fluid. Boron Carbide (B_4C) wire is located in the region between the fuel pins and heat pipes in the lattice as a thermal neutron absorber. See Fig. 2.17.

Uranium Dioxide (UO_2) has more design heritage than Uranium Nitride (N_2N_3). UO_2 with Zircaloy cladding is used in commercial reactors and has been studied with 316 stainless steel cladding in liquid metal fast breeding reactor experiments. The Russian space program used UO_2 clad in molybdenum, but information about this is not publicly available. A domestic study of UO_2 molybdenum fuel would be required to validate its use. UN/SS and UO_2/SS are limited to clad temperature less than approximately 973 K due to loss of creep strength in the fuel cladding.

2.4.8 Stirling engine system

Stirling engines are desirable for space power applications since they operate at the highest efficiency of any heat engine. This decreases fuel burnup, radiation levels, and the amount of heat that must be rejected by reactor. Heat pipes from the HOMER-15 or HOMER-25 reactor are connected to a heat exchanger that transfers heat to the heater head of the Stirling engine. The exit temperature for a stainless steel reactor is 900 K and the average temperature of the heater head is 850 K. This thermal energy is converted to mechanical energy by pistons inside the Stirling engine, which is then converted to electricity by a generator. The pistons are supported by flexure bearings in a high pressure helium working fluid.

The baseline design for HOMER-15 is a single 3 kWe Stirling engine. An alternate configuration of three 1 kWe engine, or for redundancy in case of failure, three 1.5 kWe engines could be used. In the HOMER-25 design, six Stirling engines are configured to produce 25 kWe. Only four of the six engines operate at one time, leaving two for redundancy in care of failure. This is a necessary feature since current Stirling engines technology is not suitably reliable at high temperatures.

In a single Stirling engine configuration, the unbalanced load the single piston will cause vibration in the system. This effect can be offset by attaching a counter moving mass balance at the end of the piston. In multiple Stirling engine configurations, the engines can be arranged opposite to each other to cancel forces and angular momenta from the moving pistons.

The heat exchanger serves an important structural element. The heat pipes, reactor core, control drum, and radial reflector are suspended from the heat exchanger and the Stirling engine is supported above it. In this arrangement, the heat pipes are free to expand during the warm-up to the operating conditions.

2.4.9 Heat pipe design

The heat pipes are constructed of 316 stainless steel with the same diameter as the fuel pins, but a thickness of 0.889 mm. At one end, the evaporator section of the heat pipes in the core uses an annular wick structure, as shown in Fig. 2.18 here.

The wick is composed of stainless steel and the heat transfer fluid is Sodium (Na) or Potassium (K) or reactor designs that use a Sodium–Potassium eutectic which is liquid at room temperature. The heat pipes pass 40 cm through the axial reactor shield. In this region, the heat pipes are insulated with a SS-316 vacuum thermos structure. The length of the heat pipes in the condenser section or Sodium boiler is 20 cm. Including small axial gaps at the shield core interface and the shield-boiler interface, the total length of each heat pipe is approximately 107 cm.

Stainless steel 316 is compatible with Sodium up to temperature of 1.050 k if the Oxygen (O_2) content in the Sodium is maintained below approximately ~10 ppm.

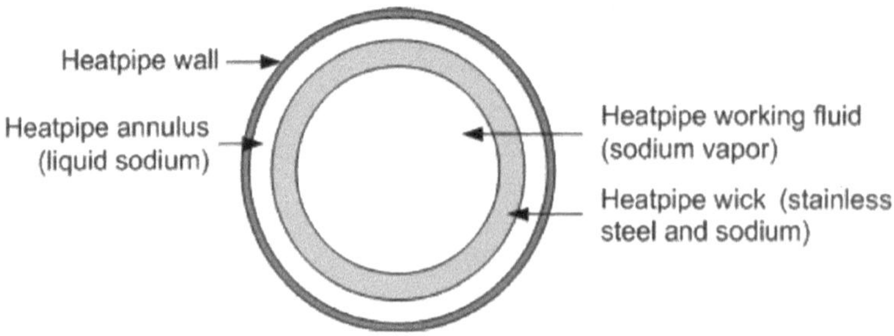

FIG. 2.18 Evaporator region of heat pipe image.

2.4.10 Nuclear reactor power system

To meet the high power requirements of the VASIMR engine, an onboard nuclear reactor will be needed to power the craft. Research has already been done on fission power in space, but it has been underfunded in the past. With proper funding and modern equipment, it should be possible to make a space hardened reactor capable of providing more than 1MWe.

The SP-100 was a joint DOD/DOE/NASA program to design a space-application nuclear reactor that operates in the 10–1000 kW range. Current systems, such as Alkali Metal Thermal-to-Electric Conversion (AMTEC)xe "Alkali Metal Thermal-to-Electric Conversion (AMTEC)" can achieve this with greatly reduced mass.

The following Figs. 2.19 and 2.20, are a few images from the initial design of the SP-100.

The SP100 space nuclear reactor was designed for use as an orbital power supply, lunar or Martian surface power station, and power supply for nuclear electric propulsion, with a saleable power range of 10's kWe to 100's kWe. The original mission was an orbital power supply for the United States (US) Strategic Defense Initiative (SDI)xe "Strategic Defense Initiative (SDI)" of the 1980s. Although the original sponsors were a consortium of the US Department of Defense, US Department of Energy, and the US National Aeronautics and

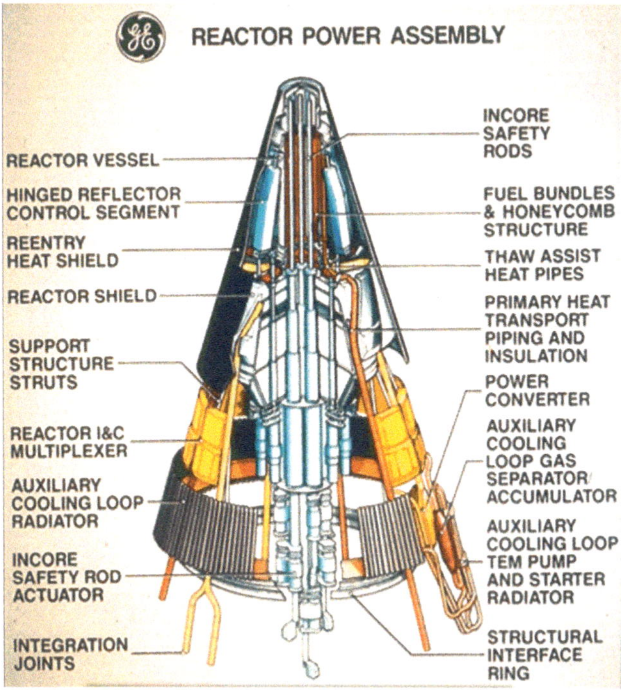

FIG. 2.19 Cross-section of SP-100 image. *Courtesy of General Electric.*

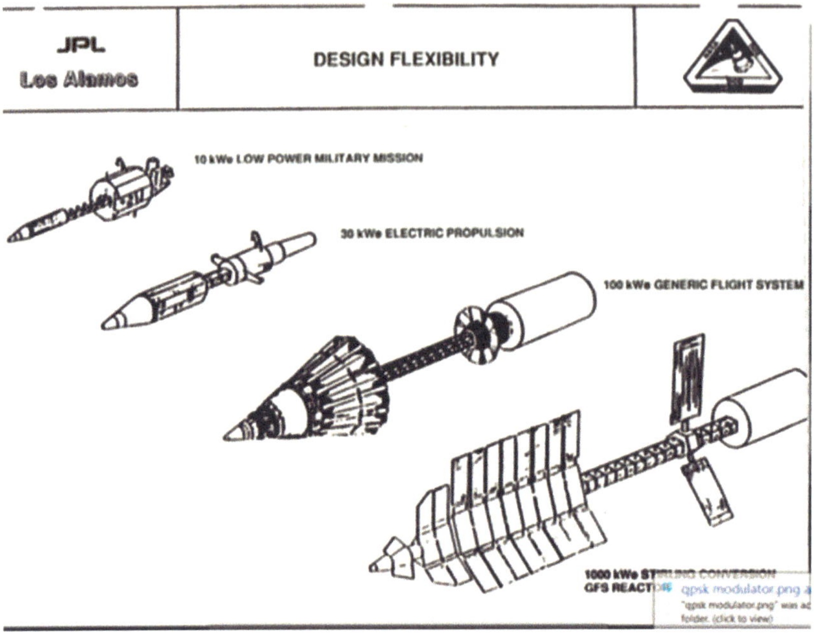

FIG. 2.20 Initial design of the SP-100.

Space Administration (NASA)xe "National Aeronautics and Space Administration (NASA)", as the SDI effort diminished with the demise of the Soviet Union the mission evolved more toward the needs of NASA. Eventually, as the grandiose missions of NASA came into question in the early 1990s and less extravagant missions became more palatable, the SP-100 program was discontinued. At the time of program discontinuation, a complex infrastructure of industry and national laboratories had been established, and approximately $1-billion invested in design and development. The SP-100 was not intended as a low cost one-time-use device; but rather, a highly flexible power supply that realized a cost advantage by being capable of multiple missions, based on a common design with the flexibility to evolve. The design and development team made major progress successfully fabricating and testing technology features that were essential to meeting the stringent safety, performance, lifetime and reliability requirements of proposed missions. The following paper attempts to describe the high level of sophistication incorporated in the SP-100 design, and the high degree of technology readiness at the time of program discontinuation. See Fig. 2.20 as images from the initial design of the SP-100.

Here is Table 2.5, which is listed of masses of current on-board nuclear reactors as well.

TABLE 2.5 Masses of current on-board nuclear reactors.

Parameter/subsystem	SP-100 SNRP System†	AMTEC-SNRP (*Low Temperature*)	AMTEC-SNRP (*Small Reactor*)	AMTEC-SNRP (*Small radiator*)
Nominal Parameter:				
- Electrical Power (kWe)	100.8	100.8	100.8	100.8
- Converter Efficiency (%)	5.3	25.4	30.1	27.8
- System Efficiency, T_{sys} (%)	4.2	22.9*	27.0*	25.0*
- Reactor Power (kWt)	2400	440	373	403
- Rx Exit Temperature (K)	1373	1127	1180	1180
- Radiator Temperature (K)	793	650	650	650
- Structure	Refractory Alloys	Super-steel alloys	Super-steel alloys	Super-steel Alloys
- Rx Cooling	Li-pumped loops	Na-heat pipes	Na-heat pipes	N-heat pipes
- Radiator	K-heat pipes	K-heat pipes	K-heat pipes	K-heat pipes
- Output voltage (V DC)	208.6	206	277	176
- Output current (A)	483	490	364	572
- Converter type	SiGe, PCAs	AMTEC	AMTEC	AMTEC
- No. of converter units	12	24	24	18
- Converter unit Power (kWe)	9.0	4.2	4.2	5.6
- System load-following	Yes	Yes	Yes	Yes
- % of Peak Power	>97%	95.0%	87.5%	94.0%
- η_{sys} at peak Power (%)	4.2	20.0*	24.0*	21.8*
- Radiator HP sonic limit (K)	$\leq$ 580 K	$\leq$ 630 K	$\leq$ 630 K	$\leq$ 630 K
- Radiator Area (m^2)	111.4	26.9	21.1	23.4
- Rad. Specific Area (kg/m^2)	7.67	7.67	7.67	7.67
- Space xink temperature (K)	250	250	250	250
- Half cone angle (*)	17	17	17	17

Continued

TABLE 2.5 Masses of current on-board nuclear reactors.—cont'd

Parameter/subsystem	SP-100 SNRP System†	AMTEC-SNRP (*Low Temperature*)	AMTEC-SNRP (*Small Reactor*)	AMTEC-SNRP (*Small radiator*)
Key Subsystems Mass:				
- Nuclear Reactor (kg)	700	477	456	466
- Radiation Shield (kg)	960	437	405	420
- Convener (kg)	450	1063	1063	798
- Heat Rejection Radiator (kg)	854	206	162	180
- Total Subsystem Mass (kg)	2964	2183	2086	1864
- Total Subsystem Mass (%)	100	73.7	70.4	62.9
SNRP System:				
- Total System Mass (kg)	4600	$\leq$ 3390#	$\leq$ 3227#	$\leq$ 2893#
- System Sp. Power (We/kg)	21.9	$\geq$ **29.7**#	$\geq$ **31.2**#	$\geq$ **34.8**#
- System Sp. Mass (kg/kWe)	45.6	$\leq$ **33.6**#	$\leq$ **32.0**#	$\leq$ **28.7**#

(*): 90% of converter efficiency. (#): Projected: (†): Marriott and Fujita [123].

TABLE 1. Design, Performance Parameters, and Mass Estimates of Na-AMTEC SNRP Systems for $T_{rad} = 650$ K.

Looking at the masses of the smaller reactors, we expect that, given a full research budget, it would be possible to develop a suitable reactor with mass 2 tons.

Defense Advanced Research Projects Agency (DARPA)xe "Defense Advanced Research Projects Agency (DARPA)", in conjunction with DOE's Office of Nuclear Energy, and NASA's Office of Aeronautics and Space Technology *are* jointly sponsoring a space nuclear reactor power system program known as the Space Power-100 (SP-100) Development Project. The program is presently in the critical technology phase. This phase, better known as technology assessment and advancement, includes mission requirements definition, system conceptual designs, and critical technology development. A ground test phase decision is scheduled for July 1985. If the decision is positive, the next phase would begin in fiscal year I986. An overriding concern in conducting this program is to ensure that nuclear safety is being properly addressed even in these early stages.

Recent and past studies have indicated that several classes of missions would be enabled or significantly enhanced by space nuclear reactor power systems, including military and civil satellite missions, manned space stations, and missions to the planets, comets, and asteroids. For systems requiring more than 25 kWe of power, the power-to-mass ratio of a nuclear reactor system can be considerably higher than that of competing solar/battery systems. The nuclear system can readily be used for deep-space applications, and it need not be oriented toward the sun. The rugged, low-cross-section configuration would enhance survival in radiation fields, reduce drag in orbit, decrease the detection cross section, and enhance maneuverability and harden ability.

A space nuclear reactor power system based on already proven technology is ready for a ground test phase now, but its performance would be limited in terms of power-to-mass ratio, total power, and life. It is likely that potential users would not find it interesting enough to employ in their missions.

To meet this challenge, DARPA, in conjunction with DOE's Office of Nuclear Energy, and NASA's Office of Aeronautics and Space Technology have combined efforts to sponsor a space nuclear reactor power system program known as the Space Power-100 (SP-100) Development Project. Three major laboratories are collaborating in this Program: NASA's Jet Propulsion Laboratory (JPL), and Lewis Research Center (LeRC)xe "Lewis Research Center (LeRC)", and DOE's Los Alamos National Laboratory (LANL)xe "Los Alamos National Laboratory (LANL)".

A significant portion of the SP-100 Project is aimed at identifying plausible missions, determining their requirements, and identifying space nuclear reactor systems that would be capable of meeting these requirements. These system identification activities are being closely followed by the development of enabling technology. To ensure that the concepts and supporting technology will not result in system designs that introduce unacceptable risk with regard to nuclear safety, a concurrent safety evaluation program is being conducted.

The remainder of this paper presents the status of the various activities in support of the SP-100 Project.

2.4.11 Material choices

The primary material for reactor components 316 stainless steel, chosen as low cost and off the shelf material. SS-316 also has material advantages when exposed to the Martian atmosphere, which is predominately carbon dioxide. Stainless steel is carburized by CO_2, increasing its emissivity. The CO_2 atmosphere also provides a good thermal conduction path in fuel pin and heat pipe connections. However, carburization tends to make stainless steel more brittle. The use of stainless steel structures makes it unnecessary to hermetically seal the reactor core, but dust buildup may cause problems with the internal components.

Ideally, power output could be increased by decreasing the diameter of the heat pipes and increasing the number of fuel pins per module, but this would increase the internal operating conditions of the reactor core and stainless steel is limited to use at relatively low temperatures compared to temperatures in the reactor core. Stainless steel cannot stress or 2/3 rupture stress for 316 stainless steel at 923 K is approximately 35 Mpa for a ten year operational lifetime.

2.4.12 Safety considerations

In case of a heat pipe failure, the operating temperatures in the fuel cladding would reach 1067 K, leading to a significant shortening of the reactor lifetime. All-important design improvement would be to reduce the maximum temperature after the failure of a heat pipe to under 970 K to minimize thermal creep. Although stainless steel has several advantages in a CO_2 atmosphere, it is incompatible with CO_2 for long term exposure at temperature above 923 K.

2.4.13 Reactor control

The core reactivity is controlled by six stainless steel clad Beryllium Oxide (BeO) cylindrical drums arranged symmetrically about the core in the radial reflectors as Shown in Fig. 2.6. Each drum contains a 1 cm thick Boron Carbide (B_4C) absorber section along a 120° arc.

The drums can be rotated so that either the B_4C face or the Beryllium face of the drum is oriented toward the reactor core. Each drum is independent and can be operated in sequence to provide the necessary reactivity conditions in the core. The radial neutron reflector is composed of six Beryllium Oxide (BeO) sections clad in stainless steel.

A server reduction in the strength and thermal conductivity of BeO in the reflectors and drums occurs at fluences above 2×10^{20} n/cm^2 due to radiation-induced micro cracking.

Fluences of 1.2 × 1021 b/cm^2 in the HOMER-15 reactor produces more severe micro cracking below 600 K. Crakes in the BeO reflectors would allow neutrons to leak to the outside, affecting the neutron properties inside the core. There is no known solution to avoid micro cracking in BeO at fluences above $\sim 1 \times 10^{21}$ n/cm^2 and temperatures below 900 K. Replacing the solid slabs of BeO with fine-grained BeO encased in stainless steel would prevent thermal cracking but the design lifetime of fine grained BeO is uncertain.

Below 770 K the impacts of neutron irradiation fluences up 1×10^{21} n/cm^2 on the properties of BeO is small. Significant swelling occurs at temperatures above 900 K, requiring extra volume within the stainless steel cladding of the reflector. The proposed radial reflector operating temperature above 873 K in BeO irradiated to greater than 1×10^{21}n/cm^2. Operation of the radial reflector below ~823 K should be suitable for the lifetime of the HOMER-15 reactor.

2.4.14 Neutron shielding

Lithium Hydride (LiH) has been used previously in neutron shielding technology in the Aircraft Nuclear Propulsion (ANP)xe "Aircraft Nuclear Propulsion (ANP)" and Space Nuclear Auxiliary Power (SNAP)xe "Space Nuclear Auxiliary Power (SNAP)" space nuclear reactor programs Of candidate shielding materials, LiH has the best neutron attenuation per unit mass due to its low density (0.775 g/mL) and high hydrogen content of 12.68 in weight percent.

Note that: The Space Nuclear Auxiliary Power (SNAP) program was begun in 1955 and terminated in 1973. During that period approximately $850 million *then-year* dollars were spent to develop nuclear power sources capable of producing from 500 W up to 1000 kWe. The technology base was broad and encompassed a wide range of materials and power conversion systems.

Hydrogen is primarily responsible for neutron moderation in LiH and it has poor thermodynamic properties such as low thermal conductivity and high coefficient of thermal expansion. LiH is difficult to fabricate and the only fabrication capability in the USA is at the Y12 National Security Complex in Oak Ridge, Tennessee.

At temperature above 700 K, hydrogen dissociation is pronounced and in hot spot, a streaming path for higher energy neutrons can be created, leading to a loss of shielding effectiveness. The upper operating temperature can be increased by enclosing the LiH in a thin walled stainless steel pressure vessel. For temperatures above 800 K, the dissociation pressure and hydrogen permeation rate through stainless steel are unacceptably high. However, at temperature below 600 K, radiation-induced swelling and cracking is significant.

In the SP-100 reactor program LiH shield material was limited to 600–700 K for the final design. Furthermore, other possible candidates for shielding include Lithium Magnesium alloy, ZrH, TiH and Hydrogenated graphite foam.

Water shielding would be the least expensive method of shielding since there is extensive terrestrial experience with stainless steel and water. The effects of irradiation on water shielding system would be easy and inexpensive to test on Earth. One concern with water shielding in that the vapor pressure of water rises considerably as shielding temperature approaches 400 K, requiring a sturdier pressure vessel and accommodation for extra volume for water vapor. At higher temperatures, the mass of the stainless steel pressure vessel could become prohibitively large. To keep the temperature under 400 K, a pontifical solution is to include Heat Pipes (HPs)xe "Heat Pipes (HPs)" bonded to the shield and connected to a radiator.

In a study by Poston et al. [57], the amount of thermal power that would be rejected from the radial reflector into the water shield is about ~2% of the total reactor thermal power. An important consideration in radiation does is that a large radiator will cause greater neutron scattering, increasing the radiation does to humans. The major drawback of using water shielding is mass. Further mission would benefit from the discovery of indigenous water to supply resources for shielding and other human requirements.

2.4.15 Reactor sitting

The reactor will need to be located at an appropriate distance from human habitation and equipment to provide another level of safety. Radiation dose drops approximately by the square of the distance away from the reactor, however increasing the distance from the reactor increases the mass of the cabling required to transmit electricity to where it is needed. An optimization between shielding mass and cabling mass would need to be determined.

Locating the reactor at a distance could also take advantage of the lunar topography. Any regolith or structures between the reactor and human habitation is beneficial for blocking radiation. The reactor could be located in a crater or on the opposite side of a ridge to provide extra safety.

To protect humans during setup, the location of the reactor could be done by moving the reactor with a vehicle or, if the reactor is launched separately from human habitation modules, a small rover on the reactor lander could transport cabling from the reactor on the habitation. A rover carrying cables would be less massive than a rover for the full reactor. A disadvantage to locating the reactor away from humans is that it becomes infeasible to transport reactor waste heat to provide habitation heat.

In locations, where the topography does not have any useful shielding features, burying the reactor in a hole is an option for shielding. The depth of the hole is relative to the distance that the reactor can be located from humans. Poston 59 found that to keep the human effective dose limit under 5 rem per year, the required hole depth, (i.e., depth of the bottom of the reactor) is approximately 1 m for a reactor 160 m away and 2 m for a reactor 40 m away.

Scattering neutrons irradiating the Stirling engine becomes a problem as the depth of the hole increases. At approximately 1.5 m, the fast neutron fluence in the Stirling alternators is minimized. For a deeper or shallower hole, the fluences increases, requiring that the Stirling engines be surrounded by thicker shielding or located at a further distance from the reactor. Either case adds system mass, so there is an optimal arrangement for shielding mass, Stirling engine configuration, and depth at which the reactor is buried. An added benefit to burying the reactor is shielding it from micro meteorites.

One important point that is not covered in regolith shielding discussion is the mass of equipment needed for burying the reactor. There are two approaches to burying the reactor manual digging and robotic digging. The gloves on space suits used on the Moon in the Apollo program were not comfortable nor suitable for extended manual labor. On Apollo 17 Astronauts Eugene Cernan and Harrison Schmitt experienced blistered knuckles and fatigued muscles from using field geology tools on their three day stay on the Moon. Manually digging would require the astronauts to dig in the radiation environment in the vicinity of the reactor or to dig a hole first and have the reactor moved automatically into the hole.

An alternative to manually digging is to have a machine dig the hole for the reactor. Through using Lunar regolith would decrease shielding mass, but the mass of a machine for burying the reactor would add extra launch mass. Both of these options are less preferable than finding a suitable indigenous topography that can be used. Machines used for burying the reactor could also be utilized for protecting human habitats from the ambient radiation environment. Mars and Moon do not have magnetic fields to protect from solar or cosmic radiation, and the Moon lacks an atmosphere that would provide some protection. For protection, humans will be required to live in environment that have radiation shielding as part of the structure.

2.4.16 Nuclear energy propulsion of aircraft (NEPA)

Beginning in 1946, the Army Air Forces (AAF)xe "Army Air Forces (AAF)" first sponsored a study on the Nuclear Energy Propulsion of Aircraft (NEPA)xe "Nuclear Energy Propulsion of Aircraft (NEPA)" project. The effort progressed over the next several years after reviews by the Atomic Energy Commission (AEC) and the Massachusetts Institute of Technology (MIT). From this emerged the joint USAF-AEC Aircraft Nuclear Propulsion (ANP)xe "Aircraft Nuclear Propulsion (ANP)" program. Ten more years of studies were undertaken by governmental laboratories and industrial firms until the Kennedy administration canceled the effort in 1961. There was, however, some useful follow-on work on high temperature materials and high performance reactors under AEC direction. Also, some of the developmental work continued under space nuclear programs. *Aircraft Nuclear Propulsion* is a comprehensive, unclassified, annotated bibliography of the US government program to develop a nuclear-powered aircraft.

Attempts by the United States government to develop nuclear-powered aircraft for military applications between 1946 and 1961 was one of the most technologically challenging programs ever undertaken. Adding to the difficulties, this development effort was a politically contentious issue throughout its fifteen-year lifetime. Because of the continuing interest in this subject, the Air Force History Support Office contracted for research of the literature and documentation in an annotated bibliography.

The origins of US attempts to power aircraft by nuclear energy began with the World War II Manhattan Project. In 1945, as the war in Asia ended with the dropping of two atomic bombs on Japan, numerous ideas for use of the awesome force of the atom were considered. Controlled use of atomic energy had been demonstrated with the nuclear reactors built for the Manhattan Project, and the concept of a nuclear-powered aircraft that could stay aloft for weeks was attractive to the military. Beginning in 1946, the Army Air Forces sponsored a feasibility study, the Nuclear Energy for Propulsion of Aircraft (NEPA) project. The NEPA project was carried out by a team of industrial firms led by the Fairchild Engine and Airplane Corporation, initially located at the Oak Ridge National Laboratory (ORNL)xe "Oak Ridge National Laboratory (ORNL)", where space was made available by the newly-formed Atomic Energy Commission (AEC). Two years later, the AEC requested that the first significant review of the NEPA project be conducted by a distinguished group of scientists and engineers organized by the Massachusetts Institute of Technology (MIT)xe "Massachusetts Institute of Technology (MIT)" and referred to as the Lexington Group. The Lexington Report, issued in September 1948, is often cited as the basis for proceeding further with development work on the nuclear-powered aircraft. The MIT group concluded that, with adequate resources and competent manpower for an intensive effort, flight of a nuclear-powered aircraft would take fifteen years and cost about a billion dollars.

By 1951, the AEC and the Air Force had established a joint program that was known as the Aircraft Nuclear Propulsion (ANP) program. Major industrial firms, including General Electric (GE)xe "General Electric (GE)", Pratt and Whitney, Lockheed, and Convair participated in the ANP program, along with ORNL and what became the Lewis Laboratory of the National Advisory Committee for Aeronautics (NACA)xe "National Advisory Committee for Aeronautics (NACA)", see Fig. 2.21.

Development of a compact, high-powered nuclear aircraft engine proved to be a formidable undertaking. Research and development was required for high temperature materials, radiation shielding, and other exceptionally difficult engineering problems. Determining the radiation effects on people and equipment was another major question. Two different approaches to the nuclear-powered engine design were pursued, GE's open cycle, air-cooled reactor, and Pratt and Whitney's closed cycle, liquid metal-cooled concept. Among the accomplishments were the operation of a series of tests run by

FIG. 2.21 The NB-36H in a test flight, shadowed by a Boeing B-50 *Superfortress*.

GE at the Idaho National Reactor Test Station, where modified J24 turbojet engines were run on nuclear power for a total of 150 h. Also, a small nuclear reactor was flown many times aboard a modified B−36 bomber to study radiation shielding, as well as radiation effects on aircraft materials. After continuing for fifteen years and spending one billion dollars, as predicted by the Lexington group, the ANP program was canceled in early 1961 by the new administration of President John F. Kennedy. Significant technological advances had been made, but the fundamental question—can a nuclear airplane be flown safely?—was never answered.

2.4.17 Project prometheus 2003

In 2002 NASA announced its Nuclear Systems Initiative for space projects, and in 2003 this was renamed Project Prometheus and given increased funding. Its purpose is to enable a major step change in the capability of space missions. Nuclear-powered space travel will be much faster than is now possible and will enable manned missions to Mars.

One part of Prometheus, which is a NASA project with substantial involvement by DOE in the nuclear area, is to develop the Multi-Mission Thermoelectric Generator and the Stirling Radioisotope Generator described in the RTG section above.

A more radical objective of Prometheus is to produce a space fission reactor system such as those described above for both power and propulsion that is safe to launch and which will operate for many years. This will have much greater power than RTGs. Power of 100 kW is envisaged for a nuclear electric propulsion system driven by plasma.

The FY 2004 budget proposal was $279 million, with $3 billion to be spent over five years. This consists of $186 million ($1 billion over 5 years) building on last year's allocation plus $93 million ($2 billion over five years) toward a first flight mission to Jupiter - the Jupiter Icy Moon Orbiter, expected to launch in 2017 and explore for a decade. Project Prometheus received $430 million in 2001 budget.

In 2003 NASA's Project Prometheus successfully tested a High Power Electric Propulsion (HiPEP)xe "High Power Electric Propulsion (HiPEP)" ion engine. This operates by ionizing xenon with microwaves. At the rear of the engine is a pair of rectangular metal grids that are charged with 6,000 V of electric potential. The force of this electric field exerts a strong electrostatic pull on the xenon ions, accelerating them and producing the thrust that propels the spacecraft. The test was at up to 12 kW, though twice that is envisaged. The thruster is designed for a 7 to 10-year lifetime with high fuel efficiency, and to be powered by a small nuclear reactor.

(In July 14, 2006 at their facility in Racine, WI) - Modine Manufacturing Company (NYSE: MOD), a world leader in designing and developing heating and cooling solutions for a diversified group of markets including electronics cooling, automotive, truck, heavy-duty and industrial, announced that it has delivered, through its wholly owned subsidiary Thermacore International Inc., high temperature titanium heat pipes to NASA Glenn Research Center in Cleveland, Ohio. These heat pipes are evaluation prototypes in support of NASA's effort to develop radiator panels for rejecting heat from the power generation systems for long duration space and planetary base missions and could be used on trips to the moon, Jupiter and points further in space.

The Thermacore heat pipes are intended for use at temperatures up to 250° Celsius. The delivered prototypes, which are 12.7 mm in diameter and 1.15 m long, are made from Titanium and use water as the internal working fluid. These devices, capable of transporting over 500 W, are produced at Thermacore's research and development facilities in Lancaster, Pennsylvania. NASA is pursuing the development of power generation systems capable of producing tens of kilowatts of electrical power for future missions to Jupiter and beyond. The Brayton cycle is the leading power conversion system being evaluated by NASA. Just like all electric generating systems, the unconverted waste heat must be rejected. In this situation, it will be done with large radiator panels that contain heat pipes operating in the temperature range of 20^0 Celsius to 250° Celsius. To minimize weight, these heat pipes were constructed from Titanium. Heat pipes have been widely accepted in military applications and have high reliability standards.

Spacecraft applications to date have been for heat pipes operating between 200 and 350°K. Consequently, a working fluid that's freezing and boiling points encompass this temperature range and has a high latent heat, a low viscosity, and high heat transport capability must be selected. Recent research study by Goddard Space Flight Center (GSFC)xe "Goddard Space Flight Center (GSFC)" has selected ammonia as an appropriate working fluid whose fluid properties meet these criteria. However, for safety reasons, the toxicity of ammonia precludes its use in manned environments such as the shuttle cabin [reference 4]. GSFC has selected aluminum alloys, such as 6061 and 6063, for the container material of the heat pipe because of their long-term compatibility with ammonia (See Tables 1–2); heritage; ability to have an extruded axial groove wick structure; ease of fabrication, shaping, and configuring; good thermal compatibility with aluminum radiators and heat sinks; and weldability characteristics.

Mahefkey and Lundberg [6] paper summarizes the envisioned, future usage of high and low temperature heat pipes in advanced Air Force spacecraft. Thermal control requirements for a variety of communications, surveillance, and space defense missions are forecast. Thermal design constraints implied by survivability to potential weapons effects are outlined. Applications of heat pipes to meet potential low and high power spacecraft mission requirements and envisioned design constraints are suggested. A brief summary of past Air Force sponsored heat pipe development efforts is presented and directions for future development outlined, including those applicable to advanced photovoltaic and nuclear power subsystem applications of heat pipes.

2.4.18 Mars one mission

British entrepreneur Richard Branson aims to populate Mars. His company Virgin Galactic has already developed a spaceship designed to provide tourists with suborbital spaceflights, but the project has fallen far behind schedule (Figs. 2.22–2.25). Elon Musk of Tesla, the developer of the electric Tesla automobile founded the space company SpaceX to send space tourists 50 million kilometers to Mars [50].

FIG. 2.22 The mars project. *Courtesy of NASA.*

FIG. 2.23 Mars one dwelling exterior [50].

FIG. 2.24 Mars one dwelling interior [50].

FIG. 2.25 Mars one greenhouse [50].

SpaceX presented its model Dragon 12 spaceship in May 2014. It has room for 7 passengers, and it is hoped that it will begin flying astronauts to the International Space Station (ISS)xe "International Space Station (ISS)" as early as 2016. The Mars One Project hopes that it will ferry the first settlers to Mars.

Mars One hopes to begin a series of unmanned cargo flights to Mars in 2018 to establish the necessary infrastructure.

In 2025 the first four out of a planned 24 settlers touch down, they will find a "reliable living environment". There will be a living unit for each along with inflatable greenhouse for lettuce, tomato and zucchini.

Water will be provided by extracting ice particles from the Martian soil and some of it will be broken down to provide Oxygen. Solar cells will be used to generate electricity [50].

Mars One hopes that $6 Billion will be enough to finance the project up to the first landing. The plan calls for most of the money to be generated by the sale of broadcasting rights. The major television channels already spend around $4 Billion for the privilege of televising the Olympic Games [50].

2.4.19 Kilopower reactor using stirling technology (KRUSTY) experiment

Kilopower (full name is Kilopower Reactor Using Stirling Technology (KRUSTY) [51] is an experimental project aimed at producing a new design for nuclear reactors for space travel [52,53]. The project started in October 2015, led by NASA and the Department of Energy's National Nuclear Security Administration (NNSA)xe "National Nuclear Security Administration (NNSA)" [54]. The Kilopower reactors will come in 4 sizes able to produce from one to 10 kW of electrical power (1−10 kWe), continuously for 12–15 years [55,56]. The fission reactor uses Uranium-235 to generate heat that is carried to the Stirling converters via passive sodium heat pipes. See Fig. 2.26.

Fig. 2.26 is presentation of Prototype NASA 1 kW Kilopower nuclear reactor for use in space and planet surfaces.

Potential applications include nuclear electric propulsion and a steady electricity supply for crewed or robotic space missions that require large amounts of power, especially where sunlight is limited or not available.

A fission reactor that uses uranium-235 in a molybdenum alloy was selected due to its suitability for the application, design simplicity, and its manufacturability at existing facilities. A uranium core has the additional benefit of avoiding uncertainty in the supply of Plutonium (Pt). The core of the reactor will be of solid, cast Uranium-235 alloy surrounded by a beryllium

FIG. 2.26 Prototype NASA 1 kW kilopower nuclear reactor. *Courtesy of NASA.*

oxide reflector. This reflector focuses neutron emissions and returns their energy back into the core to minimize nuclear gamma radiation which could impair on-board electronics.

The prototype KRUSTY 1kWe Kilopower prototype reactor weighs 134 kg and contains 28 kg of ^{235}U. The space rated 10 kW_e Kilopower for Mars is expected to mass 226 kg and contain 43.7 kg of ^{235}U

Nuclear reaction control is provided by a single rod of boron carbide which is a neutron absorber that is initially inserted, so that pre-launch radiation is negligible. Once the reactor reaches its destination, the neutron absorbing *poison rod* is removed to allow the nuclear chain reaction to start. Once the reaction is initiated, it cannot be stopped completely, although the depth of insertion provides a mechanism to adjust the heat output from the reactor core to the load demand.

Passive heat pipes filled with liquid sodium then transfer the reactor core heat to one or more Stirling enginesxe "Stirling engines", which converts heat into rotary motion that drives a conventional electric generator. The melting point of sodium is 98°C (208°F) which means that liquid sodium can flow freely at high temperatures between about 400 and 700°C (750 and 1,300°F). Nuclear fission cores typically operate at about 600°C (1,100°F).

The reactor is designed to be intrinsically safe in a wide range of environments and scenarios. Several feedback mechanisms are employed to mitigate a nuclear meltdown. The primary method is passive cooling, which requires no mechanical mechanisms to circulate coolant. The reactor design is self-regulating, through design geometry that creates a negative temperature reactivity coefficient.

NASA and the Department of Energy's National Nuclear Security Administration (NNSA) demonstrated a nuclear reactor power system that could enable long-duration crewed missions to the Moon as well as Mars and destinations beyond. A demonstration called the Kilopower Reactor Using Stirling Technology (KRUSTY) experiment was developed at the Glenn Research Center in Cleveland Ohio. The Kilopower experiment was conducted at the NASA's Nevada National Security Site (NNTSS)xe "Nevada National Security Site (NNTSS)" from November 2017 through March 2018.

"Safe, efficient and plentiful energy will be the key to future robotic and human exploration" said Jim Reuter. NASA's acting associate administrator for the Space Technology Mission Directorate (STMD) in Washington. "I expect the Kilopower project to be an essential part of Lunar and Mars power architectures as they evolve".

Kilopower is a small, lightweight fission power system capable of providing up to 10 kWe of electrical power., capable to run several average households continuously for at least 10 years. Four Kilopower units would provide enough power to establish an outpost on the Moon or Mars. On the Moon power generation from sunlight is difficult because Lunar nights are equivalent to 14 days on Earth.

The power system uses a solid cast ^{235}U reactor core, about the size of a paper towel roll. Passive Sodium Heat Pipes transfer reactor heat to high-efficiency Stirling engines, which converts the heat to electricity.

The Kilopower team conducted the experiment in four phases. The first two phases conducted without power, confirmed that each component of the system behaved as expected. During the third phase, the team increased power to heat the core incrementally before moving on to the final phase. The experiment culminated with a 28-h, full-power test that simulated a mission as illustrated in Fig. 2.27, including reactor startup, ramp to full power, steady operation and shutdown.

Throughout the experiment, the team simulated power reduction, failed engine, and failed heat pipes as well, showing that the system could continue to operate and successfully handle multiple failure. Accordingly.

Such a demonstration could pave the way for future Kilopower system that power human outputs on the Moon and Mars, including missions that rely on In-situ Resource Utilization to produce local propellants and other materials.

The KRUSTY reactor was run at full power on March 20, 2018 during a 28-h test using a 28 kg Uranium-235 reactor core. A temperature of 850°C (1,560°F) was achieved, producing about 5.5 kW of fission power. See Figs. 2.28 through 2.29.

The test evaluated failure scenarios including shutting down the Stirling engines, adjusting the control rod, thermal cycling, and disabling the heat-removal system. A Scram test concluded the experiment. The test was considered to be a highly successful demonstration.

FIG. 2.27 Kilopower reactor experimental setup.

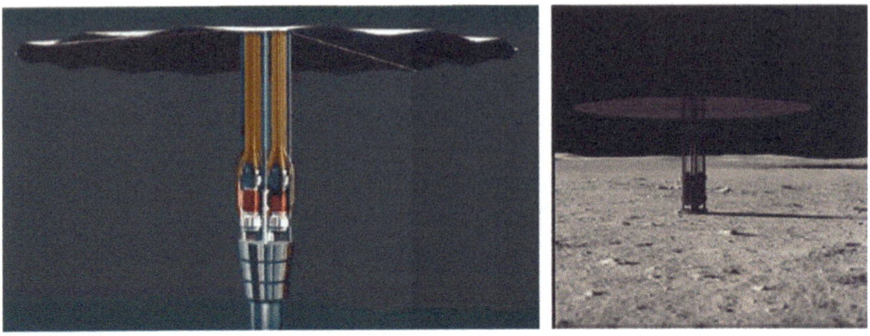

FIG. 2.28 Kilopower reactor with stirling engine in vacuum of space.

FIG. 2.29 Artistic image of kilopower reactor with a stirling engine deployment on the mars surface. *Courtesy of NASA.*

Fig. 2.28 is illustration of Kilopower reactor with a Stirling Engine uses radiators to dissipate heat in the vacuum of space.

The development of Kilopower also included an experiment called Demonstration Using Flattop Fissions (DUFFxe "Demonstration Using Flattop Fissions (DUFF") experiment, which was tested in September 2012 using the existing Flattop assembly as a nuclear heat source. When DUFF was tested at the Device Assembly Facility at the Nevada Test Site (NTS)xe "Nevada Test Site (NTS)", it became the first Stirling engine powered by fission energy and the first use of a heat pipe to transport heat from a reactor to a power conversion system.

According to David Poston, the leader of the Compact Fission Reactor Design Team, and Patrick McClure, the manager for small nuclear reactor projects at Los Alamos National Laboratory, https://en.wikipedia.org/wiki/Kilopower-cite_note-DUFF-Gibson-2 the DUFF experiment showed that "for low-power reactor systems, nuclear testing can be accomplished with reasonable cost and schedule within the existing infrastructure and regulatory environment".

2.5 Space shuttle orbiter heat pipe applications

Grumman under contract by NASA launched an investigation that was made to formulate and evaluate heat pipe applications for the space shuttle orbiter.

Of the 27 specific applications which were identified, a joint NASA/Grumman evaluation resulted in the selection of five of the most promising ones for prototype development [1].

The primary objectives of this study were to:

- Identify potential heat pipe applications for the space shuttle orbiter
- Evaluate the applications and recommend the most promising ones for further development
- Perform detailed design and analysis on the recommended applications
- Prepare design drawings with necessary material specifications to permit fabrication of prototype hardware for at least three of the recommended applications
- Prepare test plans for performance verification of the three or more prototype applications

The secondary objectives were to:

- Evaluate a general design concept employing "off-the-shelf" heat pipe components to be used in minimizing costs, in the event of an extensive commitment to heat pipe systems
- Create study plans for the development of prototype heat pipe hardware for space station, space shuttle, and common shuttle/station applications (including space radiators)

The formulation process is described in a report prepared by Grumman [1], along with the applications which evolved. The bulk of the discussion deals with the "top" five applications, namely:

- Heat pipe-augmented cold rail
- Avionics heat pipe circuit
- Heat pipe/phase change material modular sink
- Air-to-heat pipe heat exchanger
- Heat pipe radiator for compartment temperature control

The philosophy, physical design details, and performance data are presented for each concept along with a comparison to the baseline design where applicable. A sixth application, heat pipe space radiator for waste heat rejection, was also recommended for prototype development—but its development would be more efficiently handled under a separate contract.

Each of the shuttle subsystems, i.e., structure, propulsion, avionics, power and environmental control, and life support, was reviewed in detail, with possible heat pipe application areas indicated by the heat sources and sinks located throughout the shuttle vehicle. Twenty-seven initial applications were defined, from which 11 were chosen for further design and analysis. The procedure used to evaluate these 11 was based on a better than/worse than comparison with the baseline system for each of six criteria: temperature gradient, capacity margin, power requirements, control requirement, weight, and safety. Because of the lack of factual data, parameters such as cost, maintainability, reliability, durability, and development risk were only evaluated on a secondary basis.

The 11 prime contenders are briefly summarized below:

1. Isothermalization of the leading edge of the wing to lower peak temperature and to increase mission life.
2. Wheel well radiators to maintain minimum temperatures sufficient for tire survival by supplying waste heat.
3. A design similar to (2) for the air-breathing engine compartments.
4. A HP avionics circuit to collect and transfer the thermal load from electronic boxes to the heat transfer system.
5. Modular heat sinks for cooling remotely located components without the need for long extensions of the pumped coolant system.
6. An adaptation of (5) for the flight/voice recorders located in the tail.
7. A modular heat pipe heat exchanger system for adapting air-cooled commercial and military avionics to the shuttle.
8. An all-HP radiator system for waste heat rejection.
9. A modified version of (8) incorporating a pumped fluid loop header.
10. A HP-augmented cold rail capable of absorbing an order of magnitude greater local power density when compared to a simple fluid cold rail.
11. A high-temperature heat rejection system for the fuel cells.

The preliminary design studies of these prime contenders included a description of the overall system, supporting drawings showing the heat pipe systems and shuttle interfaces, and heat pipe design details including capacity requirements, working fluids, wick design, pipe lengths, and diameters (See Fig. 2.30) shuttle landing.

FIG. 2.30 The space shuttle discovery lands on kennedy space Center's runway 33 Tuesday, April 20, 2010, in cape canaveral, Florida.

Further evaluation resulted in 6 of the 11 concepts being selected for detailed design and analysis. These six are noted in Table 2.6, which summarizes the results of the evaluation process.

2.6 Heat pipe in electronics

All electronic components, from microprocessors to high-end power converters, generate heat, and rejection of this heat is necessary for their optimum and reliable operation. As electronic design allows higher throughput in smaller packages, dissipating the heat load becomes a critical design factor. Many of today's electronic devices require cooling beyond the capability of standard metallic heat sinks. The heat pipe is meeting this need and is rapidly becoming a mainstream thermal management tool [13].

Heat pipes have been commercially available since the mid-1960s. Only in the past few years, however, has the electronic industry embraced heat pipes as reliable, cost-effective solutions for high-end cooling applications. The purpose of this section is to explain basic heat pipe operation, to review key heat pipe design issues, and finally to discuss current heat pipe electronic cooling applications [14].

Perhaps the best way to demonstrate the heat pipe application to electronic cooling is to present a few of the more common examples. Currently, one of the highest volume applications for heat pipes is cooling the Pentium processors in notebook computers. Due to the limited space and power available in notebook computers, heat pipes are ideally suited for cooling the high-power chips [14].

Fan-assisted heat sinks require electrical power and reduce battery life. Standard metallic heat sinks capable of dissipating the heat load are too large to be incorporated into the notebook package. Heat pipes, on the other hand, offer a high efficiency, passive, compact heat transfer solution. Three- or 4-mm diameter heat pipes can effectively remove the high heat flux from the processor. The heat pipe spreads the heat load over a relatively large area heat sink, where the heat flux is so low that it can be effectively dissipated through the notebook case to ambient air. The heat sink can be the existing components of the notebook, from Electromagnetic Interference (EMI)xe "Electromagnetic Interference (EMI)" shielding under the key pad to metal structural components [14]. Various configurations of notebook heat pipe heat sinks are shown in Fig. 2.31.

Typical thermal resistances for these applications at 6–8-W heat loads are 4–6°C/W. High-power mainframe, mini-mainframe, server and workstation chips may also employ heat pipe heat sinks. High-end chips dissipating up to 100 W are outside the capabilities of conventional heat sinks. Heat pipes are used to transfer heat from the chip to a fin stack large enough to do the convection of the heat to the supplied airstream. The heat pipe isothermalizes the fins eliminating the large conductive losses associated with standard sinks.

FIG. 2.31 Typical notebook heat pipe heat sink [15].

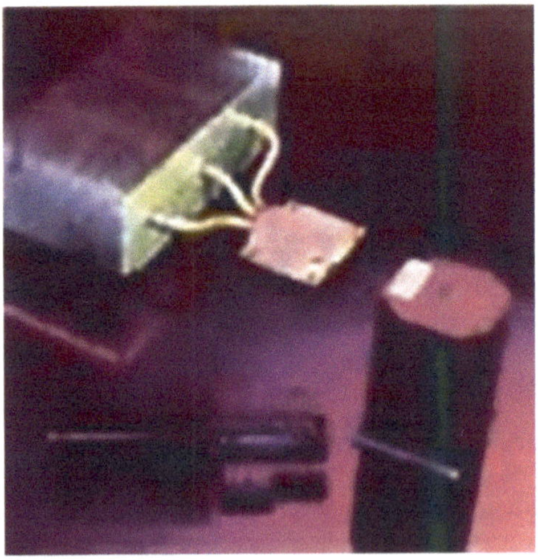

FIG. 2.32 High end CPU heat pipe heat sink [10].

The heat pipe heat sinks, shown in Fig. 2.32, dissipate loads in the 75–100-W range with resistances from 0.2 to 0.4°C/W, depending on the available airflow [15].

In addition, other high-power electronics, including Silicon-Controlled Rectifiers (SCRs)xe "Silicon-Controlled Rectifiers", Insulated-Gate Bipolar Transistors (IGBTs)xe "Insulated-Gate Bipolar Transistors (IGBTs)", and thyristors, often utilize heat pipe heat sinks. Heat pipe heat sinks similar to the

FIG. 2.33 High power IGBT heat pipe heat sink [10].

one shown in Fig. 2.33 are capable of cooling several devices with total heat loads up to 5 kW.

These heat sinks are also available in an electrically isolated version where the fin stack can be at ground potential with the evaporator operating at the device potentials of up to 10 kV. Typical thermal resistances for the high-power heat sinks range from 0.05 to 0.1°C/W. Again, the resistance is predominately controlled by the available fin volume and airflow [15].

2.6.1 Electronic and electrical equipment cooling

Miniaturization of electronic components is accompanied by increased demands on heat dissipation systems due to the increased density of the components. For example, the digital computer has evolved from a massive system that filled an entire room to a unit which can be stored in a small briefcase. However, the overheating problems associated with the dense packing of heat-generating integrated circuit chips used in the computer such as Central Processing Unit (CPU)xe "Central Processing Unit (CPU)" and Graphics Processing Unit (GPU)xe "Graphics Processing Unit (GPU)" cooling have escalated dramatically. Since the reliability of these and other types of electronic components is sensitive to their operating temperature, steps have been taken to improve heat dissipation by using heat pipes.

Other applications to electronic cooling have included rectifiers, thyristors, transistors, traveling wave collectors, audio and RF amplifiers and high density semiconductor packages.

After the introduction of the Pentium processor in 1993, the processor performance and power consumption trend has significantly increased annually. In the year 2000, the heat flux was approximately 10–15 W/cm^2, eventually reaching 120–150 W/cm^2 in 2010. The average power consumption for laptop/notebook computers is currently between 25 and 50 W, while desktops and servers consume between 80 and 150 W. Regardless of components, power level, or type of computer/processor, as electronics are packed into smaller volumes, it is important to optimize the thermal management components with higher cooling rates.

Presently, almost all laptop/notebook computers use a heat pipe Remote Heat Exchanger (RHE)xe "Remote Heat Exchanger (RHE)", where a heat pipe is used to transfer heat from the processor to the heat exchanger. Heat pipes have been an important part of laptop/notebook computer cooling systems and will remain so for the foreseeable future. Figure 1 shows a simplified typical layout of a heat pipe remote heat exchanger for laptop/notebook computers, using the heat pipe embedded inside a metal block (CPU case) over the silicon (die). For simplicity, a single heat sink is shown in Fig. 2.34 to illustrate the

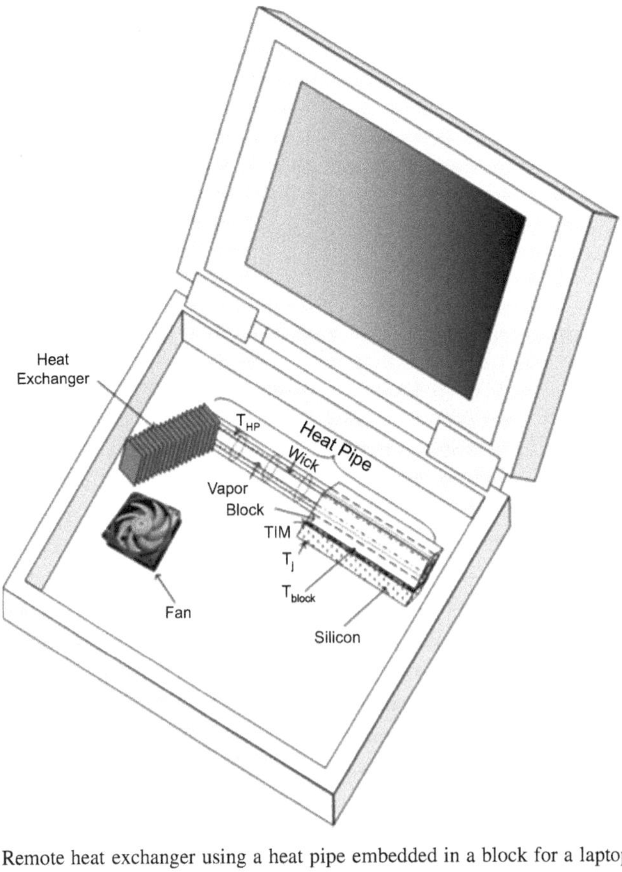

FIG. 2.34 Remote heat exchanger using a heat pipe embedded in a block for a laptop computer.

concept of thermal resistance optimization for personal computers. The overall thermal resistance of this system is presented by the following equation:

$$Q = \text{Heat Dissipation } (W) = \frac{T_j - T_\infty - T_{sys}}{R_{j-\infty}} \tag{2.2}$$

where:

$R_{j-\infty} = RS_i + RTIM + R_e + R_c + R_a + RHT_{-\infty}$
T_j = Junction temperature [°C]
T_∞ = Ambient temperature [°C]
T_{sys} = Impact of other energy sources in the system [°C]
$R_{j-\infty}$ = Thermal resistance of the CPU die to ambient [°C/W]
RS_i = Thermal resistance of silicon (microprocessor package) [°C/W]
$RTIM$ = Thermal resistance through thermal interface material (TIM) [°C/W]
R_e, R_c, R_a = The thermal resistances for the evaporator, condenser, and adiabatic section of the heat pipe, respectively.
$RHT_{-\infty}$ = Thermal resistance of heat exchanger to ambient [°C/W]

The thermal resistances of both the evaporator and condenser sections of the heat pipe are much higher than the adiabatic sections. Fig. 2.35 [61] shows a simple pie chart with fractional distributions of various thermal resistances for a heat pipe remote heat exchanger technology for laptop computers. [61] also showed that the overall thermal resistances presented in Fig. 2.35 can be reduced by 17% by reducing the thermal resistances of the evaporator and condenser sections. [62] presented the various typical CPU thermal resistances versus heat dissipation (See Fig. 2.36) by assuming a junction temperature of 100°C, outside ambient temperature of 35°C, and a system temperature rise of 10°C. Mochizuki et al. (2006) [5] presented a number of potential future

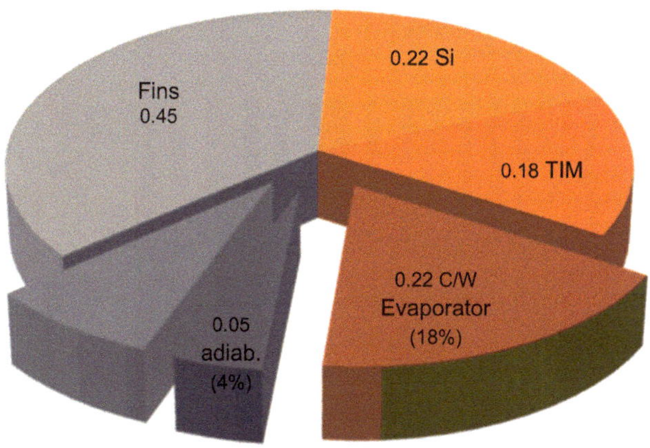

FIG. 2.35 Pie chart showing component thermal resistances of a heat pipe remote heat [64].

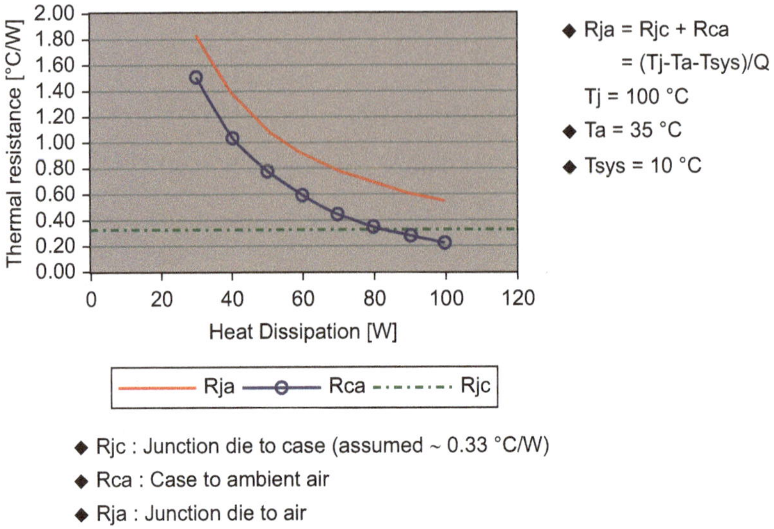

FIG. 2.36 Various typical CPU thermal resistances versus heat dissipation [65].

cooling technologies for higher power cooling chips, including vapor chambers and Loop Heat Pipes (LHPs) xe "Loop Heat Pipes (LHPs)" for desktop computers and servers.

Fig. 2.35 is depiction of Pie Chart showing component thermal resistances of a heat pipe remote heat exchanger for a laptop PC [64].

Fig. 2.37 [62] fins are soldered directly to the vapor chamber and proposed to replace the solid copper metal Interface Heat Spreaders (IHS)xe "Interface Heat Spreaders (IHS)" with a two-phase micro-channel vapor chamber. [67] proposed a cold energy storage system using wickless thermosyphons for cooling large scale data centers (8800 kW).

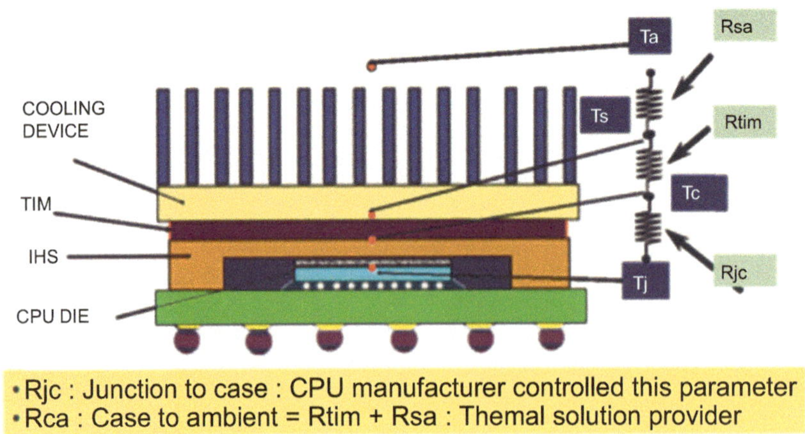

FIG. 2.37 Thermal solution integration for desktop and servers [66].

Annular heat pipes are proposed for use in the fusing units of copy machines by some copy machine manufacturers as heat drums for fast thermal response and uniform fusing [64]. The proposed annular heat pipe is composed of two unequal diameters, where the evaporation and condensation is separated in the radial direction so that the heat is transferred in the radial direction. While miniaturization is not a significant factor in large-scale electrical equipment, the application of heat pipes to this area is important due to the increase in efficiency when the components operate at lower temperatures [67–70].

Electric motors have been designed which incorporate heat pipes around the periphery of the rotor for cooling during operation. Another design replaced the solid motor shaft with an on-axis rotating heat pipe with an internal taper for cooling. These designs removed heat from the interior of the electric motor so that the bearings operated at a lower temperature, which increased efficiency. Also, since the resistance of the electric windings decreases with temperature, less power was required to maintain a particular load. Similarly, circuit breakers and electric transformers designed with heat pipe cooling benefit from the decrease in internal temperature.

Underground power transmission cables that are placed close to hot water or steam pipes are sometimes subjected to limitations in the maximum permissible current due to thermal interference. Therefore, it is advantageous to place maintenance-free cooling systems using heat pipes at these locations. Several heat pipes are placed at the hot spot along the direction of the cable, which are bent so that the condenser sections are above ground or in the soil away from the heated section [71]. The two methods of removing heat, either by natural convection to the atmosphere or by conduction to the soil, provide sufficient cooling so that higher current loads are capable of being provided to the user.

2.7 Heat pipe in defense and avionics

Dissipating heat reliably and quickly, for satellite thermal control, is a challenge that grows tougher all the time. Space electronics are steadily becoming more miniaturized, more complex, and more powerful. Increased packaging densities are also limiting the volume available for thermal management systems. The end result, the performance, flexibility, and versatility of thermal management systems, must improve to meet the growing challenge. Plus, "low-maintenance" systems are not good enough in space; thermal control must be "no maintenance." But proven heat pipe technology developed by different companies and designer can meet all these challenges. With a wide range of advanced solid conduction, heat pipe and loop heat pipe technologies are available to meet the need for high performance, system efficiency, and low mass with no moving parts to fail. Companies like Thermacore's advanced solid conduction solutions, heat pipes, and loop heat pipes are

FIG. 2.38 F-22 image.

maintenance-free, making them ideal for satellite, military, and aerospace applications. See Fig. 2.38, which is an image of F-22 Fighter in flight.

Military equipment thermal management solutions from Thermacore stand up to the toughest challenges in the world—from submarine electronic cooling and radar electronics in harsh environments to satellite thermal control management in space.

Thermal technology design, development, and manufacturing have helped deliver thousands of solutions that operate in mission-critical applications in the harshest environmental conditions. Thermacore's advanced thermal management technologies are a major source of superior solutions to allow our military equipment customers to overcome the Size, Weight, and Power (SWaP)xe "Size, Weight, and Power (SWaP)" constraints in current and future generation electronic equipment.

2.7.1 On the ground application

To get a high-performance, lightweight, and compact cooling solution for next generation transmit/receive modules used in military radar systems, prime contractors throughout the world turn to Thermacore for high-performance cold plate technology. These Thermacore solutions are based on advanced solid conduction (k-Core®)-embedded heat pipe technology, vapor chamber heat pipe technology, and vacuum-brazed liquid cooling cold plates. These thermal solutions provide high performance, high-efficiency heat transfer in this space- and weight-constrained application, with heat pipes and k-Core giving cold plates the weight of aluminum with an effective thermal conductivity more than four times that of copper. See Fig. 2.39.

Cooling solutions for military ground-based radars use our experience in sealed enclosure cooling technology, providing reliable performance in extreme temperatures, rain, sand, and snow. That includes reliable electronic cooling for communications and target acquisition in tanks and Humvees, where displays and electronics—as well as thermal solutions themselves—face sweltering heat, high humidity, shock, and vibration.

FIG. 2.39 Next generation of military radar system.

Military electronic cabinet cooling is provided by compact, rugged, and versatile cross flow heat exchangers that resist everything from shock and wind to microorganisms. These military cooling solutions provide superior performance and take advantage of our experience in power electronic heat management.

2.7.2 In the sea application

Highly efficient radar electronics and power supply cooling for next-generation naval destroyer radar systems are an area of expertise for Thermacore heat pipes and cold plates. These technologies are specially designed for high thermal performance, low mass, and rugged conditions, including thermal cycling between -40 and + 75°C. See Fig. 2.40.

Navy antenna cooling is the mission of the heat pipe assemblies used to reduce heat load on USG-2 naval Planar Array Antenna Assemblies (PAAAs) xe “Planar Array Antenna Assemblies (PAAAs)”. Able to absorb heavy shocks from weather, thermal cycling, and impact, these superior Thermacore liquid cooling systems provide peace of mind for Navy warships all over the world.

Under the sea, nuclear submarine power conversion electronic cooling is provided by embedded heat pipe assemblies and vapor chamber assemblies from Thermacore. Thermacore thermal solutions are also used to protect the power subsystems supporting naval radar platforms.

Thermacore heat exchangers are in action cooling nuclear reactor control electronics as well.

FIG. 2.40 Next generation naval destroyer radar systems.

2.7.3 In the air application

Thermacore patented encapsulated graphite k-Core heat pipes are cooling critical electronic components on the new F-35 Joint Strike Fighter (JSF)xe "Joint Strike Fighter (JSF)", providing improved thermal performance over traditional graphite conduction solutions at lower cost.

Loop heat pipes from Thermacore provide highly reliable electro-optics cooling for target acquisition systems, remote wing electronics, and navigational avionics onboard the F-16 fighter aircraft (See Fig. 2.41), standing up to extreme temperatures and up to 9 g of force. It is a performance typical of our thermal solutions for aerospace and avionics.

Superior thermal protection of radar systems for US Coast Guard helicopters is the job of Thermacore remote dissipation heat pipe assemblies, combining heat pipes with fins for better overall thermal performance. For thermal management of aircraft electronics under the most severely cold conditions (down to -70°C), designers turn to Thermacore heat pipe assembly solutions to maintain temperatures close to the optimum for avionics operation.

2.7.4 In the space application

For satellite thermal control and heat transfer, advanced solid conduction, axially grooved low-temperature heat pipes, and loop heat pipe from Thermacore offer the widest variety of high-performance thermal solutions

FIG. 2.41 F-16 Fighter aircraft.

FIG. 2.42 Satellite in space.

for space applications. Thermacore's k-Core space radiator panels and doublers provide lightweight, high-performance heat spreading. See Fig. 2.42.

Thermacore's axially grooved low-temperature heat pipes, using ammonia and ethane as working fluids, can be embedded into radiator panels for improving heat spreading within radiator panels over extended lengths. In addition, loop heat pipes from Thermacore offer high total power dissipation (>2000 W) and lightweight (a deployable radiator panel with a weight of <55 lb.). Deployable, honeycomb-bonded panel construction enables reliable.

2.8 Heat pipe as heat exchanger

The most commonly known and used thermal devices is the heat exchanger, which transfers heat from a hot fluid to a cold one. Heat exchanger is used in a variety of thermal devices ranging from the modern power generation such as nuclear reactor system to domestic heating and air conditioning. There are three basic deigns for heat exchangers, as shown in Fig. 2.43, and they are as follows:

1. Parallel flow.
2. Counterflow.
3. Cross-flow.

The effectiveness of heat exchangers ε is defined as follow:

$$\varepsilon = \frac{Q}{C_{\min}\left(T_{h_{in}} - T_{c_{in}}\right)} \tag{2.3}$$

where.

Q = The heat transfer rate
$C_{\min}$ = The heat capacity rate defined as the product of specific heat and the mass flow rate of the hot or cold fluid whichever is smaller
$T_{h_{in}}$ = Inlet temperature of the hot fluid
$T_{c_{in}}$ = Inlet temperature of the cold fluid

It can be shown theoretically that other things being equal the counter-flow heat exchanger effectiveness than either the parallel-flow or cross-flow heat exchanger.

It can be shown theoretically that other things being equal the counterflow heat exchanger effectiveness than either the parallel-flow or cross flow heat exchanger.

Because heat pipes can operate almost near an isothermal mode, i. e., with a very small temperature drop, the effectiveness of heat pipe heat exchanger can be very high. The main heat-flow resistance for a heat pipe heat exchanger is at the interfaces of the heat pipe exterior surfaces and the hot and cold fluids. In order to have reduction in heat-flow resistances, external fins may be provided at the evaporator and condenser sections of the heat pipe. Excellent heat transfer from the hot fluid to the cold fluid can then be obtained because both the hot and the cold fluids are passed through the coils of finned tubes and the heat is transferred from the hot fluid side to the cold fluid side by the "isothermal" heat pipes. A counterflow heat exchangers using heat pipes is shown in Fig. 2.44 and easily can designed and constructed without any difficulties.

Heat pipe heat exchangers can also take advantage of staggering the finned tubes. Fig. 2.45 illustrates a typical arrangement of heat pipe tubes. A large variety of heat pipe heat exchangers are commercially available at the present

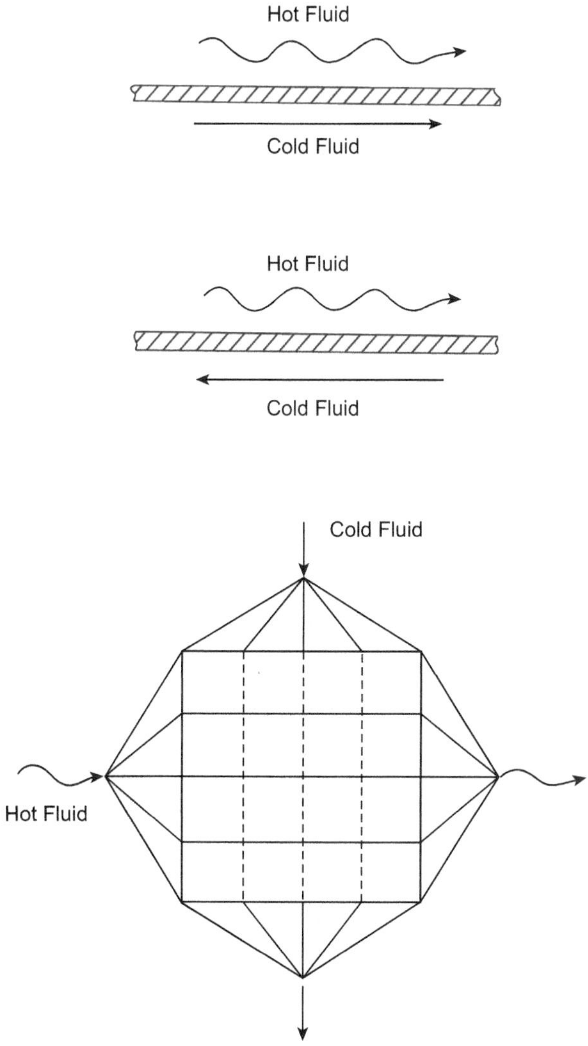

FIG. 2.43 Types of heat exchange [9].

time. Technical data provided by manufacturers for their product usually include the thermal effectiveness and the operating temperature ranges [16].

The quoted effectiveness is almost invariably based upon operation in the counterflow mode, and designer of heat pipe heat exchangers should take advantage of the counterflow between the hot and cold fluids. Design or sizing of a heat pipe heat exchanger involves both the design of the heat pipe and the heat exchanger as a heat transfer device such that it will meet the specified heat

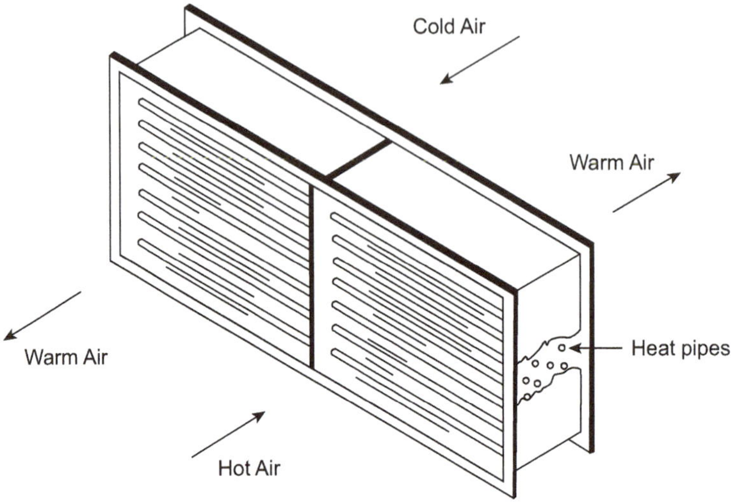

FIG. 2.44 Schematic of a counter-flow heat pipe heat exchanger [9].

FIG. 2.45 Illustration of a staging heat pipe tube [15].

duty, pressure drops on both sides, and the heat pipe will work properly in the operating design range [15].

The Heat Pipe Heat Exchanger (HPHE)xe “Heat Pipe Heat Exchanger (HPHE)” design methodology is illustrated in Fig. 2.46. This design procedure may be characterized by a *case study* method.

It is a complex procedure because of the many qualitative judgments (in addition to the quantitative calculations) that must be introduced [15].

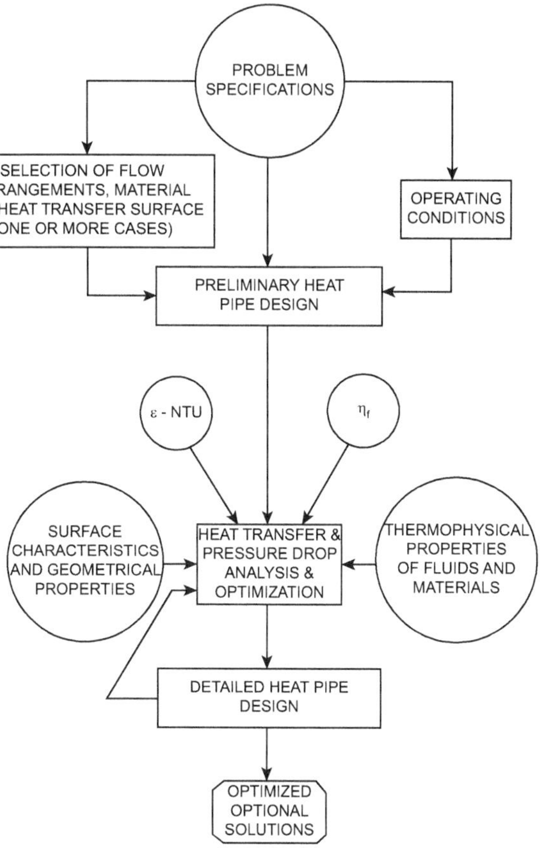

FIG. 2.46 Overall Heat Pipe Heat Exchanger Thermal and hydraulic design methodology [10].

The general characteristic of heat pipe heat exchangers can be illustrated by numerical examples.

This may help engineers at the preliminary stage of their design, although the final design should be made after the manufacturer's data are consulted and analyses are made.

2.9 Heat pipe in residential building

There so many homes built around the world, and heating, air-conditioning, and hot water supply of these homes are a challenge on annual energy consumption everywhere in the world. A typical home in the mid-region of the USA uses about 200 million BTU ($2{:}1 \times 10^{11}$ J) per heating season. Hence, there is a great incentive for home energy conservation. Several devices using heat pipes are developed, e.g., heat recovery from furnace flue gas, waste hot water, and fireplace. The temperature of flue gas at the domestic furnace for

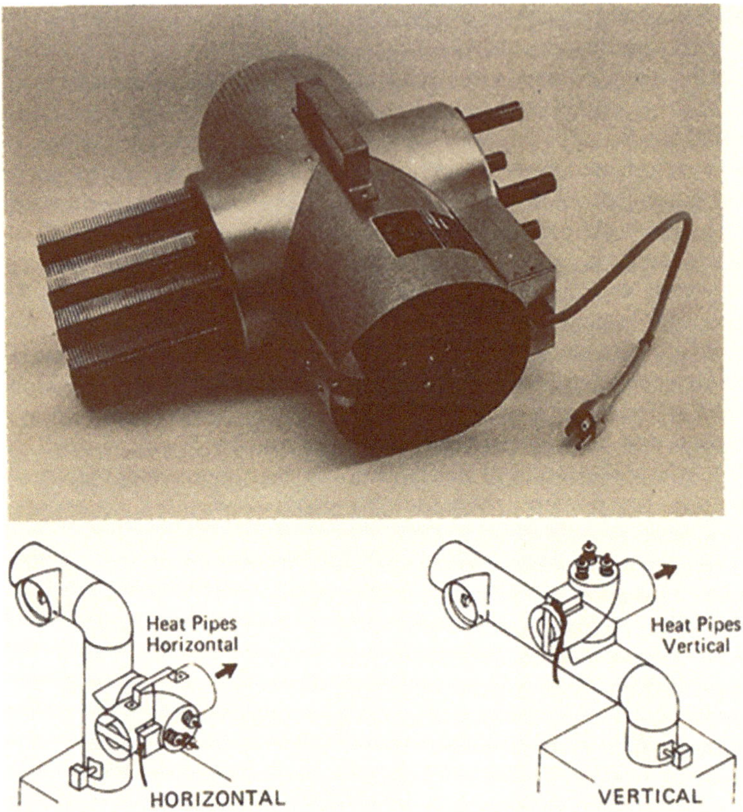

FIG. 2.47 Heat pipe heater for reclaiming heat in flue gas from residential furnaces [9].

space heating is about 500 °F (533 K), which is about 12% of the energy available from the fuel. In order to reclaim this waste heat, a device using a heat pipe has been developed by companies such as Isothermics, Inc., which is illustrated as in Fig. 2.47 and consists of a heat pipe heater that attaches to the flue pipe above the furnace. A heat made of copper and using water as working fluid are highly recommended for this application. The device uses the gas flue as a heat source; this heat then can be used to heat part or all of the home basement, recreation room, workshop, or laundry room; or a dual duct can be attached to the unit to carry the heat to otherwise chilly spaces inside the house [16].

Houses that have forced air heating can also duct the reclaimed heat to a nearby cold-air return duct in the basement so as to heat up the cold air that comes back to the furnace from the rest of the house, which in return translates to less fuel will be required to heat this air to required temperature for recirculation.

2.10 Heat pipe applications in thermal energy storage systems

Heat pipes and other thermal management systems for renewable energy applications such as in solar, wind, and geothermal can help you design more efficient, more cost-effective, and greener systems.

In Concentrated Solar Power (CSP)xe "Concentrated Solar Power (CSP)" applications, uniform heat distribution is crucial for electric power generation efficiency in applications such as Stirling engines and solar panels. Thermacore heat pipe technology, relying on the laws of thermodynamics to spread heat passively, makes uniform heat distribution possible. This eliminates problems such as nonuniform heat fluxes and associated hot spots where two solar mirrors overlap in CSP systems. In addition, Thermacore technologies remove waste heat from solar cells in Photovoltaics (PV)xe "Photovoltaics (PV)" solar systems to lower solar cell temperatures and improve PV conversion efficiencies. Fig. 2.48.

For wind power generation thermal solution, heat pipe assemblies could be ideal for cooling power generation subsystem in remote areas where maintenance is difficult.

However due to the enormous increase in the global energy demand and possible depletion of conventional energy resources such as fossil fuels, renewable energy resources turned out to be promising options to supply clean and low-cost energy.

The major issue with these sources is their intermittent nature, which causes a gap between energy demand and supply. Most of the studies about renewable energies are focused to improve their efficiency to make them

FIG. 2.48 Image of concentrated solar power panels.

appropriate replace for conventional techniques. Thermal energy storage units can be integrated to renewable power generation systems to alleviate the mentioned temporal mismatch between the energy supply and demand providing power in a dispatchable manner. employing a Phase Change Medium (PCM)xe “Phase Change Medium (PCM)” to store energy and taking the advantage of latent heat of fusion increase the energy storage density and reduce the size and cost of the system in comparison to using sensible energy storage. However, the low thermal conductivity of commercially available phase change materials has limited their performance decreasing the heat transfer rate between heat source and PCM and therefore, prolongs the melting or solidification processes. Implementing a passive heat transfer device such as heat pipe is required to enhance the heat transfer rate between the heat source and the phase change material, which increases the efficiency of the system. In this section, we will discuss the application of heat pipes in latent heat thermal energy storage systems with either high or low operating temperatures. For low-temperature applications like domestic hot water production, organic and inorganic heat storage materials, including paraffin, fatty acids, and inorganic salt hydrates with the temperature range of 0–220°C, can be used as storage media. For high-temperature latent heat thermal energy storage systems which can be integrated in Concentrated Solar Power (CSP)xe “Concentrated Solar Power (CSP)” systems, phase change materials are mostly inorganic salts or their eutectic mixtures with operating temperature up to 800°C. Each of these applications requires specific heat pipe configuration and operating conditions, which will be discussed in the following subsections.

2.10.1 Energy storage methods

Design and development of cost-effective energy storage techniques play an important role in achieving Levelized Cost Of Energy (LCOE)xe “Levelized Cost Of Energy (LCOE)” target of Department of Energy (DOE)xe “Department of Energy (DOE)” SunShot. Different forms of energy can be stored and some aspect of it are summarized here as following subsections.

2.10.1.1 Electrical storage

Pumped Hydro-Power Storage (PHPS)xe “Pumped Hydro-Power Storage (PHPS)”, Compressed Air Energy Storage (CAES)xe “Compressed Air Energy Storage (CAES)”, and flywheels are different methods in terms of mechanical energy storage. For largescale utility energy storage, PHP and CAES can be employed, while flywheels are more suitable for intermediate storage [58].

See Fig. 2.49, which is artistic schematic of the CASE approach.

When energy is available, it is used to run air compressors which pump air into the storage cavern. When electricity is needed, it is expanded through conventional gas turbine expanders. Note that some additional energy (typically natural gas) is used during the expansion process to ensure that

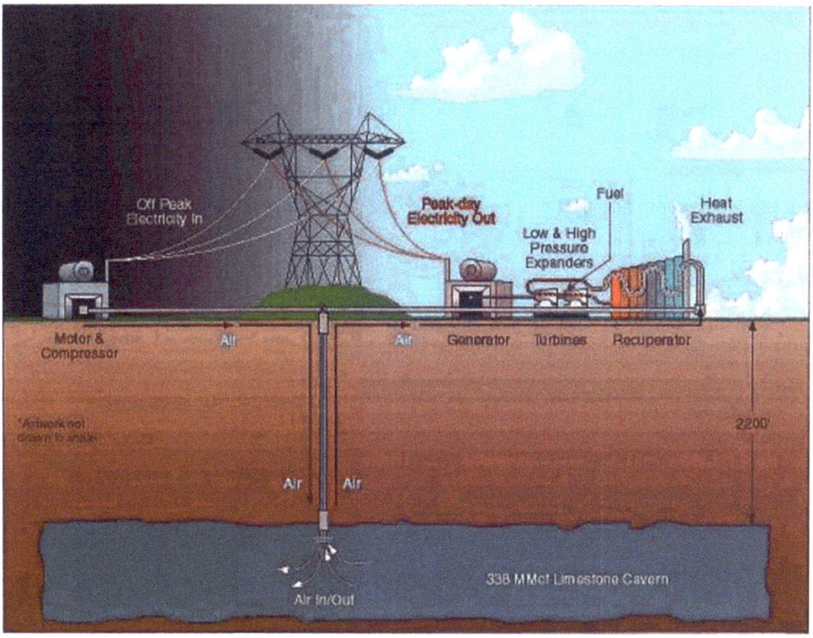

FIG. 2.49 Conceptual representation of case.

maximum energy is obtained from the compressed air (albeit as much as 67% less gas than would be used for an equivalent amount of electricity using gas turbine generators without CAES) [58].

Today, there are exit two Compressed Air Energy Storage (CAES):

1. Compressed Air Energy Storage (CAES)
2. Advanced Adiabatic Compressed Air Energy Storage (AA-CAES)

CAES plants store energy in form of compressed air. Only two plants of this type exist worldwide, the first one built over 30 years ago in Huntorf, Germany with a power output of 320 MW and a storage capacity of 580 MWh. The second one is located in McIntosh, Alabama, USA and began operation in 1991 with a 110 MW output and 2860 MWh of storage capacity. Both are still in operation [58].

2.10.1.2 Thermal energy storage

Thermal Energy Storage (TES)xe "Thermal Energy Storage (TES)" methods are described as the temporary storage of thermal energy, which occurs at high or low temperatures. Thermal energy storage can be obtained by cooling, heating, melting, solidifying, or vaporizing a material in which the energy becomes available as heat by reversing the process. Using this method provides the opportunity to mitigate environmental impacts and results in more efficient and clean energy systems. The TES methods are classified into three main groups: sensible, latent, and thermochemical.

In Sensible Thermal Energy Storage Systems (STESs)xe "Sensible Thermal Energy Storage Systems (STESs)", the energy is stored as a temperature change of the storage medium. The storage medium can be solid as soil, rock, or liquid like water. The amount of stored heat is the function of the mass of storage material and its specific heat as well as the temperature change as follows:

$$Q = \int_{T_i}^{T_f} mC_p dT \tag{2.4}$$

where Q is the amount of stored energy, m is the mass of storage medium, and C_p is the specific heat, while T_i and T_f are the initial and final temperature, respectively.

However, if the temperature range changes for the specific heat C_p at constant pressure are infinitesimal, then the simplified form of Eq. (4.4) can be presented as:

$$Q = mC_{pa}(T_f - T_i) \tag{2.5}$$

In this equation parameter C_{pa} is the average specific heat capacity at constant pressure between the initial T_i and final T_f temperatures, respectively.

Although STESs are possessing simpler design, they disadvantaged by their bigger size and not being capable of working in constant temperature in comparison to latent heat thermal energy storage [17].

2.10.2 Latent heat thermal energy storage

Latent Heat Thermal Energy Storage (LHTES)xe "Latent Heat Thermal Energy Storage (LHTES)" systems work based on absorbing and releasing heat as the storage material undergoes a phase change process. In solid–liquid LHTES, the storage capacity can be calculated as:

$$Q = \int_{T_i}^{T_m} mC_p dT + ma_m \Delta H_m + \int_{T_m}^{T_f} mC_p dT \tag{2.6}$$

where in Eq. (2.5), the symbol of Tm is the melting temperature of the storage medium, am is the fraction melted, and ΔH_m is the heat of melting per unit mass. LHTES benefits from the higher storage density and isothermal operation.

Energy storage media used in latent heat thermal energy storage systems are referred as Phase Change Materials (PCMs)xe "Phase Change Materials (PCMs)". The phase transition can occur from solid to gas, solid to liquid, and liquid to gas [17].

2.10.2.1 Thermochemical

The last approach for thermal storage is by means of thermochemical reactions. This storage method has the capability of long-term storage, but its technical complexity and high costs are not compatible with the cost-effective goal [5].

Among the abovementioned thermal heat storage techniques, LHTES is the most popular since it provides high energy storage density and has the capability of storing energy at constant temperature corresponding to the phase transition temperature of the PCM.

As it was mentioned, phase change in latent heat thermal energy storage systems can be from solid to solid, solid to liquid, and liquid to gas. Solid–solid phase change occurs when a material is transformed from one crystalline to another. These transitions are easy to handle and cost-effective, the lack of liquid material eliminates the risk of leakage, and hence there is no need for encapsulation.

The main drawback of these systems is the small heat of fusion. Solid or liquid to gas transitions have high latent heat of transition, but the large volume changes associated with them make the process complex and almost impossible. The smaller volume changes occurring in solid to liquid phase change make them economically and practically attractive as materials for TES systems despite their smaller heat of phase transition [17].

2.10.3 Latent heat thermal storage materials

The ideal PCM should have the following characteristics:

2.10.3.1 Thermal properties

- Melting temperature in the desired operating range.
- High latent heat of fusion per unit volume.
- High specific heat.
- High thermal conductivity of both phases.

2.10.3.2 Physical properties

- Small volume change during phase transformation process.
- Low vapor pressure at the operating temperature.
- Congruent melting of the PCM.
- High density.
- Kinetic properties.
- No supercooling.
- A high nucleation rate.
- An adequate rate of crystallization.

2.10.3.3 Chemical properties

- Long-term chemical stability
- A completely reversible freeze—melt cycle
- Compatibility with the construction materials
- No corrosion influence on the construction materials
- Nontoxic, nonflammable, and nonexplosive to ensure safety

The last important feature is that the large quantity of the PCM should be available at low cost [5].

2.10.4 Phase Change Material (PCM) classification

A large number of Phase Change Materials (PCMs)xe "Phase Change Materials (PCMs)" are available, which are classified and presented in Table 2.5.

It is clearly understood that lower overall costs are a key factor to make renewable energy technologies competitive with traditional energy sources. Energy storage technology is one path to increase the value and reduce the cost of all renewable energy supplies.

Concentrated Solar Power (CSP)xe "Concentrated Solar Power (CSP)" technologies have the ability to dispatch electrical output to match peak demand periods by employing Thermal Energy Storage (TES)xe "Thermal Energy Storage (TES)". In addition, TES can reduce the Levelized Cost of Energy (LCOE)xe "Levelized Cost of Energy (LCOE)" for CSP plants. In order to achieve this, energy storage technologies require efficient materials with high energy density.

To store thermal energy, sensible and latent heat storage materials are widely used. Latent heat TES systems using Phase Change Material (PCM)xe "Phase Change Material (PCM)" are useful because of their ability to charge and discharge a large amount of heat from a small mass at constant temperature during a phase transformation like melting-solidification. PCM technology relies on the energy absorption/liberation of the latent heat during a physical transformation. Unlike vapor-liquid transformations, solid-liquid transformations produce large enthalpy changes without large density changes, and due to this, salts and metallic alloys are good candidates for PCMs. Ideally these materials should have a specific melting point and high heat of fusion and offer favorable characteristics such as high working temperatures (over 500°C), low vapor pressure, good thermal and physical properties, low corrosivity and toxicity, and, of course, low cost.

Because high-melting-point PCMs have large energy density, their use can reduce energy storage equipment and containment cost by decreasing the size of the storage unit. The optimum input and output temperature of the energy storage equipment is determined by the melting point of the PCM, while the heat capacity of the TES system is determined by the PCM latent and sensible heats.

Knowledge of the thermal properties of a particular molten salt is important for its selection as a thermal storage material. In order to decrease the LCOE, operation of the CSP system must be matched to the operational characteristics of the TES system. For example, PCM TES can be optimized for use with a sensible HTF by using a cascade of materials with equally spaced melting points and uniform thermal properties. However, the information on the thermal properties of molten salt systems at high temperatures is currently incomplete or contradictory.

2.10.4.1 PCMs for different thermal storage applications

As it can be seen in Table 2.6, the PCMs with the application for thermal energy storages are organic compounds, inorganic salts, and their eutectics. Organic compounds used for PCM are generally paraffin waxes, esters, acids, and alcohols.

The salt hydrates, eutectics of inorganic salts, and metals and their eutectics are categorized as inorganic. Those PCMs belonging to organic group usually have low melting points and can only be used for low-temperature applications such as domestic hot water production, direct or heat-pump-assisted space heating, greenhouse heating, solar cooling, etc. [20]. For high-temperature thermal storage, molten salts can be implemented as phase change materials. High-temperature PCM can be used in solar power plants or industrial waste heat recovery systems [21]. Metallic materials have not been popular as other PCMs due to their elevated weight. However, they are good candidates if volume is a priority. The most notable characteristic of metallic materials is their high thermal conductivity, which eliminates the use of thermal conductivity enhancement techniques.

Generally, the materials with the melting temperature below 220°C are considered as low-temperature materials, while melting temperatures up to 420°C as medium temperature materials, and melting points greater than 420°C are classified as high-temperature materials [21]. Some detailed data from common phase change materials are shown in Tables 2.7−2.10.

Although PCMs possess great benefits, they suffer from different problems and technical difficulties which make the performance fall short out of the expectations and limit their widespread use in practical applications. One of the main drawbacks is the low thermal conductivity of the commercially available PCM which limits the heat transfer rate between the heat source and the PCM. This will result in prolonged melting and solidification processes and the overheating of the heat transfer surface. To address the mentioned issue, different techniques have been suggested. These methods include the use of finned tubes [22,23], the use of high thermal conductive small particles dispersed in the PCM [24,25], the microencapsulation of the PCM [26], and the use of highly porous PCM-filled conductive material 29. An alternative approach is embedding of heat pipe inside the PCM to spread the heat provided by heat source throughout the PCM. Latent heat thermal energy

TABLE 2.6 Shuttle heat pipe applications evaluation [14].

Original 27 candidates	11 Prelim. design studies	Six detail designs
1. TPS leading edge		
2. Landing gear		
3. Avionics HP circuit		
4. Modular sinks		
5. ATR equipment		
6. Flight and voice recorders		
7. HP radiator W/HP header		
8. HP radiator with integral HP/fluid header	TPS leading edge	
9. ECS cold rail	Landing gear	
10. HP radiator for fuel cell	Avionics HP circuit	
11. Air-breathing engine compartment	Modular sinks	
	ATR equipment	HP radiator for compartments
12. OMS LH_2 boil-off	Flight and voice recorders	HP radiator, waste heat[a]
13. High-intensity lights	HP radiator with HP header	HP circuit, avionics[b]
14. Battery	HP radiator with integral HP/fluid header	HP-augmented cold rail[b]
15. Tracking radar		Modular sink (voice rec)[b]
16. Fluid evaporator	HP-augmented cold rail (high capacity)	Air-cooled equip rack
17. Fuselage TPS, interference heating		
	HP radiator for fuel cell	
18. TPS panel	Air-breathing engine compartment	
19. Control surface pivots		
20. OMS LO_2 boil-off		
21. Main LO_2 tank boil-off		
22. C-band directional antenna		
23. Electrical wiring		
24. Hydraulic actuators		
25. APU		
26. LO_2 natural recirculating system		
27. Water chiller		

[a]Selected for study under separate contract
[b]Selected for prototype fab dwgs and test plans

More details can be found in Alario and Prager's report [1].

TABLE 2.7 Phase change material (PCM) classification diagram [19].

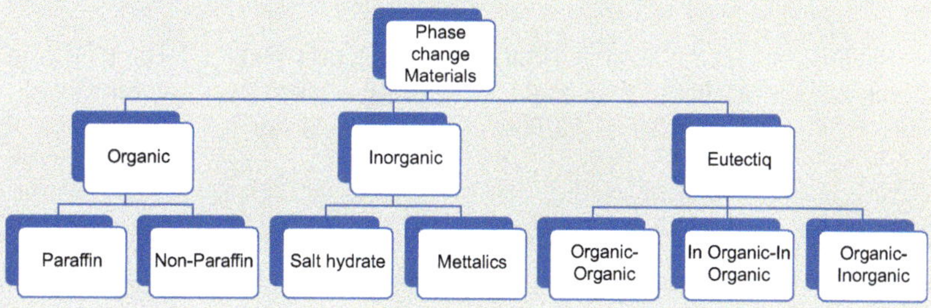

TABLE 2.8 Inorganic PCMs with melting temperature between 100°C and 280°C.

Compound (wt.%)	Melting point (°C)	Latent heat (kJ/kg)	Density (kg/m^3)	Energy density (kJ/m^3)	Thermal conductivity (W/m K)
$MgCl_2$–$6H_2O$	117	168.6	1450 (liquid,120 °C)	244,470 (liquid, 120°C)	0.570 (liquid, 120°C)
			1569 (solid, 20°C)	264,533 (solid, 20°C)	0.694 (solid, 90°C)
$NaNO_3$–KNO_3 (50:50)	220	100.7	1920	193,344	0.56
KCl–$ZnCl_2$ (68.1:31.9)	235	198	2480	491,040	0.8
LiCl–LiOH (37:67)	262	485	1550	751,750	1.10

storage systems assisted by heat pipes can be implemented in different applications [28–32].

Corrosivity it is a critical consideration in a Thermal Energy Storage (TES)xe "Thermal Energy Storage (TES)" system designed to run at high temperature for three decades. The high-temperature container materials that are able to resist the aggressive chemical behavior of the molten salts used in NGNP are basically high-temperature alloys (some stainless steels, Inconel, and Hastelloy-N), graphite, and ceramics (Williams 2006) [45]. The author

TABLE 2.9 Inorganic PCMs with melting temperature between 280°C and 400°C.

Compound (wt	Melting point (°C)	Latent heat (kJ/kg)	Density (kg/m^3)	Energy density (kJ/m^3)	The miai conductivity (W/m K)
$ZnCl_2$	280	75	2907	218,025	0.5
$NaNO_3$	308	199	2257	449,143	0.5
NaOH	318	165	2100	346,500	0.92
KNO_3	336	116	2110	244,760	0.5
NaCl–KCl (58:42)	360	119	2084.4	248,044	0.48
KOH	380	149.7	2044	305,987	0.5

TABLE 2.10 Inorganic PCMs with melting temperature above 400°C.

Compound (wt.%)	Melting point (°C)	Latent heat (kJ/kg)	Density (kg/m^3)	Energy density (kJ/m^3)	Thermal conductivity (W/m K)
$MgCl_2$–NaCl (38.5:61.5)	435	351	2480	870,480	N/A
Na_2CO_3–Li_2CO_3 (56:44)	496	370	2320	858,400	2.09
NaF–MgF_2 (75:25)	650	860	2820	2,425,200	1.15
$MgCl_2$	714	452	2140	967,280	N/A
LiF–CaF_2 (80.5:19.5)	767	816	2390	1,950,240	1.70 (liquid)
					3.8 (solid)
NaCl	800	492	2160	1,062,720	5.0
Na_2CO_3	854	275.7	2533	698,348	2.0
K_2CO_3	897	235.8	2290	539,982	2.0

reported See this author book on Small Modular Reactors as Renewable Energy Sources on Amazon.com, published by Springer Publisher) that some important factors should be considered for molten salts as coolants for NGNP [59]:

(i) Corrosive behavior can occur at high temperatures (e.g., halides of Fe and Zn are usually highly corrosive);
(ii) Heavy halide salts containing bromine and iodine usually have poor heat-transfer and high cost; and;
(iii) Mixed-halide salts with dissimilar halide anions (mixtures of chlorides, fluorides, bromides, and iodides) are considerably more complicated systems to prepare and understand.

Major issues involve new materials that can withstand higher temperatures, higher burn-ups and neutron doses, corrosive coolants; reactor designs that eliminate severe accidents; and development of fuel cycles for waste minimization and elimination of proliferation risks. Fast Reactors are expected to be commercially available from 2040 [59].

So far nuclear power has primarily been used to produce electricity, but it can also be used for process heat applications [13,14]. Currently, LWRs are already being used to a limited extent for some lower temperature applications (200°C), such as district heating and desalination of seawater. Existing designs of High-Temperature Reactors (HTR)xe "High-Temperature Reactors (HTR)" that can reach 800°C can be deployed in the coming decades and Very-High Temperature Reactors (VHTR)xe "Very-High Temperature Reactors (VHTR)" that can reach gas coolant temperatures beyond 1 000°C are being studied as a Generation-IV concept for later deployment. Process heat applications include petroleum refinery applications (400°C), recovery of oil from tar sands (600–700°C), synthetic fuel from CO [2] and hydrogen (600–1000°C), hydrogen production (600–1000°C) and coal gasification (900–1200°C). Small reactors that can be inherently safe and used to support specific high energy applications and often in remote areas are another very interesting application that is receiving more attention, in particular in the IAEA INPRO Initiative.

The management of radioactive waste, as either spent fuel or ultimate waste, depending on the national strategy, is a key issue for public acceptance of nuclear energy. There is scientific consensus that geological disposal is the only safe long-term solution for the management of ultimate waste. After a long period of intensive research and development coupled with in-depth political and social engagement, the world's first deep geological repositories for nuclear waste will be in operation in Sweden and Finland by 2020, with France following a few years later, demonstrating that practical solutions exist for the safe long-term management of hazardous waste from the operation of nuclear power plantsxe "Nuclear Power Plants". Though there will also be ultimate waste from Generation-IV fast reactor fuel cycles after reprocessing, the volumes and heat loads will be greatly reduced thereby facilitating disposal operations and optimizing use of space in available geological repositories [59].

2.10.5 Latent heat thermal energy storage systems assisted by heat pipes

In one of the very first studies, Abhat [33] considered the idea of utilizing heat pipe to enhance the PCM melting and solidification rates for solar heating application (See Fig. 2.50). He studied the performance of a latent heat thermal energy storage system assisted by a heat pipe with paraffin PCM. Annular fins were attached to the outer surface of the heat pipe to better spread the heat through the PCM [33].

In another study by Abhat [32], the performance of a modular latent heat thermal energy storage equipped with a finned heat pipe as presented in Fig. 2.51 and it was studied during charging, discharging, and simultaneous charging and discharging modes. An axially grooved copper–water heat pipe was selected to transport heat of 1000 W over the 3-m length with the maximum temperature difference of 10 K.

The module consists of a container with square cross section with heat pipe located at the center. Section B is filled with the PCM and sections A and C are in contact with the heat transfer fluids [33].

Liu and Ma [60] experimentally studied LHTES with paraffin as PCM ($T_m = 52.1°C$) which was assisted by a gravity-driven heat pipe as shown in Fig. 2.51.

The effects of Heat Transfer Fluid (HTF)xe "Heat Transfer Fluid (HTF)" inlet temperature and flow rate on the charging only, discharging only, and simultaneous charging/discharging processes were studied (Fig. 2.52).

Sharifi et al. [32] develop a numerical model to examine the melting of the phase change material in the vertical cylindrical container which is heated by the heat pipe located vertically at the center of the container. The heat

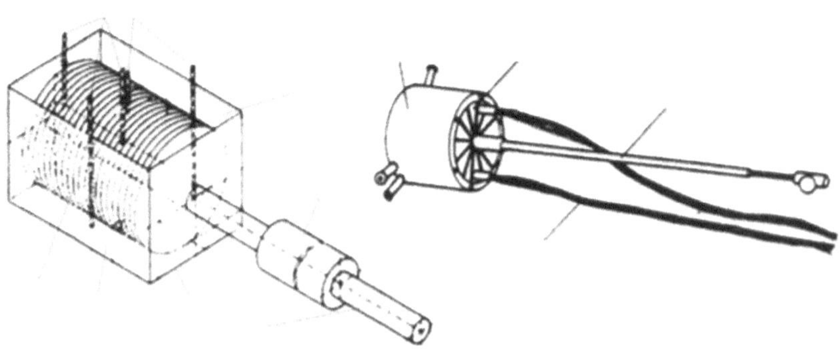

FIG. 2.50 Latent heat thermal storage unit with heat pipe [32].

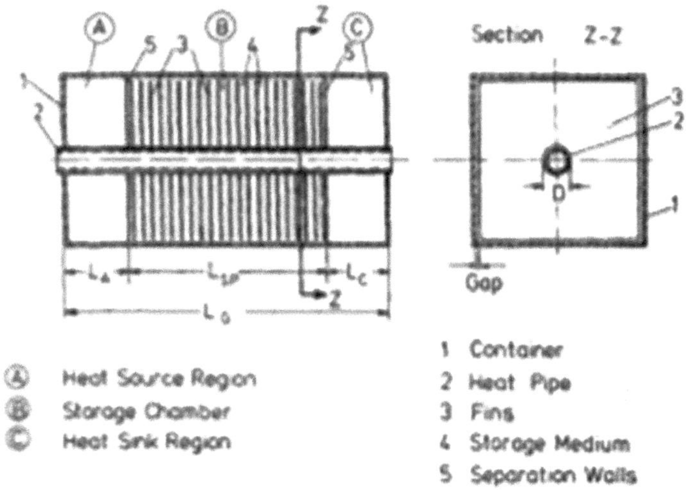

FIG. 2.51 Latent heat thermal energy storage with heat pipe heat exchanger [32].

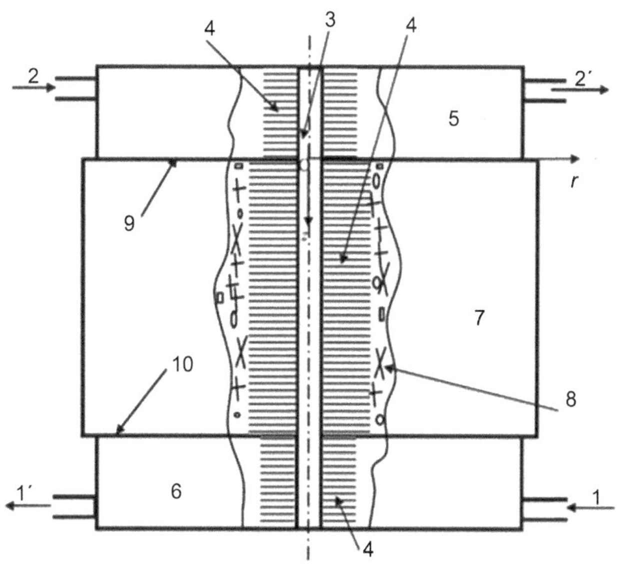

FIG. 2.52 A heat pipe heat exchanger with latent heat thermal storage.

pipe condenser is surrounded by the Phase Change Material (PCM)xe "Phase Change Material (PCM)" (See Fig. 2.52). The performance of the heat pipe integrated system is assessed by comparing the melting fraction at each time to those of assisted by isothermal surface or hot concentric rod or tube. Sodium nitrate ($NaNO_3$, $T_m = 580K$) was chosen as the PCM, and the heat pipe working fluid was selected to be potassium (K, with p_{Sat} at $T_{Sat} = T_m = 580$ 580K).

In Fig. 2.52,: the systematic configuration (1) hot fluid in, (10) hot fluid out, (2) cold fluid in, (20) cold fluid out, (3) heat pipes, (4) annular fins, (5) cold fluid-flow passage, (6) hot fluid-flow passage, (7) PCM chamber, (8) PCM, (9) upper separation, and (10) lower separation.

The heat pipe wick and wall material as well as the rod and tube was stainless steel which satisfies the compatibility issue associated with heat pipes.

Later on, Sharifi et al. [32] benefited from embedding the PCM within an aluminum foil matrix to enhance the melting and solidification rate by approximately 200 and 600% relative to the case with heat pipe and no metal foam. N-Octadecane was chosen as the PCM which was coupled with a copper–water heat pipe.

Liu et al. [60] presented a novel HP-LHTES in which composite granular solid–liquid PCM was piled up as the porous-medium layer. The PCM was composed of a mixed metal salt and inorganic bentonite which melts over temperature range of 200–250°C. Naphthalene was selected as heat pipe working fluid. In another study, they used composite granular solid–liquid PCMs compounded by RT100 and high density polyethylene with phase change temperature of 100°C. To be compatible with the phase change temperature of the PCM, water was chosen as the working fluid of the heat pipe.

Fig. 2.53 shows the schematic of fabricated experimental setup.

Tiari et al. [61,62] conducted a numerical investigation to study the effects of heat pipes on the melting and solidification rate of the PCM enclosed in a square cavity as shown in Fig. 2.54. The effects of heat pipe spacing and fin length and numbers and the influence of natural convection on the thermal performance of the LHTES unit were examined (See Fig. 2.55).

They have also studied the effect of embedded circular finned heat pipes on the melting of a eutectic mixture of potassium nitrate and sodium nitrate as the PCM in a vertical cylindrical container (See Fig. 2.56). Different arrangements of heat pipes were considered.

Fig. 2.57 illustrates the effect of heat pipe configuration on PCM liquid fraction evolution during the time at different elevation. As it can be seen, case 3 has provided better spread of heat hence more uniform melting of the phase change material and higher melting rate. Robak et al. [45] experimentally evaluated the effectiveness of latent heat thermal energy storage systems assisted by heat pipes. To obtain this, the experiment was conducted for (a) a system assisted by heat pipes, (b) a system using fins instead of heat pipes, and (c) a system without heat pipes or fins (benchmark case) for melting and solidification processes.

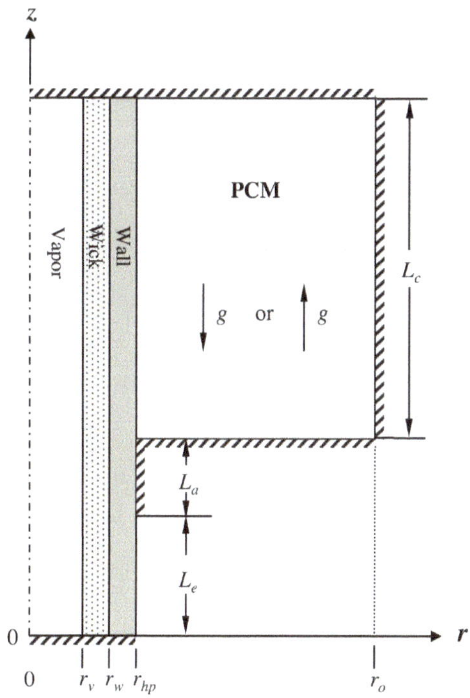

FIG. 2.53 Physical model and Computational Domain for the Heat Pipe and PCM.

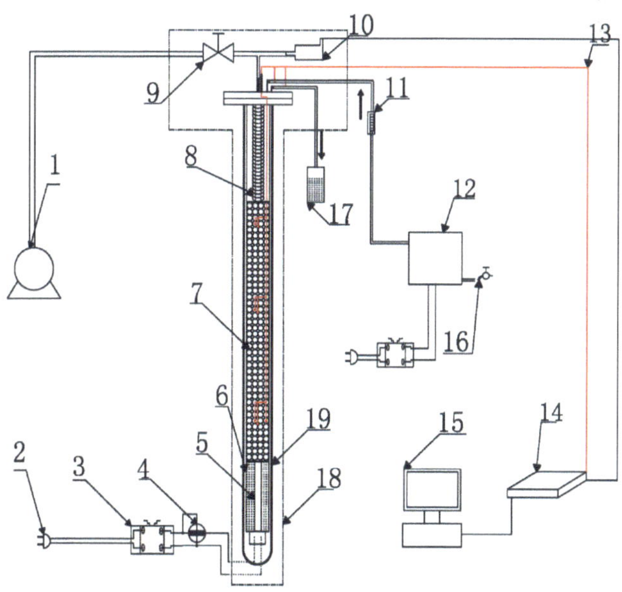

FIG. 2.54 Schematic view of the experimental setup 1. Vacuum pump, 2. Power, 3. Voltage regulator, 4. Power meter, 5. Electric heating rod, 6. Water, 7. PCM, 8. Cooling coil, 9. Vacuum valve, 10. Pressure sensor, 11. Rotor flow meter, 12. Electric heater, 13. Thermocouple, 14. Data acquisition unit, 15. Computer, 16. Tap, 17. Water tank, 18. Rubber insulation cotton and 19. Vacuum Glass Tube [41].

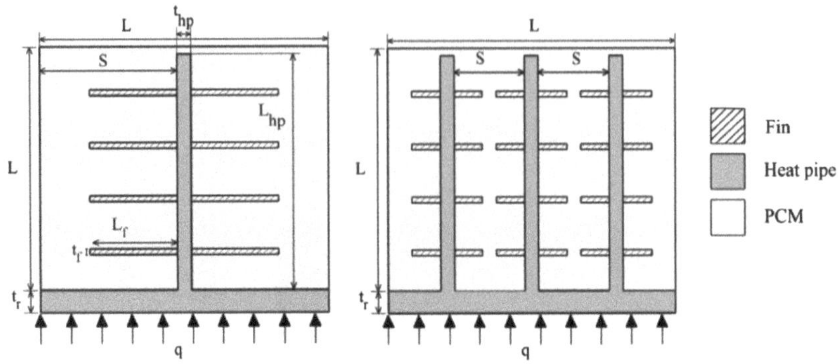

FIG. 2.55 Schematics of the LHTES units with one heat pipe (left) and with three heat pipes (right).

The effectiveness of each scenario was defined as the ratio of stored or released energy in cases with heat pipes or fin to the benchmark case:

$$\varepsilon_{HP} = \frac{E_{HP}}{E_{BM}} \quad \varepsilon_F = \frac{E_F}{E_{BM}} \tag{2.7}$$

where E is the stored or released energy. Also see Fig. 2.58.

A paraffin, n-octadecane ($C_{18}H_{38}$, Tm = 27.5°C) of 99% purity, was used as the PCM which was enclosed in a vertical cylindrical container. During the charging process, the heat was transferred from Heat Transfer Fluid (HTF)xe "Heat Transfer Fluid (HTF)" to the heat pipe evaporators using a heat exchanger, placed underneath the container. The schematic of the container with embedded heat pipes is shown in Fig. 2.59.

The result showed an increased melting rate of 70% and 50% for the heat pipeassisted charging comparing to the cases with no fins or heat pipes and the case with fins, respectively. During the discharging process, the solidification rate was doubled by implementing the heat pipes. It was also reported that the maximum effectiveness associated with the assisted heat pipe unit was around 1.6, while the fin-assisted case only achieved the value of 1.1.

Nithyanandam and Pitchumani [27] developed a numerical procedure utilizing the thermal resistance network model to study the performance of the heat pipeassisted thermal energy storage system during charging and discharging processes. The effects of the heat pipe and the LHTES system geometry and the LHTES operational conditions were the focus of this study. The system configuration implemented in their study is shown in Fig. 2.60 which includes an array of tubes which are enclosed in a shell. Four heat pipes are mounted on the outer surface of the tube walls (two horizontally oriented and two vertically oriented heat pipes).

Two different arrangements can be considered regarding the relative locations of the PCM and the HTF. In case 1, the HTF flows inside the tubes

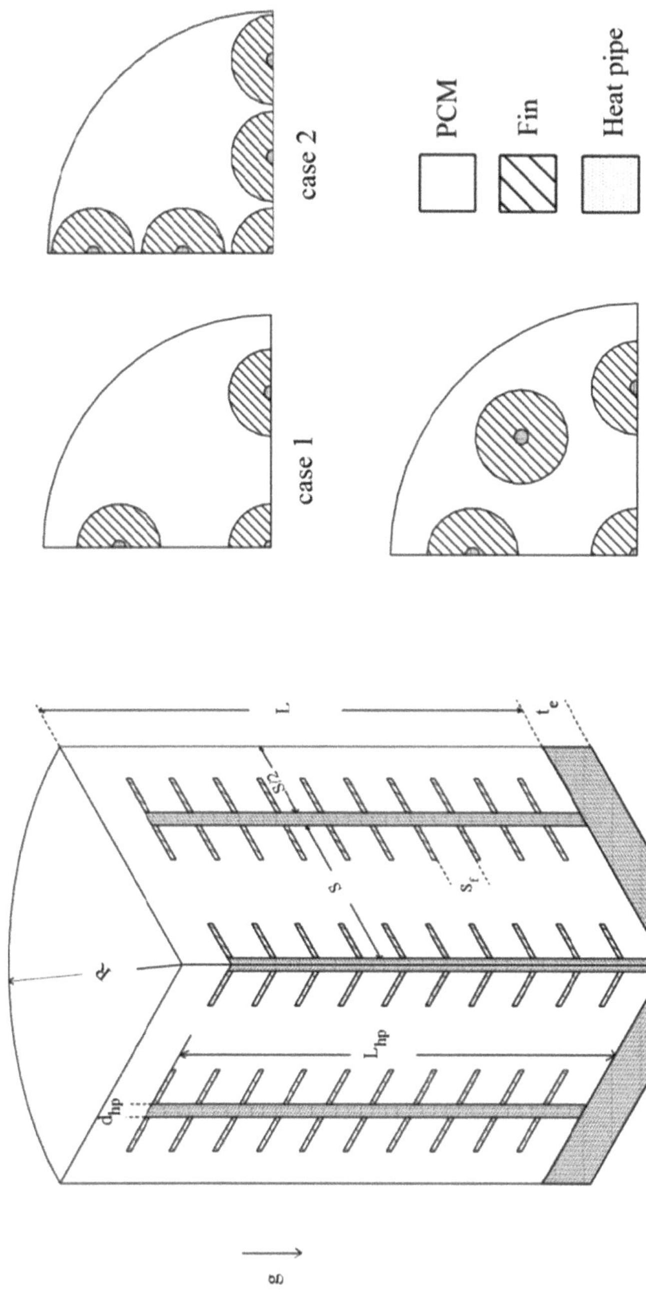

FIG. 2.56 Schematics of heat pipe-assisted thermal energy storage unit and different arrangements of heat pipe.

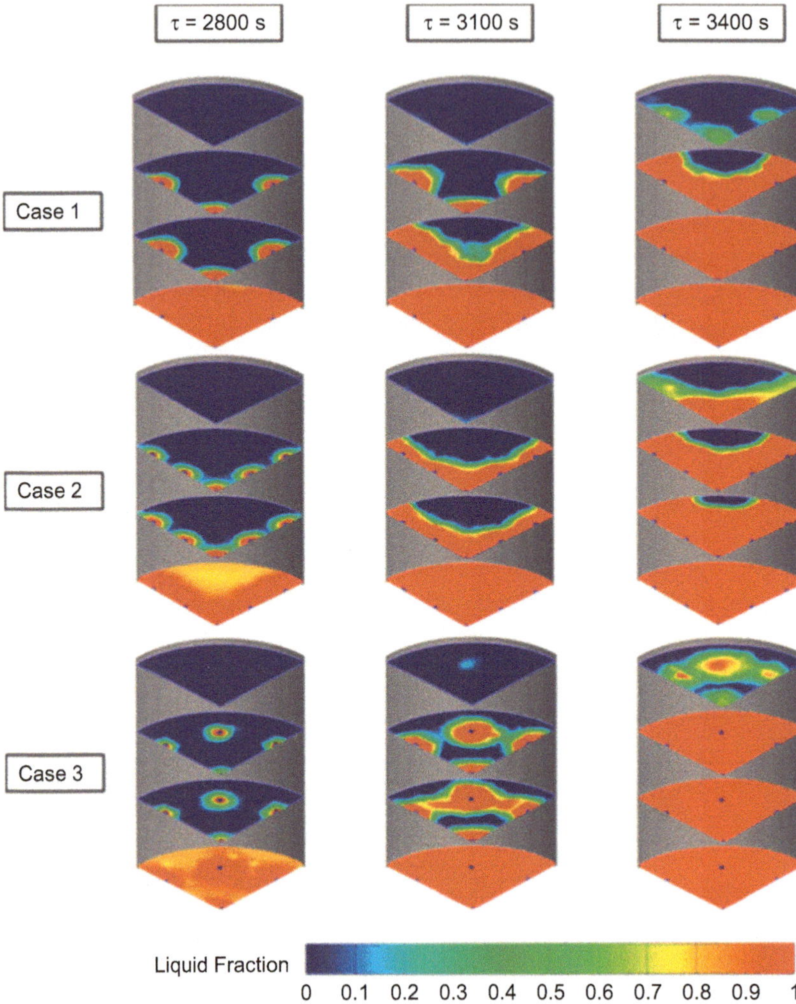

FIG. 2.57 The PCM liquid fraction for studied heat pipe Configurations at different elevations [44].

that are encompassed by the PCM, while in case 2 the PCM is placed in the pipes, and the HTF is flowing over the tubes in transverse direction.

The transient operation of the system was investigated applying the thermal resistance network model along with quasi-steady approximation. The developed thermal resistance network is illustrated in Fig. 2.61. The thermal elements E_1 to E_6 belong to heat pipe which are composed of radial conduction for the evaporator (wall, E_1; wick, E_2), condenser (wall, E_3, wick, E_4), and the axial conduction in the adiabatic section (wall, E_5; wick, E_6). The elements E_7 to E_{11} represent the PCM melt (solid) front adjoining the heat pipe, and E_{12} to E_{17} correspond to those of adjoining tube.

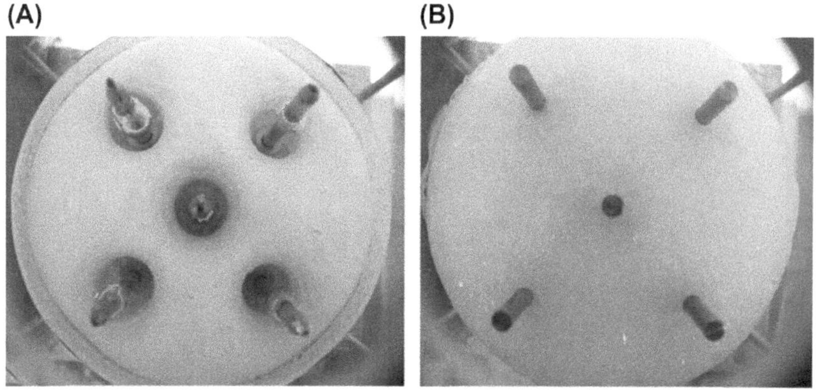

FIG. 2.58 Charging process of TES unit testing by Robak et al. [18]. (A) Heat Pipe Assisted. (B) Fin Assisted [45].

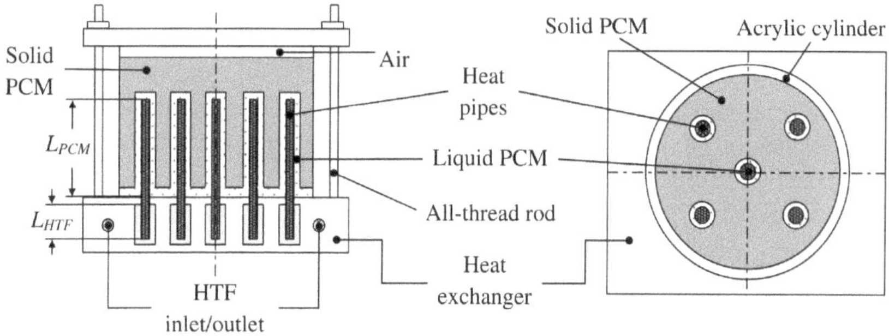

FIG. 2.59 Schematic of the PCM container with embedded heat pipes [45].

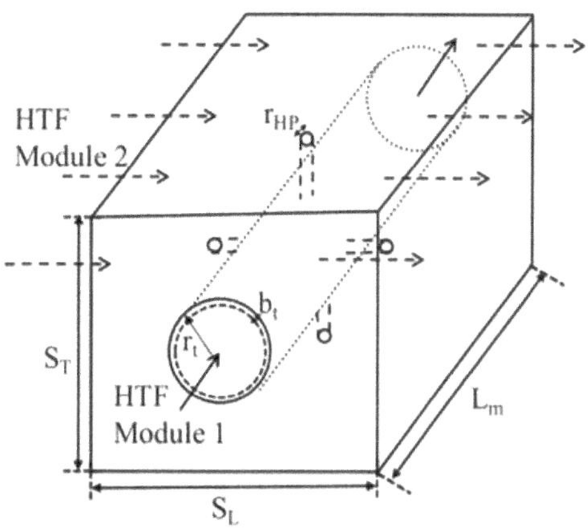

FIG. 2.60 Schematic of the flow Configuration in module 1 and module 2 of the LHTES system [46].

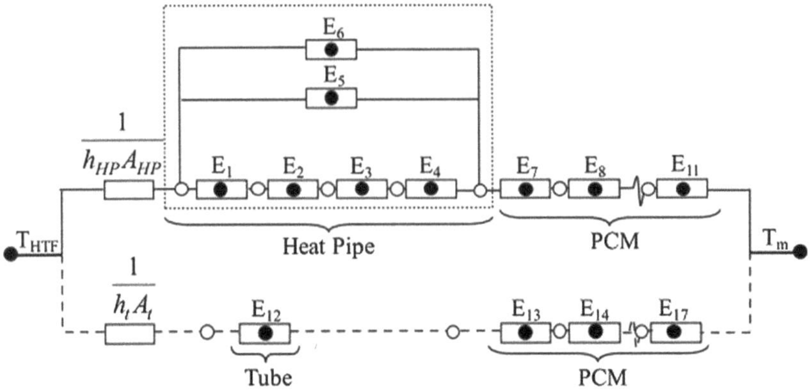

FIG. 2.61 The thermal resistance network of LHTES [29].

The thermal resistances associated with vapor flow and evaporation and condensation in the heat pipes are neglected.

The HTF temperature change between the tube inlets and its outlets is negligible. The radius of the melt front at the time is assumed to be uniform; however, its advancement is included by the enhancement of thermal conductivity of the liquid PCM. The thermal energy transfer through the heat pipes (the solid lines of the network) and the tube (the dashed lines of the network) occurs between the HTF at temperature T_{HTF} and the liquid (solid) PCM at temperature T_m.

For any thermal element, in the network, the energy balance can be written as:

$$\rho_i C_{p,i} V_i \frac{dT_i}{dt} = \left[\frac{T_{i,1} - T_i}{R_{i,1}} - \frac{T_i - T_{i,2}}{R_{i,2}} \right] \tag{2.8}$$

In Eqs. (2.7), T_i refers to surface temperatures at the middle of each element, while $T_{i,1}$ and $T_{i,2}$ are denoting the temperatures at either end of the heat conductor element. The thermal resistances for the heat pipes are explained Zohuri book Chapter 3 [59].

Surface energy balance is utilized to track the solid–liquid interface of the PCM adjacent to the heat pipe or tube's external surfaces as follows:

$$\rho \Delta H_f A_{11} \frac{ds_{HP}}{dt} = \pm 2\pi L_{11} k_{11} (T_{11} - T_m) \Big/ \ln\left(1 - \frac{L_{11}}{2S_{HP}}\right) \tag{2.9}$$

$$\rho \Delta H_f A_{17} \frac{ds_{HP}}{dt} = \pm 2\pi L_{17} k_{17} (T_{17} - T_m) \Big/ \ln\left(1 - \frac{L_{17}}{2S_{HP}}\right) \tag{2.10}$$

In above two equations variable S is the position of the melt (solid) front and T_m is the melting temperature of PCM. The positive and negative sign should be adopted for charging and discharging, respectively. The physical model was combined with a numerical optimization technique to maximize the effective rates during charging and discharging. The selected decision variables were HTFxe "Heat Transfer Fluid (HTF)" outer radius of the tube r_t length of evaporator and condenser section L_e and L_c of the heat pipe vapor core r_V and the thickness of the wick of the heat pipes T_W.

In general, the obtained result of this study shows the increase in the HTF mass flow rate, module length and tube radius decreases the effectiveness of the heat pipe, whereas the longer sections of evaporator and condenser of the heat pipe leads to higher effectiveness. For case 1, the effectiveness of the heat pipes in enhancing the thermal performance of the LTES is significantly higher while case 2 shows higher effective charging (discharging) rate of the PCMxe "Phase Change Materials (PCMs)" [46].

The effects of heat pipe configuration were investigated later by combining the simplified thermal resistance network model for heat pipe with three-dimensional numerical model of the HTF and melting/solidification of the PCM. The enthalpy-porosity technique is applied to model the melting and solidification processes within the PCM. Five different arrangements of heat pipe were studied as shown in Fig. 2.62. For the case with two heat pipes, they can be oriented vertically or horizontally. Two and one arrangements are considered for the system with three and four heat pipes, respectively [47].

In order to better understand the effect of heat pipe configurations, the contours of molten PCM during charging are presented here in Fig. 2.63 The arrangement with no heat pipe is served as the basic configuration for comparisons. The solid PCMxe "Phase Change Materials (PCMs)" is colored as the light shaded color while the molten PCM is represented by the white color. One of the points that should be noted is that as the results of the natural convection within the molten PCM, faster melting of PCM at the top parts are observed compared to the bottom sections. The figure illustrate that the case with 4-HPs represent higher rates of melting for both modules [47].

In another configuration proposed by Nithyanandam and Pitchumani, the heat pipes are embedded inside a PCMxe "Phase Change Materials (PCMs)". During charging process, the hot HTFxe "Heat Transfer Fluid (HTF)" flows into the bottom channel from right ($x = 0$) and it exchanges heat with the heat pipe evaporator hence the PCM enclosed in the container (See Fig. 2.64). While the discharging occurring, the cold HTF enters the channel located on the top from the right $x = l_d$. In order to provide enough spacing for volumetric expansion of the PCM, the air gap is supplied [48].

The conservation of energy for the HTF along the channel can be written as:

$$\rho_f \Delta H_f \frac{\partial T_f}{\partial t} + \rho_f c_f U_f \frac{\partial T_f}{\partial x} = \dot{Q}_d + N_{HP} \dot{Q}_{HP} \quad (2.11)$$

Where the f refers to the HTF and U_f is the velocity of HTF through a single channel. The heat transfer between PCM and HTF can occur either through the

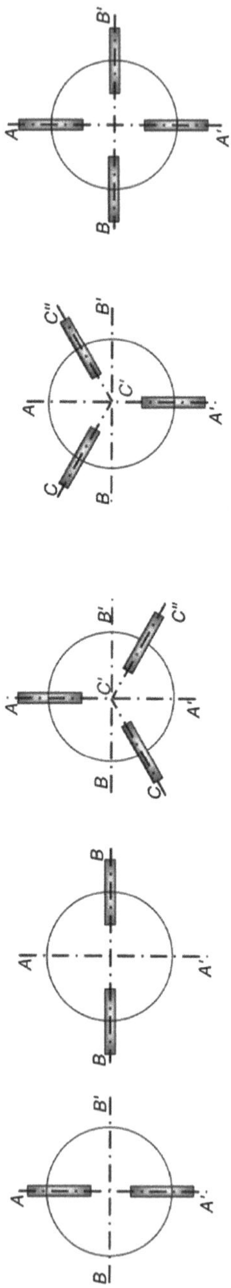

FIG. 2.62 Schematic illustration of the different arrangement of heat pipes [47].

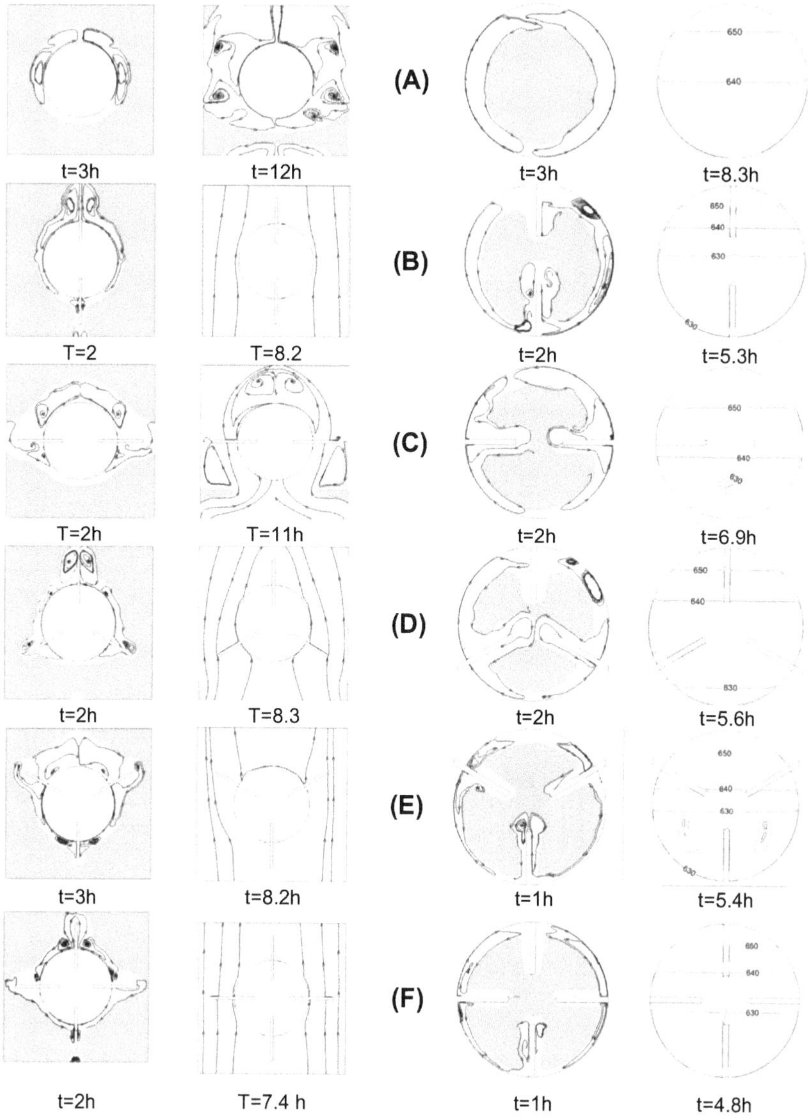

FIG. 2.63 contours of molten PCM during charging [47].

channel wall Q'_d or the heat pipe wall Q'_{HP} in Fig. 2.65. In Eqs. (2)−(11), N_{HP} represents the number of heat pipes in each control volume. Similar to previous work, the heat exchange between the PCM and HTF through channel walls or heat pipes is described using the thermal resistance network analogy.

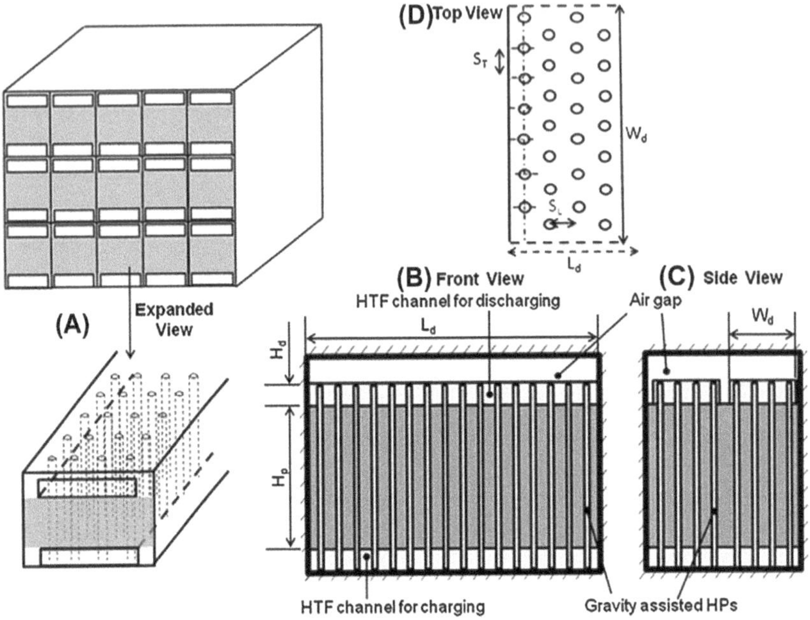

FIG. 2.64 Gravity heat pipes embedded latent thermal energy storage system [48].

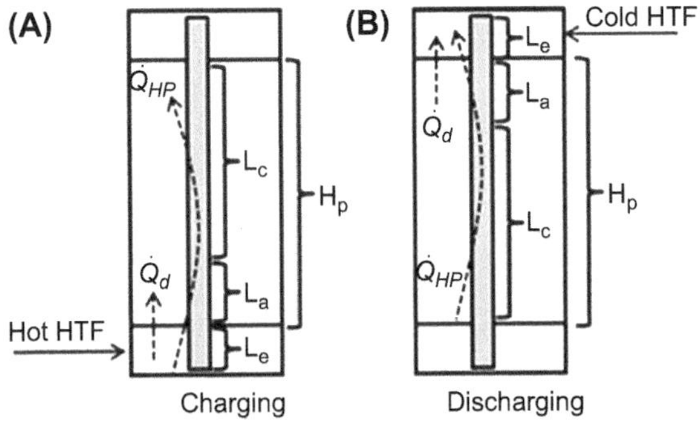

FIG. 2.65 Heat transfer pathways during [47]. (A) Charging and, (B) Discharging of HP-TES System.

In this work the energy balance equation for each thermal element mentioned earlier is modified to include the PCM melting.

$$\rho_i C_{p,i} V_i \frac{dT_i}{dt} + p_i h_{sl,i} V_i \frac{d\gamma}{dt} = \left[\frac{T_{i,1} - T_i}{R_{i,1}} - \frac{T_i - T_{i,2}}{R_{i,2}} \right] \tag{2.12}$$

Where γ is the melt fractions of the PCMxe "Phase Change Materials (PCMs)" and the h_{sl} is the latent heat of fraction for the PCM. It should be noted that this term only presents for the PCM thermal elements. The obtained results confirmed that there is an optimal longitudinal spacing between the heat pipes. This can be justified noting that the closer spacing leads to higher storage cost while larger spacing cause limited heat exchange rate between the PCM and the heat transfer fluid. This study provided the maximum discharging time, exergetic efficiency and the minimum storage cost as a function of different operating and the design parameters applied by US Department of energy SunShotxe “US Department of energy SunShot” [48].

Using thermal resistance network, Khalifa et al. evaluated the performance of the high temperature latent heat storage unit incorporating finned heat pipes. The finned heat pipes were placed in suspension arrangement adjacent to the heat transfer channel to enhance the overall heat conductance of the (PCM). A unit model cell of length L_m is shown in Fig. 2.66 [49].

Another advantage of latent heat thermal energy storage system is their near-isothermal operation which highly qualifies them for power generation systems which require isothermal input such as Stirling engine. Qiu et al. designed and tested the Thermal Energy Storage (TES)xe “Thermal Energy Storage (TES)” system integrated with concentrated solar power. The primary goal of their work was to show the feasibility of the concept. The system was composed of a 3kw Stirling engine power generator and thermal energy storage system as shown in Fig. 2.67. A different configuration of heat pipe is utilized in order to enhance the system performance, which provided the simultaneous charging and discharging of the phase change material. To aim this, a network of heat pipes was embedded inside the PCMxe “Phase Change Materials (PCMs)” consists of a primary heat pipe and an array of secondary

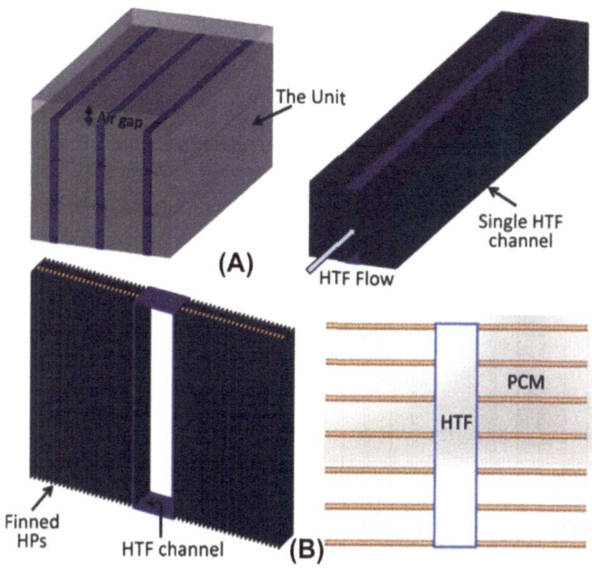

FIG. 2.66 Lhtes unit: [49]· (A) The proposed design. (B) The unit model cell.

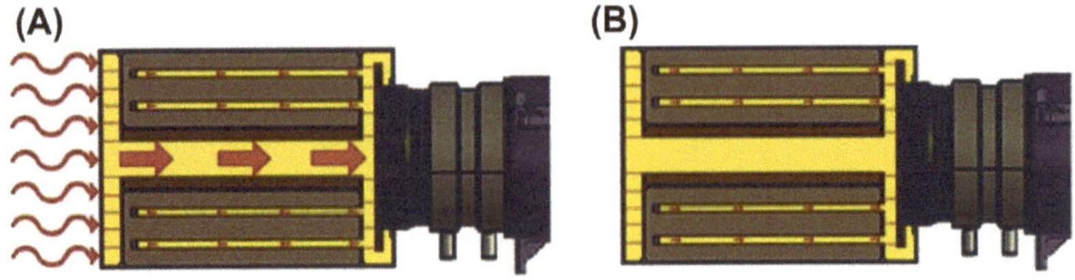

FIG. 2.67 Thermal energy storage module tested by Qui et al. [50] in (A) Charging Mode. (B) Discharging Mode.

heat pipes. As shown in Fig. 2.66 a, during the charging process, the solar energy, which impinges on the evaporator surface of the primary heat pipe, turns the liquid sodium to vapor, which moves toward the heat engine heat receptor where it condenses and releases heat. The excess heat is delivered via the array of secondary heat pipe to charge the phase change material. During the discharging mode the PCM will solidifies and the released latent heat will be transferred to the heat engine. The testing of the module was performed in two phases, in phase 1, the TES module was charged and discharged while the helium working fluid of the heat engine was evacuated and therefore the engine was inoperative. In phase 2, the engine charged and put in place and it was running in parallel with charging of the phase change material. The test results revealed that the system performance fell short of the expectations.

The shortfall was very likely due to the fact that the condensed liquid could not be returned properly to the evaporator by the heat pipe wick. Other reasons could be the secondary heat pipe configurations such as the heat pipe spacing and their number in addition to primary heat pipe geometrical features [50].

In order to improve the performance of the system, Mahdavi et al. focused on the hydrodynamics and thermal characteristic of the heat pipe. The purpose of their study was to understand the detailed effects of physical and geometrical features, especially secondary heat pipe arrangements on the overall performance of the heat pipe. It was also of interest to find the configurations that provides more uniform heat rejection in the primary condenser to the heat engine which improves the engine efficiency and more uniform phase change of PCMxe "Phase Change Materials (PCMs)" for efficient thermal energy storage. The effects of heat input, the heat transfer to secondary heat pipe in percentage of heat input, the adiabatic section radius, main condenser thickness, and secondary heat pipe arrangement were investigated on the temperature distribution of the primary condenser and secondary heat pipes.

The results showed that due to the high expansion ratio from adiabatic to primary condenser, the vapor flow leaving the adiabatic section acts similar to a jet which impinges on the main condenser surface and spreads out in radial direction. As the vapor flow proceed in radial direction, due to interaction of the viscose and inertial forces, several recirculation zones are formed which make the heat pipe performance convoluted.

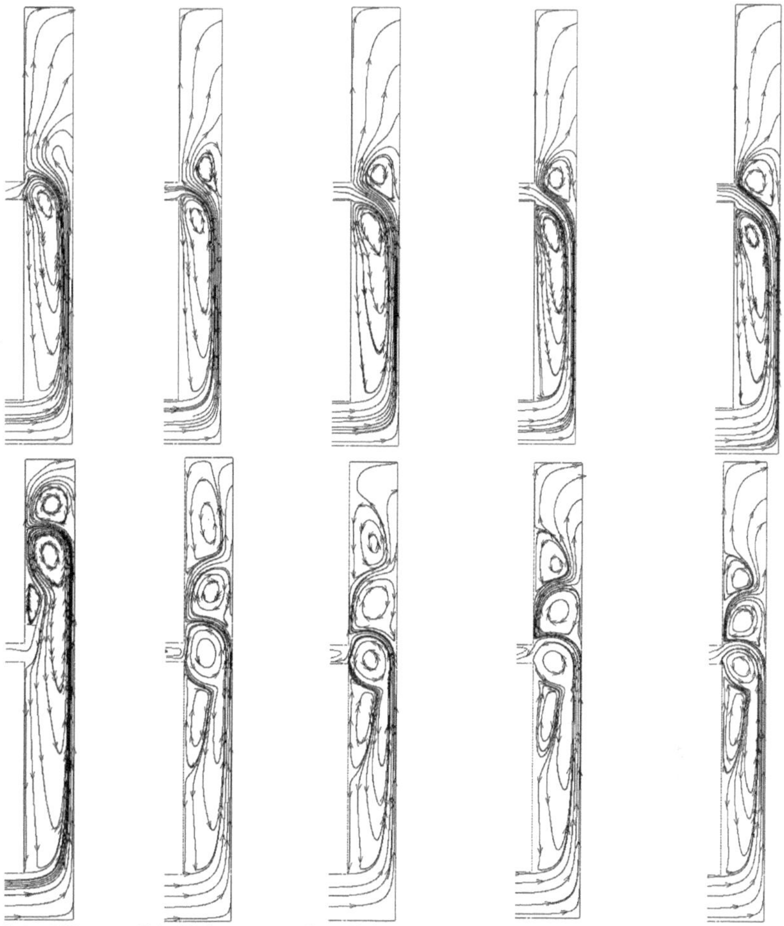

FIG. 2.68 The effect of heat input on the main condenser streamline patterns. (A) $Q_e = 1000$ W, (B) $Q_e = 3000$ W.

As one can see from Fig. 2.68, the recirculation zone plays a significant role in determining the temperature distribution in main. condenser and secondary heat pipe. In order to obtain more uniform temperature distribution, the configuration should be modified to diminish the size and quantity of the vortices [51,52]. They also studied the effects of condensers entrance shapes and the adiabatic section location on the uniformity of the temperature distributions of the main condenser. It was concluded that moving the adiabatic section radially outward to the surrounding has significant effect on reducing the size and quantity of the recirculation regions and leads to more uniform temperature distribution [54,55].

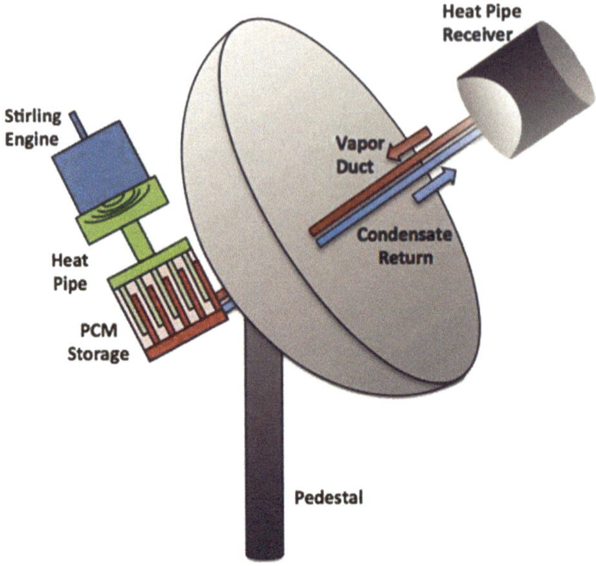

FIG. 2.69 The proposed Configuration of heat pipe assisted LHTES unit with a dish stirling system.

Andraka et al. [55] from Sandia national laboratories proposed a configuration heat pipe assisted thermal energy storage system integrated with dish-Stirling, shown in Fig. 2.69. In the proposed configuration, the latent heat thermal storage unit and the engine are placed on the rear of the dish instead of being at the focus which enables the power conversion unit to support higher phase change material mass. The thermal energy is transferred from the solar receiver to the storage media, which can be a salt or metallic PCMxe "Phase Change Materials (PCMs)", via a pumped heat pipe. A second heat pipe isothermally transfers the heat from the PCM to Stirling engine which decreases the exergetic losses of the system and also leads to a higher efficiency [55].

Shabgard et al. [56] numerically studied the configuration proposed by Andraka et al. [55]. to predict the transient response of the storage unit and thermal energy provided for Stirling engine. In order to reduce the computational cost, the HP-PCM unit was simplified to a two-dimensional numerical model. Fig. 2.70 shows the three-dimensional and simplified two-dimensional views of thermal energy storage unit. Sodium Chloride (NaCl) with melting. temperature of 800°C was considered as the PCM and heat pipe working fluid was sodium. Three operation stages were considered for the study based on time-varying solar irradiance. In the first stage which starts at sunrise and lasts until 9 a.m., all the input thermal energy is used to charge the PCM. The second stage of operation which is simultaneous charging and discharging of

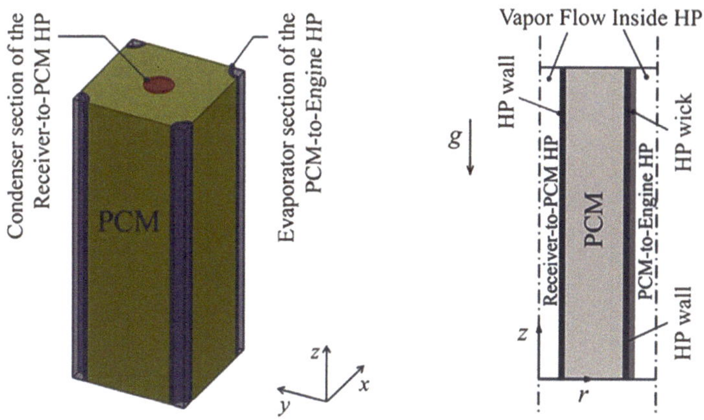

FIG. 2.70 (Left) three-dimensional schematic of the storage unit showing a receiver to PCM heat pipe and four-quarter sections PCM to engine heat pipes, (right) the simplified two-dimensional model of the system [56].

the storage unit starts at 9 a.m. and ends at sunset (approximately 7 p.m.). The third stage starts after sunset, during which the stored thermal energy is retrieved to generate power. It was found that the LHTESxe "Latent Heat Thermal Energy Storage (LHTES)" xe "Latent Heat Thermal Energy Storage (LHTES)" unit damps the time-dependent variations of solar irradiance and supplies Stirling engine with a relatively smooth thermal energy. It was also shown that the heat pipe spacing plays a significant role in thermal performance of the LHTESxe "Latent Heat Thermal Energy Storage (LHTES)" system and the system with the minimum heat pipe spacing provides the maximum exergy efficiency.

Malan et al. [57] experimentally and numerically analyzed a Concentrated Solar Power (CSP)xe "Concentrated Solar Power (CSP)" system integrated with a heat pipe-assisted latent heat thermal energy storage unit. For the experimental study, a low temperature storage unit with paraffin as the PCM was built and tested. For the numerical modeling a high temperature LHTES unit with a eutectic mixture of KCl and KF ($T_m = 605^\circ$C) was analyzed which is adapted to simulate a solar tower facility in Meerendahl, South Africa which is part of a project called the *Helio 100*. The experimental study of the test module gave an indication of how the PCMxe "Phase Change Materials (PCMs)" behaves during melting and solidification processes in the specific geometry used for the system. On the other hand, the developed numerical model analyzed the thermal performance of the thermal energy storage system which, numerical model analyzed the thermal performance of the thermal energy storage system which gave a more detailed picture about system design and operation modes. They also concluded that additional study is required to deal with corrosion issue on heat pipes and fins surface. See Fig. 2.71.

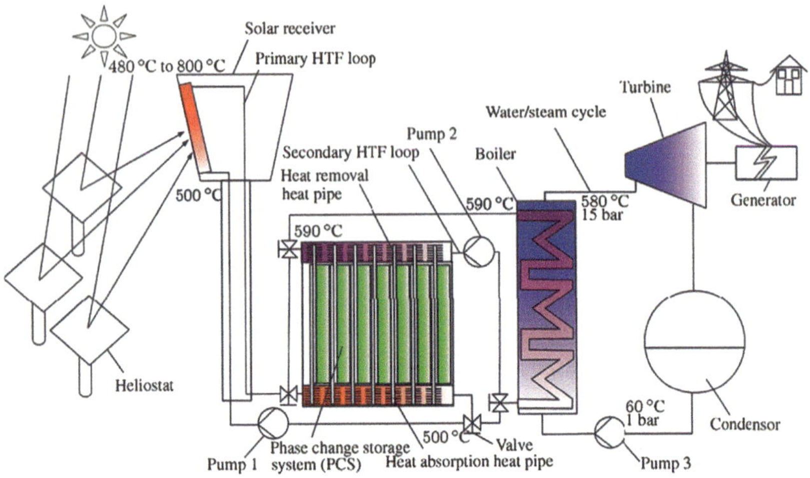

FIG. 2.71 The solar thermal system used for the numerical modeling by Malan et al. [57].

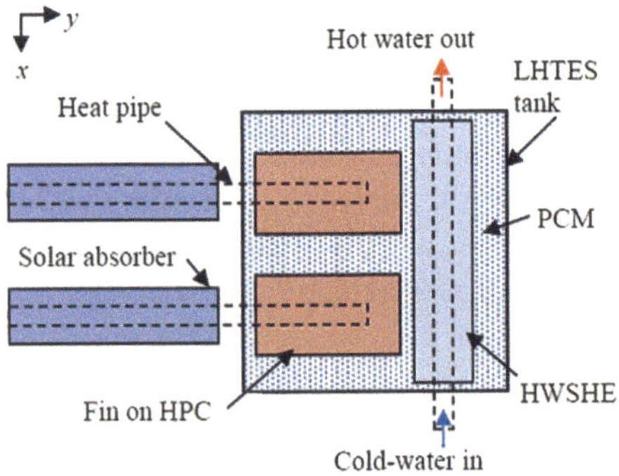

FIG. 2.72 Evacuated tube heat pipe solar collectors [58].

Naghavi et al. [58] presented a theoretical model for a solar hot water system consisting of an array of Evacuated Tube Heat Pipe Solar Collectors (ETHPSC)xe "Evacuated Tube Heat Pipe Solar Collectors (ETHPSC)" connected to a common manifold filled with phase change material as shown in Fig. 2.72. The heat pipe evaporator is exposed to solar radiation and the condenser section is inserted in into finned sockets welded into the manifold. Therefore, the energy received form sun will be transferred to the PCM via the heat pipe. The heat pipe condenser is equipped with fins to spread heat through the PCM. Domestic cold water will be heated as it flows through the LHTES tank. Fig. 2.73 illustrates three different heat transfer processes during charging and discharging.

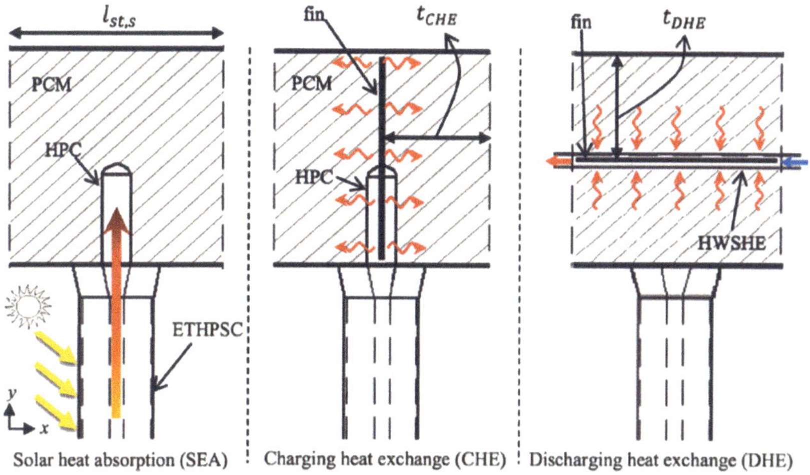

FIG. 2.73 Heat transfer mechanisms in LHTES [58].

2.11 Passive thermal technical discipline lead (TDL) for luna lander

In response to an announcement of opportunity from NASA's Science Mission Directorate (SMD) Discovery Program, the Southwest Research Institute in collaboration with the Aerospace Corporation and the NASA Johnson Space Center (JSC) proposed a lunar lander science mission.

The Moon Age and Regolith Explorer (MARE)xe "Moon Age and Regolith Explorer (MARE)" would use a lunar lander to reach a young, nearside lunar lava flow for the collection and analysis of the lunar soil. This would be used for the determination of the impact history of the inner solar system and the evolution and differentiation of the interiors of one-plate planets. The lunar lander proposed was based on the NASA JSC Morpheus lander vehicle. The thermal environments for the proposed mission were both challenging and unprecedented, since survival of multiple lunar day/night cycles at the south-west region of the Aristarchus plateau were required. Other thermal design challenges included the need for a low mass, robust and reliable thermal management system that would assure the success of the proposed mission. As part of the proposal effort and leveraging on existing NASA Small Business Innovative Research (SBIR), a completely passive thermal control system concept was used to meet mission requirements. The thermal management system proposed uses a novel type of hybrid grooved and sintered wick variable conductance heat pipe and a high conductivity heat spreader; the thermal management systems tested were developed by Advanced Cooling Technologies, Inc. (ACT)xe "Advanced Cooling Technologies, Inc. (ACT)" in Lancaster, Pennsylvania. This publication will present the design, analysis and prototype component and system level performance testing done at NASA JSC and presented by Alvarez-Hernandez et al. [61].

In their work they show that, there will be some differences between the MARE thermal control system design proposed and the actual test article built and tested. The end goal of the effort remains the same: the proof of concept of a completely passive thermal management system for a mid-sized lunar lander spacecraft. This effort also addressed technical gaps identified in the NASA technology roadmaps. In this paper and per the definition of the NASA Engineering and Safety Center (NESC)xe "NASA Engineering and Safety Center (NESC)" Passive Thermal Discipline, the authors define a passive system as an architecture without the use of a mechanically pumped fluid loop. This effort, led by the NASA Johnson Space Center's (JSC)xe "Johnson Space Center's (JSC)" Passive Thermal Technical Discipline Lead (TDL) and the Technology Development team in the thermal design branch, had an additional objective; to demonstrate an initiative to reduce cost and infuse existing research into needed NASA applications by leveraging on existing Small Business Innovation Research (SBIR) contracts. A variation of the hybrid wick VCHP and the high conductivity, HiK™, plates used in this work were also tested in the ISS as part of the technology experiment suite, under the JSC Advanced Passive Thermal experiment project.

The MARE (Moon Age and Regolith Explorer) Discovery Mission concept targets delivery of a science payload to the lunar surface for sample collection and dating with the objective of determining the impact history of the inner solar system and the evolution and differentiation of the interiors of one-plate planets. The mission science is within a 100-m radius region of smooth lunar Maria terrainxe "Maria terrain" near Aristarchus crater, 23.7 °N and 47.4 °W. The lunar lander proposed for this mission was based on the NASA JSC Morpheus lander vehicle. Fig. 2.74.

FIG. 2.74 Mare lander configuration. *Courtesy of NASA.*

For *Proposed Thermal Design*, the MARE Thermal Control System (TCS) xe "Thermal Control System (TCS)" design proposed is intended to ensure that all spacecraft and science components are maintained within operational/survival limits throughout all mission phases. The lunar surface operation phase presents the most challenging thermal environment for heat rejection given the imposed thermal loads and heat rejection sink temperatures. Lunar surface temperatures can vary drastically and range from -173°C at lunar night to 87°C during lunar day. In addition, the long-duration, cold, lunar night drives heater power requirements to keep components within their temperature limits. The MARE thermal control architecture proposed consists of a simple passive thermal control system, making this a reliable, cost and mass-effective approach. This system is composed of two thermally controlled zones on the top deck of the spacecraft to maintain science and vehicle components within operational and survival limits. Cooling is achieved with two high conductivity heat spreaders connected via hybrid Variable Conductance Heat Pipes (VCHPs)xe "Variable Conductance Heat Pipes (VCHPs)" to two high conductivity radiators. In the proposal phase, two high conductivity aluminum encapsulated Annealed Pyrolitic Graphite (APG)xe "Annealed Pyrolitic Graphite (APG)" heat spreaders were assumed as the heat spreader plate and the radiators. Driven by cost reduction, and the opportunity to leverage on existing SBIR contracts, these were replaced for testing for 2 HiK™ plate heat spreaders, one HiK™ plate radiator and one Gas Charged Heat Pipe (GCHP)xe "Gas Charged Heat Pipe (GCHP)" radiator. The thermal links provide a means to control the heat leak from the heat spreaders, depending on the set-point. They conduct heat to the radiators when the heat spreader surface temperature is above the upper temperature limit and turn off by heating the non-condensable gas reservoir, disconnecting the ammonia flow within the VCHP, when the heat spreaders are below the lower temperature limit. Figs. 2.75 and 2.76 shows the original thermal management concept and technical definitions presented during the proposal.

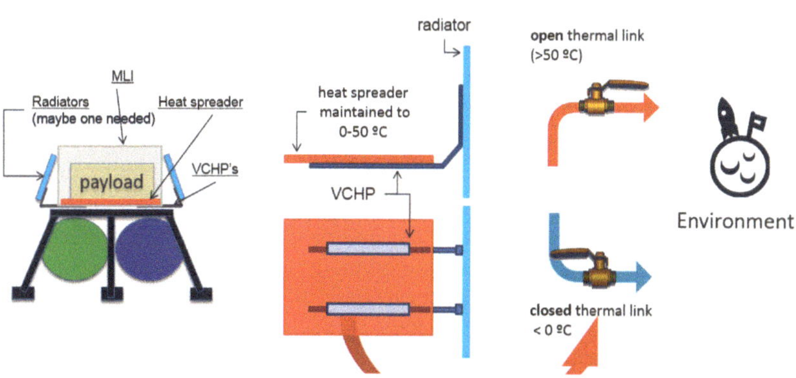

FIG. 2.75 Mare thermal concept. *Courtesy of NASA.*

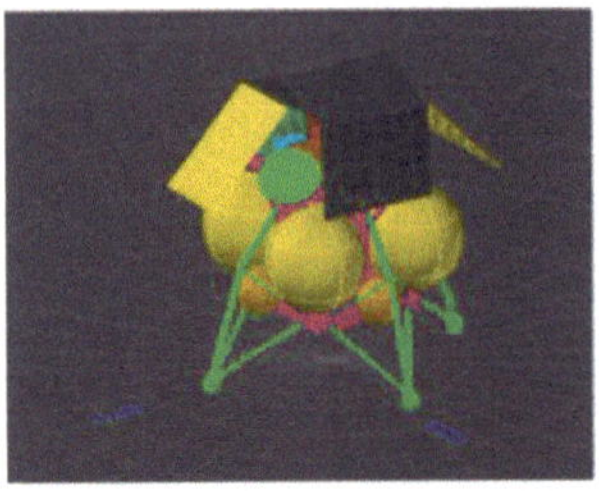

Key Capabilities / Features

Attribute	Description/parameters
Thermally controlled zones	2 controlled zones: Zone 1: temp range -10 – 40°C Zone 2: temp range -35 – 65°C
Radiators	2 High conductance plate radiator. Being resized with new thermal dissipation numbers.
Thermal links	Variable conductance heat pipes. 2 per radiator.
Heat Spreader	High conductivity plate (heat spreaders with embedded heat pipes).

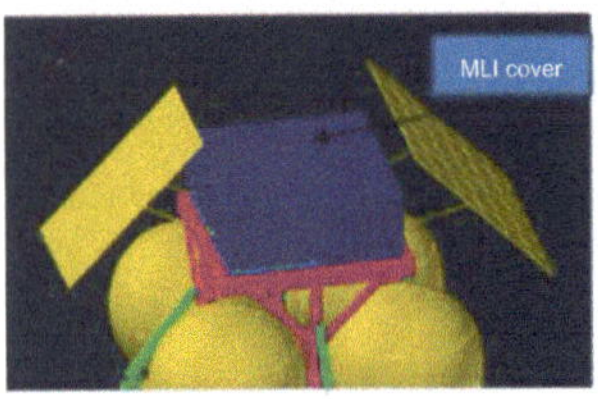

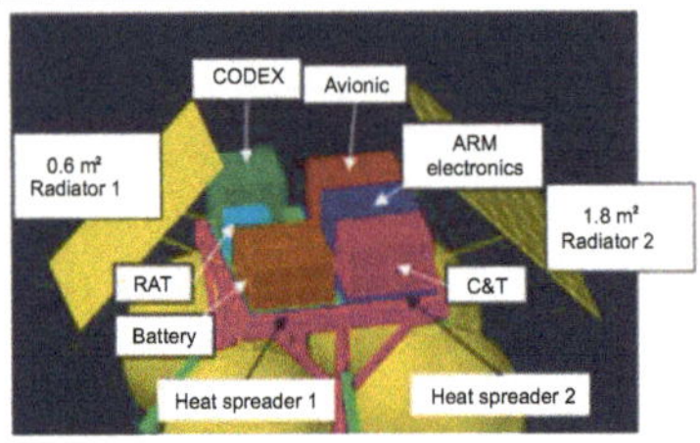

FIG. 2.76 Mare thermal concept description. *Courtesy of NASA.*

As part of the lunar operations power requirements, the vehicle would land with both radiators perpendicular to the sun angle. The radiators proposed assumed an AZ-93 surface coating with solar absorptivity/emissivity of 0.19/0.89 to reduce the amount of incident thermal load and to reject heat in the infrared spectrum. An assessment of dust degradation was not included, and the effects were expected to be covered with the design margin. The lunar surface IR thermal load on the radiator is minimized by positioning the radiators at a 45° angle from the local vertical. The first radiator size is 0.6 m^2 and is located on the south facing side of the spacecraft.

This radiator rejects a maximum waste heat of 88W produced by zone 1 (CDEX, RAT and batteries) as shown in Fig. 2.77. The second radiator is 1.8 m^2 and is located on the north facing side of the spacecraft. This radiator rejects a maximum waste heat of 235 W produced by the spacecraft avionics, telecom and lander electronics in zone 2.

In order to survive the lunar night, thermostatically controlled Kapton heaters, directly fed from the primary battery input, provide the necessary heat to keep electrical components within operational and survival thermal limits. The survival heater power needed is approximately 55 W. Multi-Layered Insulation (MLI)xe "Multi-Layered Insulation (MLI)" blankets will shroud all top deck components to reduce the amount of heat leak paths to the cold lunar environment and space environment. A thermal math model was developed for the proposed MARE TCS architecture, and thermal analysis was performed for the worst cold and hot mission environments. The results show that the Thermal Control System (TCS)xe "Thermal Control System (TCS)"

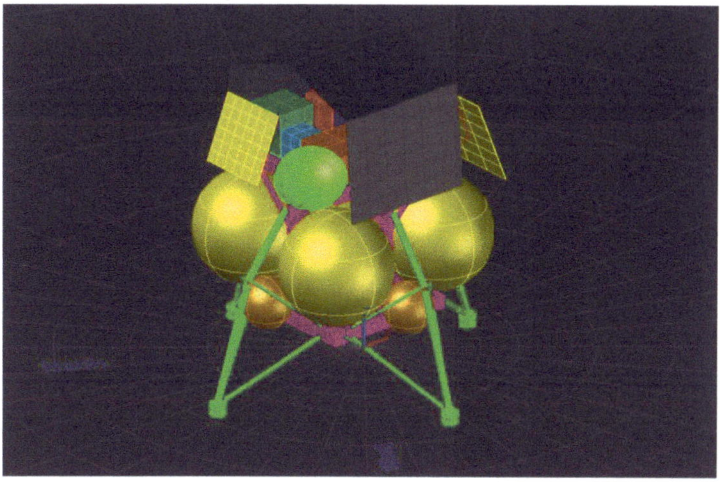

FIG. 2.77 Mare thermal mathematical model.

subsystem will protect the MARE payload from worst case hot mission environment with at least a 13°C margin for all components. Results also showed at least 12°C margin for all components for a worst case cold mission environment.

In conclusion of this mission and conceptual study for purpose of "Design Analysis and Performance testing of Novel Passive Thermal Management System for Future Exploration Mission" [60], the temperature data was used to calculate the percentage of heat transferred by the thermal control system and the percentage of heat lost to the environment through the MLI system and thermal spacers.

The publication by Alvarez-Hernandez et al. [61] presents all the work performed, from the proposal phase to the proof of concept, of a fully passive thermal management system for a mid-sized lunar lander. This testing successfully demonstrated the initiative to reduce cost and infuse existing research into needed NASA applications by leveraging on existing Small Business Innovation Research (SBIR)xe "Small Business Innovation Research (SBIR)" contracts. A series of benchtop and component level testing has been performed on the HiK™ plates and hybrid wick VCHPs, including on-orbit operations inside the Advanced Passive Thermal Experiment (APTx)xe "Advanced Passive Thermal Experiment (APTx)" [60] suites. This publication documents the integration of these technologies in a full system level thermal performance test. The results showed the hybrid VCHP to be the driving component during survival conditions, allowing a reduction of survival heater power of approximately 100 W compared to the baseline design.

As we have stated in Chapter One of this book, a VCHP is similar to a conventional heat pipe, but has a reservoir and a controlled amount of

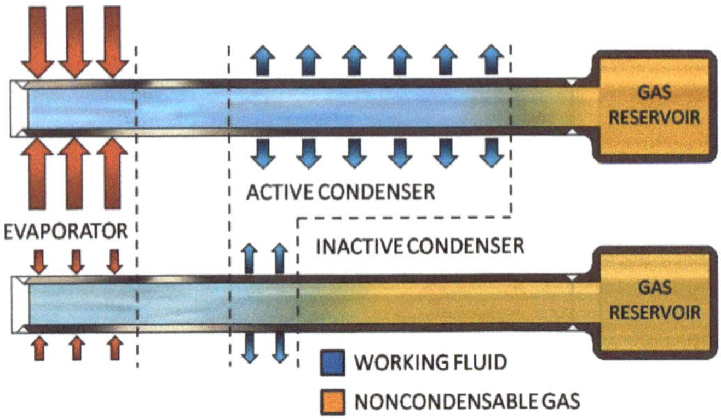

FIG. 2.78 Operation of a VCHP.

Non-Condensable Gas (NCG)xe "Non-Condensable Gas (NCG)" inside the reservoir. This NCG is used to regulate the overall thermal conductance of the heat pipe. By regulating the thermal conductance, the temperature at which the VCHP operates can be controlled even with variations in the heat load.

When the VCHP is operating, the NCG is swept toward the condenser end of the heat pipe by the flow of the working fluid vapor. The NCG then blocks the working fluid from reaching a portion of the condenser. The VCHP works by varying the amount of condenser available to the working fluid for heat transfer. As the evaporator temperature increases, the vapor temperature (and vapor pressure) rises, the NCG compresses (Fig. 2.78, top) and more of the condenser is exposed to the working fluid. This increases the effective conductivity of the heat pipe and drives the temperature of the evaporator down. Conversely, if the evaporator cools, the vapor pressure drops and the NCG expands (Fig. 2.78, bottom). As the NCG begins to fill the condenser, the heat pipe effective conductivity decreases, and the evaporator temperature decrease is minimized.

In Fig. 2.78, as heat load increases, the temperature dependent saturation pressure of the working fluid increases and the NCG is compressed into the reservoir (top). As heat input decreases, the working fluid saturation pressure decreases, and the non-condensable gas is allowed to expand into the condenser (bottom).

For the simple VCHP illustrated in Fig. 2.78, the reservoir is cold-biased in most spacecraft applications. Electric heaters on the reservoir are then used to control the temperature to within range ±2°C. During the long Lunar day, the thermal management system must be capable of removing the waste heat from the Warm Electronic Box (WEB)xe "Warm Electronic Box (WEB)" and ensuring the WEB does not get too warm. During the long Lunar night, the variable thermal link for the WEB must limit the amount of heat that is removed from the WEB and radiated to space. This will keep the electronics and battery warm with minimal power, even with the very low temperature (100 K) environment. A variable thermal link (typically a VCHP) is needed which can operate during large tilts, over a wide sink temperature swings on the Lunar surface and shut down during the long Lunar night.

In contrast to the standard cold, electrically-heated reservoir at the end of the condenser, the hybrid wick VCHP has a warm reservoir located adjacent to the evaporator. This eliminates the 1–2 Watts required to keep the reservoir at the correct temperature. This is a necessity for Lunar applications, where it is estimated that supplying 1 W over the 14 day long Lunar night requires 5 kg of solar cells, batteries, etc. Note that a warm reservoir does not provide the same tight temperature range as an electrically heated reservoir, but this is not a problem for the envisioned applications.

2.12 Heat pipe driving home energy system

With home heating costs increasing, more attention has been focused on the use of heat pipes to collect solar energy [72] and [73]. A relatively simple design incorporates a bank of inclined thermosyphons exposed to the south side of a residence. Solar energy is absorbed and transported into the living space, where it is convicted to the interior air or stored in a water tank for later use. During the night, the thermosyphons essentially act as thermal diodes, since the only way heat can be transferred from the interior to the outside is by axial conduction through the pipe walls. A similar design can be used for desalinating sea water using solar energy. In this application, however, a heat pipe would be positioned at the focal point of a trough-shaped parabolic reflector in order to generate the high temperatures and heat fluxes necessary for desalination. Singh et al. [74] presented the design and characteristics of various energy conservation systems and renewable energy systems utilizing heat pipes as the thermal control mechanism. A wide range of energy systems including data center cooling, agricultural products cold storage, bakery waste heat recovery and automotive dashboard cooling were discussed. It was argued that zero emission and economic advantages can be achieved by using thermosyphon and capillary pumped loop.

Concentrating Photovoltaic (PV)xe "Photovoltaic (PV)" systems use low-cost optical systems such as the Fresnel lens, a mini-reflecting mirror that can concentrate solar intensity from 200 to 1000 suns. The concentrated solar energy delivered from the solar cell is from 20 to $100 W/cm^2$. Part of the energy is directly converted to electricity, while the remainder is removed as waste heat. Heat pipe cooling systems (Fig. 2.79) were developed to passively remove the high heat flux at the PV cell and reject it to the ambient by natural convection [75–77].

Heat transfer devices and methodologies for novel Thermal Energy Storage (TES)xe "Thermal Energy Storage (TES)" systems for various applications including Concentrated Solar Power (CSP) generation systems using latent heat

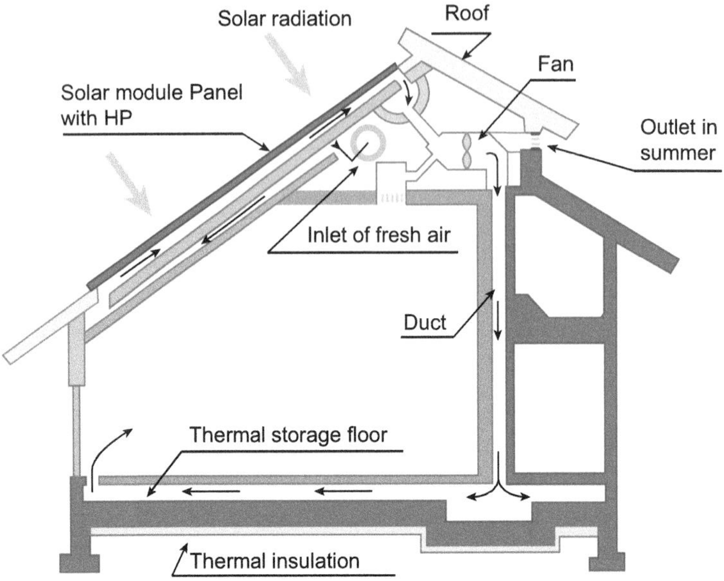

FIG. 2.79 Solar house which has an energy collection mechanism [76].

Phase Change Materials (PCMs)xe "Phase Change Materials (PCMs)" are in great demand in the energy field. However, latent heat thermal energy storage PCMs, despite their great potential, suffer from low thermal conductivity. Faghri [78,79] invented two methods (Figs. 2.80 and 2.81) to significantly increase the thermal conductivity of PCMs by embedding micro or conventional heat pipes in the PCM for applications in thermal energy storage systems and heat exchangers.

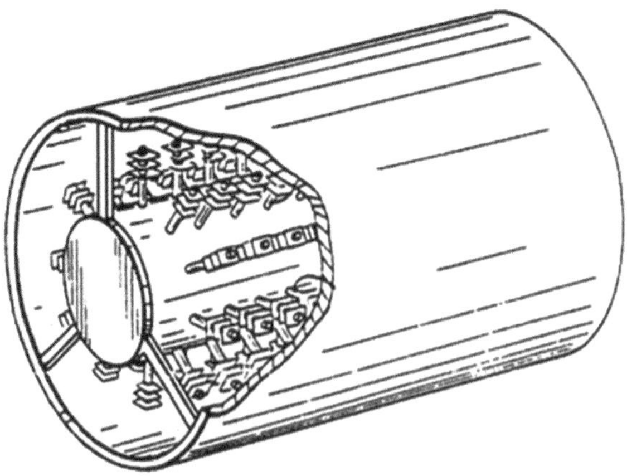

FIG. 2.80 Thermal energy storage system [78].

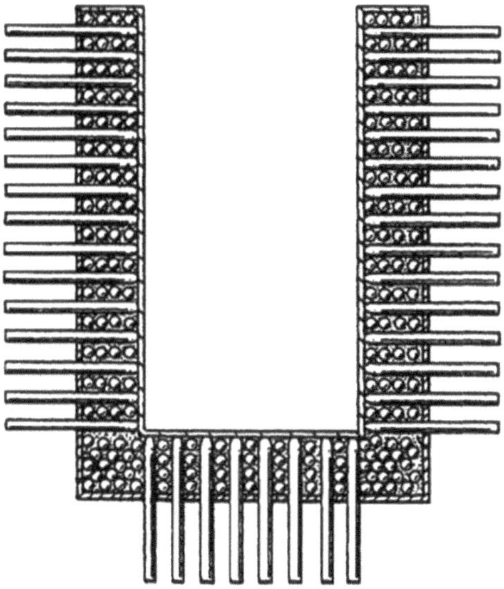

FIG. 2.81 Micro heat pipe exchanger [79].

The effect of heat pipes to improve the performance of latent heat thermal energy storage systems has been investigated experimentally and theoretically [80,83,84].

The experimental results of Robak et al. [83] showed 60% and more than 100% increase in the heat transfer rates during melting and solidification, respectively, compared to the fin-assisted storage systems as well as non-fin, non-heat pipe configurations. Shabgard et al. [81] developed a thermal network modeling approach for computationally-efficient simulation of such heat pipe-assisted latent heat thermal energy storage systems. The modeling results also showed substantial heat transfer enhancements during charging and discharging of latent heat thermal energy storage systems equipped with heat pipes. The developed thermal network model was later extended to simulate a large scale heat pipe-assisted cascaded latent heat thermal energy storage system for Concentrated Solar Power (CSP) applications [82].

Fuel Cells are versatile energy conversion devices with numerous potential applications: large electrical plants, stationary electricity generation, vehicle propulsion and small portable power [82,83]. In recent years, Proton Exchange Membrane Fuel Cells (PEMFCs)xe "Proton Exchange Membrane Fuel Cells (PEMFCs)", Direct Methanol Fuel Cells (DMFCs)xe "Direct Methanol Fuel Cells (DMFCs)", Solid Oxide Fuel Cells (SOFCs)xe "Solid Oxide Fuel Cells (SOFCs)" and Molten Carbonate Fuel Cells (MCFCs)xe "Molten Carbonate Fuel Cells (MCFCs)" have become the major types of fuel cell technologies that draw the most attention for commercialization. Heat pipe technology is beneficial for fuel cell development in two main ways: using heat pipes in fuel cells as thermal management components and using the heat pipe concept in the fuel cell systems to achieve passive high-effective fluids-thermal management. Most fuel cell systems produce electrical energy at high efficiencies that may range from 40% to 60% based on the Lower Heating Value (LHV)xe "Lower Heating Value (LHV)" of the fuel. For example, a fuel cell operating at 1.0 kW and 50% efficiency generates 1.0 kW of waste heat. This heat may be dissipated by convection, conduction, radiation or phase change. The heat generated in a fuel cell stack may be dumped to the atmosphere, but often, it is used in other system components that require heat. In some cases, the heat is used to run a thermodynamic cycle for additional power generation. Heat pipes can be utilized in fuel cell systems for thermal management purposes which allow for effective use of the fuel cell byproduct, heat, leading to a substantial increase in heat transfer and overall system efficiency [82,83].

The matter of heat generation in fuel cell stacks presents challenges for thermal management. Stacks operating at 40%–60% efficiency generate heat at the same rate to more than twice the rate of electric generation. Due to changes in mass concentration, temperature gradients, and in some cases, phase change throughout the stack, the heat generation is not uniform.

This non-uniform heat generation further increases thermal gradients in the stack. Increasing the mobility of the heat is a challenge that, if met, leads to three main benefits:

(1) The risk of stack failure due to overheating is reduced,
(2) The stack operates more closely to its design temperature, resulting in better power density and efficiency, and
(3) The heat can be reused, perhaps for reactant preheating, pre-vaporization, combined cycle operation, or cogeneration.

For example, the heat pipe embedded with a bipolar plate [83] is an innovative approach that would increase heat transfer in fuel cell stacks while requiring significantly smaller thermal gradients and much smaller volumes and weights than alternative methods. As conceptualized in Fig. 2.82, a bipolar plate is fabricated with holes, into which micro heat pipes are inserted and bonded. The micro heat pipes that are shown in Fig. 2.82 are embedded in an axial and transverse direction.

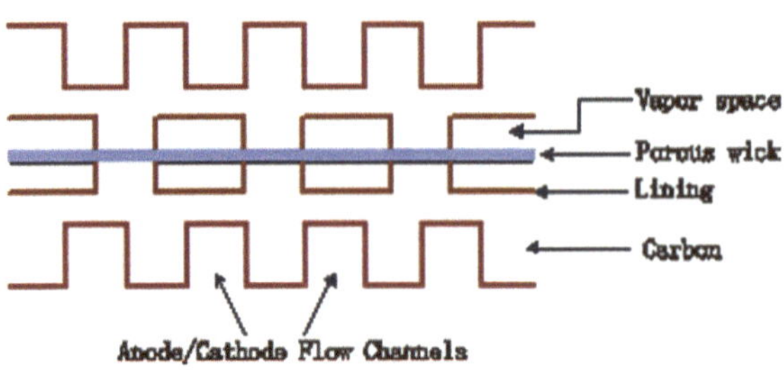

FIG. 2.82 Bipolar plate integrated heat pipe [83].

Another method for thermal control in the fuel cell stack is presented with the integrated bipolar plate flat heat pipe [83]. For example, as conceptualized in Fig. 2.83, a carbon bipolar plate contains a flat heat pipe. A challenge for developing this component is sealing the heat pipe. Due to the permeability of the carbon, two-halves of the bipolar plates are coated with a liner–silver Activated Brazing Alloy (ABA)xe "Activated Brazing Alloy (ABA)". Also, in the enclosure is a porous wick of metal foam, felt, or porous carbon. The two-halves are then sealed together, perhaps with brazing in an inert gas. The structure of the interior is designed to allow good electrical conductivity, thermal conductivity to the wick, and structural support, while the working fluid is typically at a pressure different than the surroundings.

Passive Direct Methanol Fuel Cell (DMFC)xe "Direct Methanol Fuel Cell (DMFC)" technology uses various capillary approaches to manage methanol and water without the need for a complex micro-fluidic subsystem (see Fig. 2.84), [82,84–86].

At the core of this new technology is a unique passive system that uses the heat pipe concept for fuel delivery. Furthermore, the fuel cell is designed for both passive water management and effective carbon dioxide removal.

The passive components that are critical to the fuel cell design are the fuel delivery, and air-breathing and water recirculation systems. The passive fuel delivery system stores pure methanol, which can be mixed with water in situ without the use of a pumping system and can be passively supplied to the fuel cell anode at an optimal concentration. Since water is needed in the anode for the methanol reaction to occur, the water created in the cathode can be passively supplied to the anode. This water recirculation, in conjunction with passive methanol fuel delivery, can dramatically extend the operation time of the fuel cell per refueling. The passive mass transfer concept (wick structure) developed in heat pipe technology is an effective approach for mass transfer management in various fuel cell technologies.

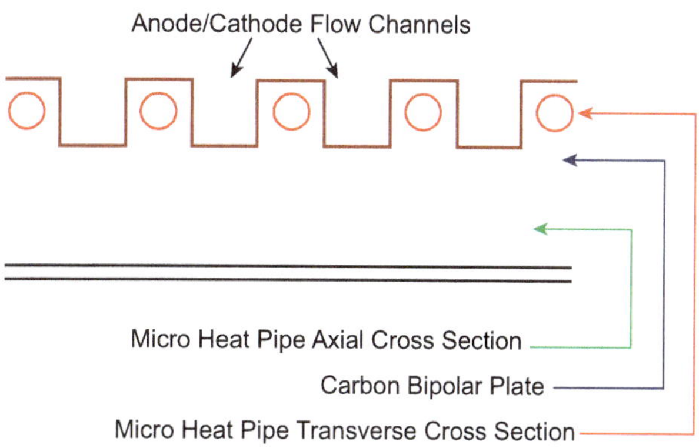

FIG. 2.83 Micro heat pipe embedded bipolar plate [83].

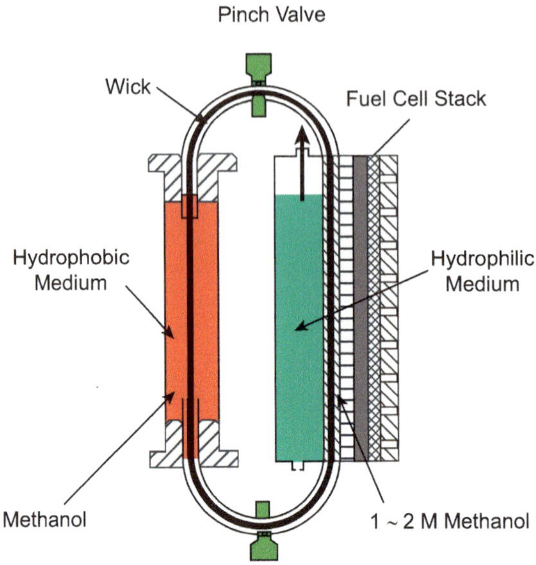

FIG. 2.84 Passive direction methanol fuel cell using heat pipe concept [83].

The proposed Direct Methanol Fuel Cell (DMFC)xe "Direct Methanol Fuel Cell (DMFC)" technologies developed were operated passively, without moving parts, which resulted in a highly reliable system. Due to their significantly longer charging life, passive miniature DMFC system are seriously being considered for replacing the battery for applications such as cell phones, digital cameras and laptops.

Current dish/Stirling systems use directly illuminated receiver design in which the concentrated solar radiation is absorbed by tubular receiver which also serve as the heater tubes of the Stirling engine. Replacing the directly illuminated receiver design with spherical heat pipe receivers improves the performance by about 20% [87]. Two primary reasons contribute to this improvement;

(a) In the heat pipe receiver, the temperature difference between the output gas temperature and the receiver peak temperature is much less than the directly illuminated receiver due to nearly isothermal operation of the heat pipe. This allows higher operating temperatures resulting in higher conversion efficiencies, and

(b) In heat pipe receiver heat is transferred to the Stirling engine by condensation on the entire surface of the heater tubes, whereas in directly illuminated receivers only one side of the tubes is irradiated. Hence the "dead volume" of the engine is decreased and higher compression ratios can be achieved which result in greater system efficiency [88].

2.13 Heat pipe driving heat exchangers and heat pumps

Increases in the cost of energy have promoted the use of heat pipe technology in industrial applications. Due to their high heat transfer capabilities with no external power requirements, heat pipes are being used in heat exchangers for various applications. In the power industry, heat pipe heat exchangers are used as primary air heaters on new and retrofit boilers. The major advantages of heat pipe heat exchangers compared to conventional heat exchangers are that they are nearly isothermal and can be built with better seals to reduce leakage. Heat pipe air heaters should also be cheaper than conventional tubular heat exchangers, as they are smaller and can be shipped in a small number of modules. Heat pipe heat exchangers can serve as compact waste heat recovery systems which require no power, a low pressure drop and are easy to install on existing lines. Heat pipe heat exchangers can be categorized into gas-gas, gas-liquid, and liquid-liquid type heat units. Among these three, gas-gas heat pipe heat exchangers have the widest application in industry. A gas-gas heat pipe exchanger consists of a group of externally finned heat pipes which reclaim waste heat [89,90]. These units eliminate cross-contamination due to the solid wall between the hot and cold gas streams. Also, the heat pipe design is totally reversible (heat can be transferred in either direction). Gas-gas energy recovery units typically fall into three categories: heat recovery in air-conditioning systems (low temperatures), recovery of excess process heat for space heating (moderate temperature), and recovery of waste heat from high temperature exhaust streams for reuse in the process (preheating of combustion air, e.g.,). The units for these applications vary in size and construction depending on the specific application, but many commercial models are now available that implement this heat pipe design. Gas-liquid heat pipe exchangers are less commonly available than gas-gas models due to the fact that the present design of waste heat boilers is very efficient. In the past, exhaust heat from boilers was simply dispersed to the atmosphere. Fig. 2.85 shows schematics for waste heat recovery with liquid-gas and gas-gas heat exchangers. Faghri [91] invented an innovative design for a centrifugal heat pipe vapor absorption heat pump (Fig. 2.86). The heat pipes in this heat pump system are disk-shaped, with one face partially or completely being the evaporator and the opposite face partially or completely being the condenser [92]. The wick is designed such that the centrifugal force aids in the delivery of the condensate to the evaporator. This design will significantly improve the heat and mass transfer characteristics of the rotating components of the vapor-absorption heat pump by increasing the heat pumping capacity that can be packaged in a given value. This results in a more efficient and compact vapor-absorption heat pump system.

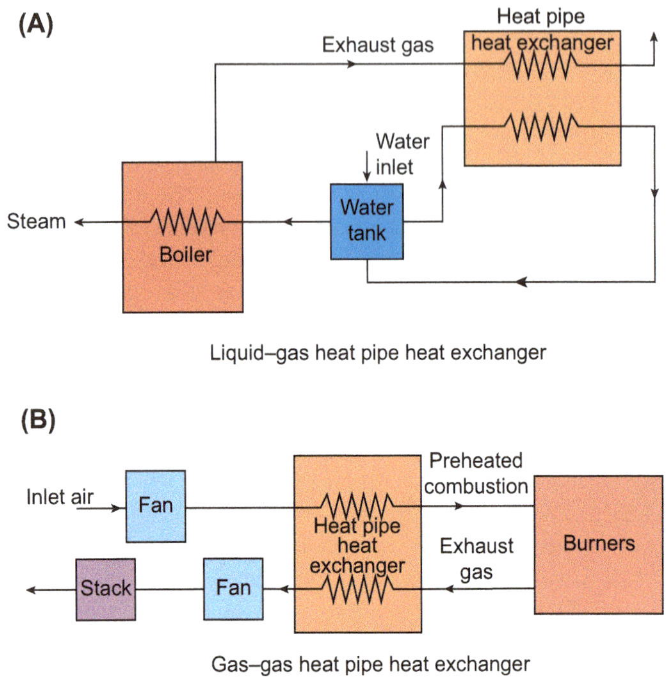

FIG. 2.85 Waste heat recovery using heat pipe heat exchangers.

2.14 Gas turbine engines and the automotive industry

The temperature limitation is one of the most crucial limiting factors related to the efficiency of a gas turbine aircraft engine or power gas turbine. An increased turbine inlet temperature decreases both the specific fuel and air consumption, while increasing efficiency. This desire for a high turbine inlet temperature, however, is often in conflict with materials available to withstand the high temperature. As a result, innovative cooling systems for hot-gas-path components are required. Among the hot-gas-path components, the first-stage rotor blade and nozzle guide vane require the most challenging cooling consideration. In addition to improving energy utilization efficiency, effective cooling could also drastically improve the reliability of high-speed rotating components. It has been observed that the creep life of turbine blades is reduced by half with every 10−15°C rise in metal temperature. Therefore, the temperature of the turbine blade must be kept within certain tolerable limits.

The primary cooling technology in use today for turbine blades and Nozzle Guide Vanes (NGVs) is the film cooling technology. Since 1995, a number of efforts have been initiated to utilize the concepts of miniature, radially rotating, high-temperature heat pipes, for gas turbine blades and disk cooling. This also includes a series of experimental and numerical studies [93–96].

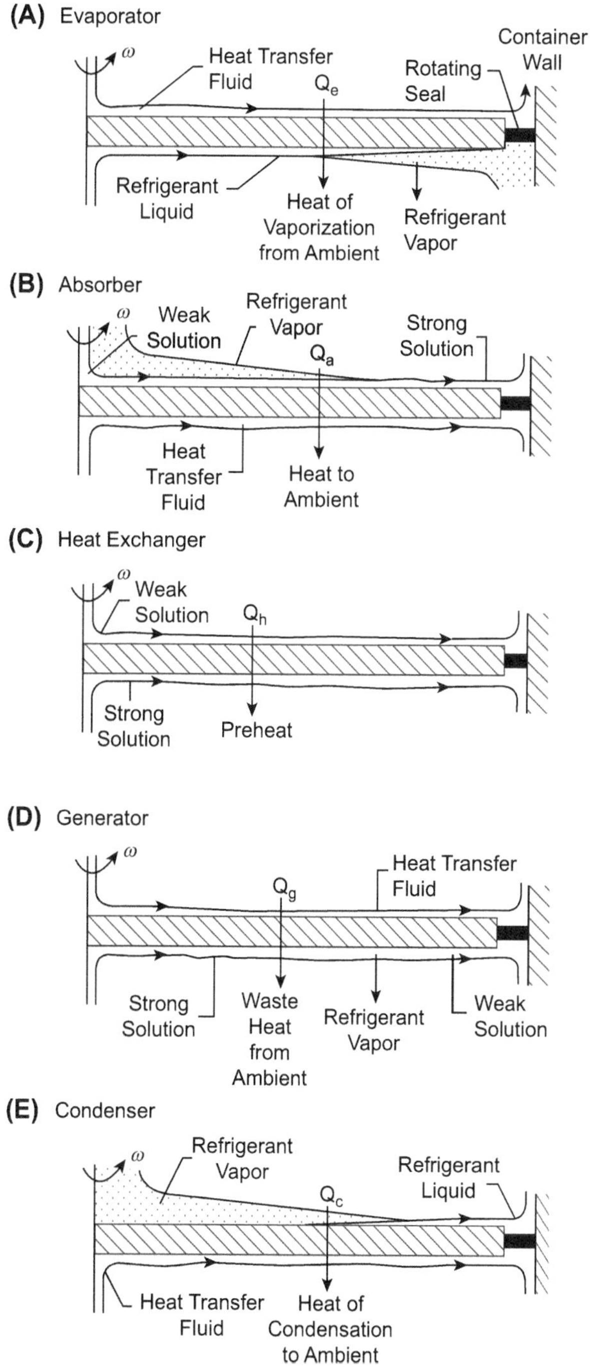

FIG. 2.86 Schematic of centrifugal heat pipe heat pump vapor-absorption cycle [91].

High temperature heat pipe cooling is a promising cooling technology for gas turbine hot components, such as first stage rotor blades, nozzle guide vanes, and rotor disks. This has the potential to significantly reduce the temperature of these hot components, allowing for a much higher gas turbine inlet temperature, while also reducing the consumption of high-pressure compressor air.

A Stirling engine heated by sodium heat pipes has been constructed and tested [97–99]. This engine operates at highest efficiency when the thermal energy is supplied to it at a very constant temperature and high heat fluxes. Helium is heated to increase its pressure, which is used to drive pistons and a crankshaft for the generation of power from thermal energy. The sodium heat pipes deliver heat from a molten salt heat storage system to the gaseous helium which is used to drive the engine. The coupling of a Stirling cycle engine with sodium heat pipes can also be used in the direct conversion of solar energy.

Nguyen et al. [100] proposed a thermosyphon Rankine engine for power generation using solar and other available waste heat energy sources. The basis of the engine (Fig. 2.87) is the modification of a heat pipe to incorporate a turbine and thereby, making the system into a Rankine cycle engine to convert thermal energy to electrical energy.

Another potential application of heat pipes has been toward the cooling of vehicle brake systems [101] https://www.thermalfluidscentral.org/encyclopedia/index.php/Heat_Pipe_Applications-cite_note-52. Normally, the brake pads and rotors of a conventional automobile brake system suffer greatly from the effects of elevated temperatures. Thermal cycling induces stresses and causes increased amounts of corrosion, which shorten the working lives of the rotors.

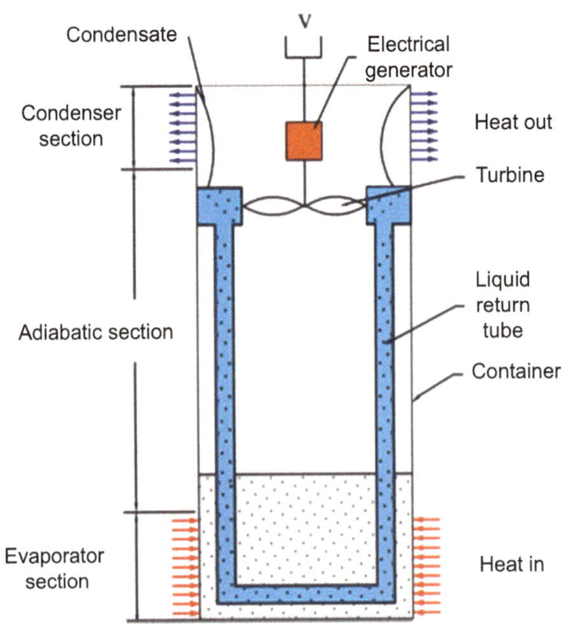

FIG. 2.87 Experimental thermosyphon rankin engine [100].

High temperatures at the rotor-pad interfaces cause embrittlement of the brake pad material, which also reduces the service life. Rotating and stationary heat pipes have been proposed to cool both the rotor and brake pads, which would result in lower, more uniform temperatures. The use of this development could also increase the service life of brake pads so that 'lifetime' pads, lasting the entire life of the automobile, may become a reality.

Vibration/shock-tolerant capillary two-phase loops are developed to meet the cooling requirements of the high performance electronics of future military vehicles which undergo intense shocks during operation [102] https://www.thermalfluidscentral.org/encyclopedia/index.php/Heat_Pipe_Applications-cite_note-53. The shock tolerant feature is achieved by utilizing an in-situ sintered wick which also reduces the manufacturing costs.

2.15 Heat pipes driving production tools

An important application of heat pipes is in the field of die casting and injection molding [103,104]. The most obvious use of heat pipes in this field is the removal of heat during the solidification process; however, heat pipes are also useful in minimizing thermal shock in the dies. Die casting involves introducing a material in molten form into a closed die. The material is cooled until it solidifies, the part is removed, and the process repeated. An important consideration is the time required for the material to cool into solid form, so most dies are water cooled. It is often difficult to cool inaccessible parts of the die, however, so heat pipes are used to cool these sections. These heat pipes can also be used to preheat the die to assure the continuous flow of molten material due to the reversible nature of heat pipes.

2.16 Medicine and human body temperature control via heat pipe

Another application is of heat pipes is related to human physiology. A surgical probe incorporating a cryogenic heat pipe is being used to destroy tumors in the human body [105]. This type of surgery, where the tissue is frozen instead rather than irradiated, is beneficial because the surrounding tissue sustains practically no damage. Also, the surgery results in very little bleeding or pain. The cryoprobe is a hand-held device with a reservoir of liquid nitrogen and a 12-inch heat pipe extension, which is maintained at approximately 77 K for about one-half hour. Fletcher and Peterson [106] invented a catheter using micro heat pipes that provide precise temperature control for treating diseased tissue.

Another application related to human physiology concerns the control of body temperature [107,108]. In regions or occupations where humans are exposed to extreme temperatures, such as workers in polar regions or foundry workers, adverse health effects from these environments are evident.

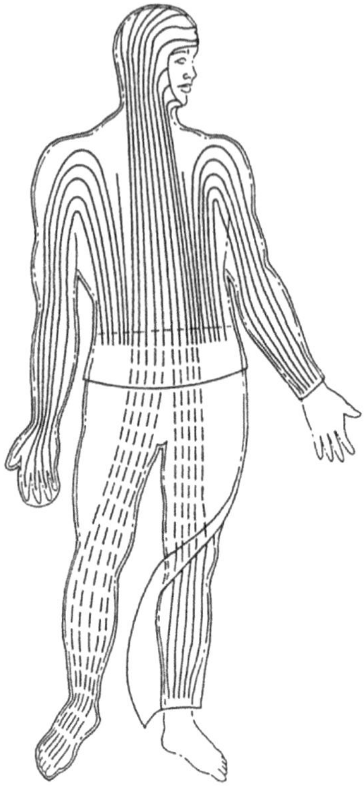

FIG. 2.88 Temperature regulation system for the human body using heat pipes [108].

Frostbite on the extremities in cold regions and heat exhaustion in warm climates are very serious problems which must be handled with extreme care. These problems can be avoided with the use of gloves, socks, and suits in which heat pipes are placed in order to transfer heat either to or from parts of the body (Fig. 2.88).

In cold climates, heat pipes could bring heat from the torso to the extremities such as the fingers, toes, and ears to prevent frostbite. Fig. 2.89 shows a conceptual design for cold weather hand wear with heat pipes, where body heat is transferred from the forearm to the fingers. In very hot environments, such as those experienced by fire fighters, a cold suit employing heat pipes could be developed which would be lighter and less bulky than the suits presently worn. This type of suit would also be beneficial in the respect that the wearer would be kept cooler, resulting in more time available for extracting people from a burning building, for instance.

Similar garments or blankets using heat pipes [113] could also be applied to medical patients whose own body temperature regulation system is impaired or not functioning. Patients with spinal cord injuries or people with Multiple

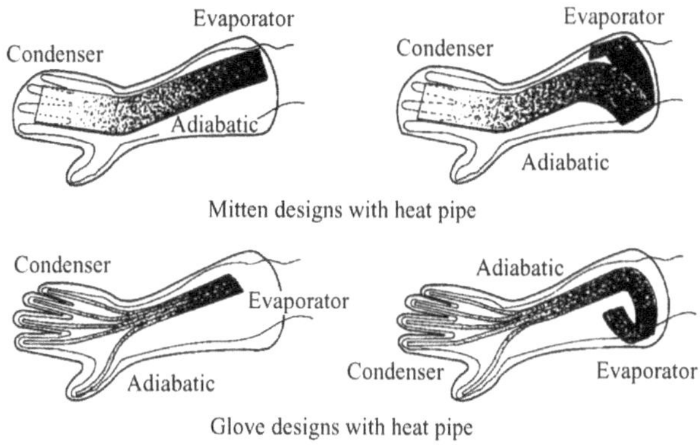

FIG. 2.89 Models of cold weather hand ware with heat pipes [108].

Sclerosis often become overheated during hot weather because of the inability of the nervous system to control their body temperature. A heat pipe suit equipped with an external heat exchanger could significantly improve the quality of life of these patients by allowing them to be more exposed to the elements. Likewise, patients who are bedridden often become overly warm or cold. In some cases, a blanket with a single-phase fluid loop is placed on the patient for heating or cooling. Since this type of blanket is quite heavy due to the fluid loop, a heat pipe blanket has been proposed, which would decrease the weight and improve the heat transfer and temperature uniformity.

It has been found experimentally that when the temperature of the cauterizing forceps used in surgical operations increase above 80°C, the tissue under operation tends to stick to the forceps. Small-diameter heat pipes are employed to precisely control the forceps temperature by removing heat from the forceps tip as well as the tissue [109]. The heat generated during cauterization evaporates the working fluid of the heat pipe. The vapor moves toward the cold section of the heat pipe where it dissipates its thermal energy content and condenses. Capillary action then returns the condensed liquid to the forceps tip. This technology represents the smallest known mass-produced heat pipe assembly (with a diameter of 2.34 mm) and makes it possible to passively control the temperature of forceps tips with diameters ranging from 0.5 to 4 mm.

2.17 Heat pipes driving ovens and furnaces

One of the first applications of heat pipes (actually, two-phase closed thermosyphons) was in baking ovens [110]. Normal flame-heated baking ovens warmed the firebrick in the firebox, which then conducted the heat to the baking chamber so that the baked food was not contaminated by the combustion products. With the advent of the heat pipe, heat was transferred to the baking chamber by the evaporation and condensation of the working fluid within the array of tubes. This arrangement was not only more efficient, saving up to 25% of t(he fuel normally needed, but also resulted in a more uniform oven temperature. A similar concept is the gas-fired restaurant griddle, which is a flat plate heat pipe separating the gas fire from the cooking surface [105]. This provides a surface with an extremely uniform temperature, even with heavy food loading. It also has the benefit of a fast warm-up time from a cold start, and the efficiency is comparable to the above-mentioned baking oven. Other food processing equipment has been designed using heat pipe technology such as braisers, kettles, saucepans, and deep fat fryers [111].

High temperature furnaces have been developed for heat-treating, sintering, and other applications [112,113]. These furnaces are actually annular heat pipes with liquid-metal working fluids. The isothermal behavior of the interior of the annular heat pipe makes it ideal for use as a furnace. The temperature gradient across the length is very small, which is always a significant problem with conventional furnaces. Due to the operation of the heat pipe, only a small portion of the outer surface of the annular heat pipe needs to be heated. Upon insertion of a cold charge, the area in contact with the cold charge becomes the condenser. Annular heat pipe furnaces are also being used in the calibration of temperature measurement devices [114,115]. Thermocouples, RTD's and thermistors have been calibrated using a variable conductance heat pipe furnace; the temperature of which was measured with a traceable platinum Resistance Temperature Detector (RTD)xe "Resistance Temperature Detector (RTD)". Optical pyrometers, which measure surface temperature from the emitted radiation, have also been calibrated with heat pipe blackbody sources.

2.18 Heat pipes driving Permafrost Stabilization

Permafrost stabilization is an important matter in designing the foundations of buildings or other structures in arctic regions. During the summer months, heat is conducted into the permafrost, which tends to melt the water trapped in the soil. Normally, this thawing would allow the foundation to sink or shift, thereby damaging the building. To avoid this problem, engineers responsible for the Trans-Alaskan pipeline placed heat pipes around the pilings of the pipeline supports (See Fig. 2.90) [116] https://www.thermalfluidscentral.org/encyclopedia/index.php/Heat_Pipe_Applications-cite_note-67. The heat from the soil adjacent to the pilings was dissipated to the ambient by the heat pipes, which maintained a permafrost bulb around the pilings throughout the year.

Mashiko et al. [117] developed an artificial permafrost storage facility in Hokkaido, Japan (See Fig. 2.91), where the evaporator of the heat pipe is buried underground, and the condenser is exposed to the ambient. The heat pipe transfers the ground heat to the ambient in winter which freezes the soil. When the temperature rises in the spring, the heat is not transmitted from the ambient to the ground (thermal diode), so the frozen soil is kept throughout the year. The main feature of this system is that no power is required, and the temperature is kept constant for a long period of time.

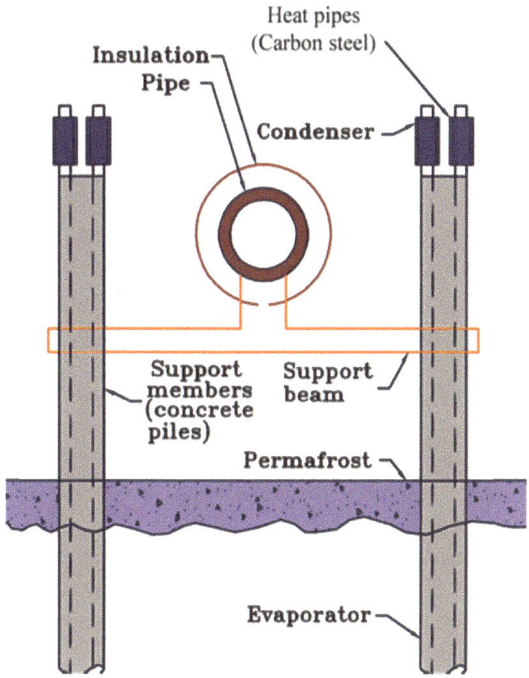

FIG. 2.90 Application of heat pipes for permafrost stabilization of structural supports on the Trans-alaskan pipeline.

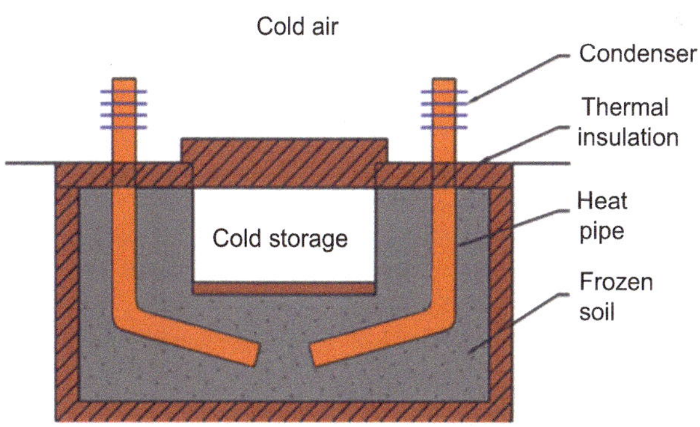

FIG. 2.91 Cold energy storage using heat pipes.

2.19 Heat pipes driving transportation systems and deicing

Several innovations using heat pipe technology have been proposed which could improve the safety and reliability of transportation systems. Heat pipes have been used to melt the ice and snow on roadways, bridges, and aircraft runways by transporting geothermal heat stored in the ground to the pavement [118,119]. Shiraishi et al. [120] presented several newly developed snow removal and deicing methods using heat pipes such as prevention of snow damage to support wires for telephone poles, a snow melting system for pavement, and a large scale snow melting system for roads in Japan. Corrugated, long heat pipes are utilized for these applications.

People working on ships at sea during the winter months are plagued by the problems associated with icing. Anything exposed to the elements, such as the decks and handrails, are constantly icing over during the winter. The usual method of removing the ice is to use an axe or hatchet, which is extremely dangerous when the seas are rough. The potential for severe cuts or falling overboard is always present. To solve this problem, heat pipes have been incorporated into the decks and handrails to transport waste heat from the engine [121]. This heat constantly melts the ice before large deposits are formed. Another similar deicing problem addressed by the use of heat pipes concerns navigation buoys [122]. Off the east coast of Canada, navigation buoys are frequently subjected to ocean spray and freezing temperatures, which results in the buoys capsizing due to the weight of the ice. Researchers have constructed a prototype buoy which was, in essence, a large ammonia thermosyphon heated from the sea. It was experimentally determined that the superstructure of the thermosyphon was kept free from ice, but that the auxiliary float seriously iced up.

References

[1] Str. Bisericii Romane no. 27, Transterm Web Site on "Cheaper Heating by Recovering", January 10, 2002, p. 500068 (Brasov, ROMANIA).

[2] R. Daniel, Koenig "Heat Pipe Nuclear Reactor for Space Power", in: LA-UR-76-998, 11th Inter society Energy Conversion Engineering Conference, Sahara Tahoe, state Line, Nevada, September 12-17, 1976, Los Alamos Scientific Laboratory, 1976.

[3] W.M. Phillips, W.C. Estabrook, T.M. Hsieh, Nuclear Thermionic Power System for Spacecraft, Jet Propulsion Laboratory, 1976.

[4] M. Ragheb, "Space Power Reactors", University of Illinois, Urbana, Illinois.

[5] J.H. Ambrose, H.R. Holmes, (Lockheed Missiles and Space Co., Sunnyvale, CA), AB (Lockheed Missiles and Space Co., Sunnyvale, CA), in: AIAA, Thermophysics Conference, 26th, Honolulu, HI, June 24-26, 1991, NASA-supported research, June 1991, p. 7.

[6] T. Mahefkey, R.R. Barthelemy, Heat Pipe Applications for Future Air Force Spacecraft, in: American Institute of Aeronautics and Astronautics, Thermophysics Conference, 15th, Snowmass, Colo., July 14–16, 1980, 1980, p. 9.

[7] W.A. Ranken, L.B. Lundberg, High Temperature Heat Pipes for Terrestrial Applications, in: International Heat Pipe Conference, 3rd, Palo Alto, Calif., May 22–24, 1978, 1978. Technical Papers. (A78-35576 14-34) New York, American Institute of Aeronautics and Astronautics, Inc., 1978, pp. 283–291. ERDA sponsored research.

[8] Uranium Information Centre Serving the Web since 1995, Now Part of Australian Uranium Association.

[9] S.W. Chi, Heat Pipe Theory and Practice, Hemisphere Publishing Corporation a McGraw Hill Book Company, 1976.

[10] R.K. Shah, A.D. Giovannelli, Heat pipe heat exchanger design theory, in: R.K. Shah, E.C. Subbarao, R.A. Mashelkar (Eds.), Heat Transfer Equipment Design, Hemisphere Publishing, 1988.

[11] D. Scott., P.E. Garner, Thermacore Inc., 780 Eden Road, Lancaster PA 17601, USA.

[12] Most of the information on subject of Nuclear Reactor in Space comes from this site and user should refer to this site for further consultation. http://www.world-nuclear.org/info/inf82.html.

[13] J. P. Alario, R. C. Prager "Space Shuttle Orbiter Heat Pipe Applications", Volume 1 Synopsis, Grumman Aerospace Corp. Final Report, 6 July 1971 – 30 April, 1972 NASA CR-128497.

[14] G.M. Grover, T.P. Cotter, G.F. Erickson, Structures of very high thermal conductance, J. Appl. Phys. 35 (6) (1964) 1990–1991.

[15] G.M. Grover, J. Bohdansky, C.A. Busse, The Use of a New Heat Removal System in Space Thermionic Power Supplies, 1965. European Atomic Energy Community – EURATOM report EUR 2229.

[16] L.L. Vasiliev, Heat pipes in modern heat exchangers, Appl. Therm. Eng. 25 (1) (2005) 1–19.

[17] A. Sharma, V.V. Tyagi, C.R. Chen, D. Buddhi, Review on thermal energy storage with phase change materials and applications, Renew. Sustain. Energy Rev. 13 (2) (2009) 318–345.

[18] K. Pielichowska, K. Pielichowski, Phase change materials for thermal energy storage, Prog. Mater. Sci. 65 (2014) 67–123.

[19] A. Abhat, Low temperature latent heat thermal energy storage: heat storage materials, Sol. Energy 30 (4) (1983) 313–332.

[20] B. Xu, P. Li, C. Chan, Application of phase change materials for thermal energy storage in concentrated solar thermal power plants: a review to recent developments, Appl. Energy 160 (2015) 286–307.

[21] A. Hoshi, D.R. Mills, A. Bittar, T.S. Saitoh, Screening of high melting point phase change materials (PCM) in solar thermal concentrating technology based on CLFR, Sol. Energy 79 (3) (2005) 332–339.

[22] A. Sciacovelli, F. Gagliardi, V. Verda, Maximization of performance of a PCM latent heat storage system with innovative fins, Appl. Energy 137 (2015) 707–715.

[23] W.-W. Wang, L.-B. Wang, Y.-L. He, Parameter effect of a phase change thermal energy storage unit with one shell and one finned tube on its energy efficiency ratio and heat storage rate, Appl. Therm. Eng. 93 (2016) 50–60.

[24] S. Mat, A.A. Al-Abidi, K. Sopian, M.Y. Sulaiman, A.T. Mohammad, Enhance heat transfer for PCM melting in triplex tube with internal-external fins, Energy Convers. Manag. 74 (2013) 223–236.

[25] Y. Jin, Q. Wan, Y. Ding, PCMs heat transfer performance enhancement with expanded graphite and its thermal stability, Procedia Engineering 102 (2015) 1877–1884.

[26] D.H. Choi, J. Lee, H. Hong, Y.T. Kang, Thermal conductivity and heat transfer performance enhancement of phase change materials (PCM) containing carbon additives for heat storage application, Int. J. Refrig. 42 (2014) 112–120.

[27] N. Calvet, X. Py, R. Olivès, J.-P. Bédécarrats, J.-P. Dumas, F. Jay, Enhanced performances of macro-encapsulated phase change materials (PCMs) by intensification of the internal effective thermal conductivity, Energy 55 (2013) 956–964.

[28] D. Zhou, C.Y. Zhao, Experimental investigations on heat transfer in phase change materials (PCMs) embedded in porous materials, Appl. Therm. Eng. 31 (5) (2011) 970–977.

[29] S. Tiari, S. Qiu, M. Mahdavi, Numerical study of finned heat pipe-assisted latent heat thermal energy storage system, Bull. Am. Phys. Soc. 59 (2014).

[30] S. Tiari, M. Mahdavi, S. Qiu, Analysis of a heat pipe-assisted high temperature latent heat energy storage system using a three-dimensional model, in: First Thermal and Fluids Engineering Summer Conference, New York, USA, 2015.

[31] N. Sharifi, A. Faghri, T.L. Bergman, C.E. Andraka, Simulation of heat pipe-assisted latent heat thermal energy storage with simultaneous charging and discharging, Int. J. Heat Mass Transf. 80 (2015) 170−179.

[32] A. Khalifa, L. Tan, A. Date, A. Akbarzadeh, A numerical and experimental study of solidification around axially finned heat pipes for high temperature latent heat thermal energy storage units, Appl. Therm. Eng. 70 (1) (2014) 609−619.

[33] E.G. Jung, J.H. Boo, Thermal analytical model of latent thermal storage with heat pipe heat exchanger for concentrated solar power, Sol. Energy 102 (2014) 318−332.

[34] A. Abhat, Performance studies of a finned heat pipe latent thermal energy storage system, in: F.D. Winter, M. Cox (Eds.), Sun: Mankind's Future Source of Energy, Pergamon, 1978, pp. 541−546.

[35] A. Abhat, Performance investigation of a long, slender heat pipe for thermal energy storage applications, J. Energy 6 (6) (1982) 361−367.

[36] Z. Liu, Z. Wang, C. Ma, An experimental study on heat transfer characteristics of heat pipe heat exchanger with latent heat storage. Part I: charging only and discharging only modes, Energy Convers. Manag. 47 (7−8) (2006) 944−966.

[37] Z. Liu, Z. Wang, C. Ma, An experimental study on the heat transfer characteristics of a heat pipe heat exchanger with latent heat storage. Part II: simultaneous charging/discharging modes, Energy Convers. Manag. 47 (7−8) (2006) 967−991.

[38] N. Sharifi, S. Wang, T.L. Bergman, A. Faghri, Heat pipe-assisted melting of a phase change material, Int. J. Heat Mass Transf. 55 (13−14) (2012) 3458−3469.

[39] N. Sharifi, T.L. Bergman, M.J. Allen, A. Faghri, Melting and solidification enhancement using a combined heat pipe, foil approach, Int. J. Heat Mass Transf. 78 (2014) 930−941.

[40] Z. h. Liu, B.-c. Zheng, Q. Wang, S.-S. Li, Study on the thermal storage performance of a gravity-assisted heat-pipe thermal storage unit with granular high-temperature phase-change materials, Energy 81 (2015) 754−765.

[41] B. w. Hu, Q. Wang, Z.-H. Liu, Fundamental research on the gravity assisted heat pipe thermal storage unit (GAHP-TSU) with porous phase change materials (PCMs) for medium temperature applications, Energy Convers. Manag. 89 (2015) 376−386.

[42] S. Tiari, S. Qiu, M. Mahdavi, Numerical study of finned heat pipe-assisted thermal energy storage system with high temperature phase change material, Energy Convers. Manag. 89 (0) (2015) 833−842.

[43] S. Tiari, S. Qiu, M. Mahdavi, Charging and discharging process of a high temperature latent heat thermal energy storage system Assisted by heat pipe, in: ASME Power & Energy 2015 − ASME 2015 9th International Conference on Energy Sustainability, San Diego, California, USA, 2015.

[44] S. Tiari, S. Qiu, Three-dimensional simulation of high temperature latent heat thermal energy storage system assisted by finned heat pipes, Energy Convers. Manag. 105 (2015) 260−271.

[45] C.W. Robak, T.L. Bergman, A. Faghri, Enhancement of latent heat energy storage using embedded heat pipes, Int. J. Heat Mass Transf. 54 (15−16) (2011) 3476−3484.

[46] K. Nithyanandam, R. Pitchumani, Analysis and optimization of a latent thermal energy storage system with embedded heat pipes, Int. J. Heat Mass Transf. 54 (21−22) (2011) 4596−4610.

[47] K. Nithyanandam, R. Pitchumani, Computational studies on a latent thermal energy storage system with integral heat pipes for concentrating solar power, Appl. Energy 103 (2013) 400–415.

[48] K. Nithyanandam, R. Pitchumani, Design of a latent thermal energy storage system with embedded heat pipes, Appl. Energy 126 (2014) 266–280.

[49] A. Khalifa, L. Tan, A. Date, A. Akbarzadeh, Performance of suspended finned heat pipes in high-temperature latent heat thermal energy storage, Appl. Therm. Eng. 81 (2015) 242–252.

[50] S. Qiu, R. Galbraith, M. White, Phase change material thermal energy storage system design and optimization, in: ASME 2013 7th International Conference on Energy Sustainability, Minneapolis, Minnesota, USA, 2013.

[51] M. Mahdavi, S. Qiu, Mathematical modeling and analysis of steady state performance of a heat pipe network, Appl. Therm. Eng. 91 (2015) 556–573.

[52] M. Mahdavi, S. Qiu, S. Tiari, Numerical investigation of hydrodynamics and thermal performance of a specially configured heat pipe for high-temperature thermal energy storage systems, Appl. Therm. Eng. 81 (2015) 325–337.

[53] M. Mahdavi, S. Qiu, Numerical Analysis and Optimization of a Complex Geometry, High Temperature Heat Pipe, First Thermal and Fluid Engineering Summer Conference, TFESC, New York City, USA, 2015.

[54] M. Mahdavi, S. Tiari, S. Qiu, Numerical Investigation of the effect of adiabatic section location on thermal performance of a heat pipe network with the application in thermal energy storage systems, Bull. Am. Phys. Soc. 60 (2015).

[55] C.E. Andraka, K.S. Rawlinson, N.P. Siegel, Technical Feasibility of Storage on Large Dish Stirling Systems, Sandia National Laboratories, Albuquerque, NM, USA, 2012.

[56] H. Shabgard, A. Faghri, T.L. Bergman, C.E. Andraka, Numerical simulation of heat pipe-assisted latent heat thermal energy storage unit for dish-Stirling systems, J. Sol. Energy Eng. 136 (2) (2013), 021025-021025.

[57] D.J. Malan, R.T. Dobson, F. Dinter, Solar thermal energy storage in power generation using phase change material with heat pipes and fins to enhance heat transfer, Energy Procedia 69 (2015) 925–936.

[58] M.S. Naghavi, K.S. Ong, I.A. Badruddin, M. Mehrali, M. Silakhori, H.S.C. Metselaar, Theoretical model of an evacuated tube heat pipe solar collector integrated with phase change material, Energy 91 (2015) 911–924.

[59] B. Zohuri, Heat Pipe Design and Technology, Modern Applications for Practical Thermal Management, Springer Publishing Company, New York, NY, 2016.

[60] M.M. Tiari, S. Qiu, Analysis of a Heat Pipe-Assisted High Temperature Latent Heat Energy Storage System Using a Three-Dimensional Model, in: First Thermal and Fluids Engineering Summer Conference, New York, USA, 2015.

[61] A.R. Alvarez-Hernandez, S. Ortega, J.T. Farmer, S. Breeding, C. Tarau, M.T. Ababneh, G. William, Anderson, Design Analysis and Performance Testing of Novel Passive Thermal Management System for Future Exploration Mission, in: 48th International Conference on Environmental Systems, 8–12 July 2018, 2018. Albuquerque, New Mexico, ICES-2018-209.

[62] S. Machiroutu, B. Kluge, M. Kuroda, H. Pokharna, The study of the effect of the gap between blower and heat exchanger in Notebook PC thermal module design, in: Proceedings of the 8th International Heat Pipe Symposium, Kumamoto, 2006, pp. 46–50.

[63] M. Mochizuki, T. Nguyen, Y. Saito, F. Kiyooka, V. Wuttijumnong, Advanced micro-channel vapor chamber for cooling high power processors, in: Proceedings of the ASME InterPack Conference, IPACK 2007, Vancouver, Canada, 2007, pp. 695–702.

[64] M. Mochizuki, Y. Saito, F. Kiyooka, T. Nguyen, High Power Cooling Chips by Heat Pipes and Advanced Heat Spreader, in: 8th International Heat Pipe Symposium, Kumamoto, Japan, 2006, pp. 214—221.

[65] X. Ping Wu, M. Mochizuki, K. Mashiko, T. Nguyen, T. Nguyen, V. Wuttijumnong, G. Cabusao, R. Singh, A. Akbarzade, Cold Energy Storage Systems Using Heat Pipe Technology For Cooling Data Centers, vol. 2, 2011, p. 013005, https://doi.org/10.5098/fhp.v2.1.3005. Available at: www.ThermalFluidsCentral.org. Frontiers in Heat Pipes (FHP).

[66] A. Jalilvand, M. Katsuta, K. Saito, J. Toyonaga, M. Mochizuki, The study of the thermal performance of concentric annular heat pipe with the application in the heat roll of fusing unit of copy machine, Annals of the Assembly for International Heat Transfer Conference 13 (2006).

[67] O. Oslejsek, F. Polasek, Cooling of Electrical Machines by Heat Pipes, in: Proceedings of the 2nd International Heat Pipe Conference, Bologna, Italy, 1976, pp. 503—514.

[68] F. Giessler, P. Sattler, F. Thoren, Heat Pipe Cooling of Electrical Machines, in: Proceedings of the 6th International Heat Pipe Conference, Grenoble, France, 1987, pp. 557—564.

[69] T. Momose, H. Ishimaru, T. Murase, S. Tanaka, S. Ishida, Application of Heat Pipe to Electron-Position Collider TRISTAN, in: Proceedings of the 6th International Heat Pipe Conference, Grenoble, France, 1987, pp. 798—799.

[70] G. Alonso, R. Perez, Cooling of Power Transformers with Two-phase Thermosyphons, in: Proceedings of the 7th International Heat Pipe Conference, Minsk, 1990.

[71] Z. Iwata, S. Sakuma, S. Koga, Heat Pipes for Local Cooling of Underground Power Cables, in: Proceedings of the 5th International Heat Pipe Conference, Tsukuba, Japan, 1984, pp. 153—155.

[72] W. Bienert, Heat Pipes for Solar Energy Collectors, in: Paper No. 12-1, Proceedings of the 1st International Heat Pipe Conference, Stuttgart, Federal Republic of Germany, 1973.

[73] C. Roberts, A Non-tracking Moderately Focusing Heat Pipe Solar Collectors, in: Paper No. 78-402, Proceedings of the 3rd International Heat Pipe Conference, Palo Alto, CA, 1978.

[74] R. Singh, M. Mochizuki, T. Nguyen, A. Akbarzadeh, Applications of heat pipes in energy conservation and renewable energy based systems, Frontiers in Heat Pipes (FHP) 2 (2011) 033003. https://doi.org/10.5098/fhp.v2.3.3003.

[75] A. Akbarzadeh, T. Wadowski, Heat pipe-based cooling systems for photovoltaic cells under concentrated solar radiation, Appl. Therm. Eng. 16 (1) (1996) 81—87. https://doi.org/10.1016/1359-4311(95)00012-3.

[76] K. Gi, S. Maezawa, Study on solar cell cooling by heat pipe, in: Proceedings of the 8th International Heat Pipe Symposium, Kumamoto, Japan, 2006, pp. 295—299.

[77] W.G. Anderson, P.M. Dussinger, D.B. Sarraf, S. Tamanna, Heat Pipe Cooling of Concentrating Photovoltaic Cells, in: 33rd IEEE Photovoltaic Specialists Conference, San Diego, CA, 2008.

[78] A. Faghri, Thermal Energy Storage Heat Exchangers, U.S. Patent No. 4976308, 1990.

[79] A. Faghri, Micro Heat Pipe Energy Storage Systems, U.S. Patent No. 5000252, 1991.

[80] H. Shabgard, T.L. Bergman, N. Sharifi, A. Faghri, High temperature latent heat thermal energy storage using heat pipes, Int. J. Heat Mass Transf. 53 (15—16) (2010) 2979—2988.

[81] H. Shabgard, C.W. Robak, T.L. Bergman, A. Faghri, Heat transfer and exergy analysis of cascaded latent heat storage with gravity-assisted heat pipes for concentrating solar power applications, Sol. Energy 86 (3) (2012) 816—830.

[82] A. Faghri, Z. Guo, Challenges and opportunities of thermal management issues related to fuel cell technology and modeling, Int. J. Heat Mass Transf. 48 (19–20) (2005) 3891–3920.

[83] A. Faghri, Z. Guo, Integration of heat pipe into fuel cell technology, Heat Transf. Eng. 29 (3) (2008) 232–238.

[84] Z. Guo, A. Faghri, Development of planar air breathing direct methanol fuel cell stacks, J. Power Sources 160 (2) (2006a) 1183–1194.

[85] Z. Guo, A. Faghri, Miniature DMFCs with passive thermal-fluids management system, J. Power Sources 160 (2) (2006b) 1142–1155.

[86] A. Faghri, Z. Guo, Vapor Feed Fuel Cells with a Passive Thermal-Fluids Management System, U.S. Patent No. 7625649B1, 2009.

[87] C.E. Andraka, K.S. Rawlinson, T.A. Moss, D.R. Adkins, J.B. Moreno, D.R. Gallup, P.G. Cordeiro, S. Johansson, Solar heat pipe testing of the stirling thermal motors 4-120 stirling engine," SAND96-1424C, Proceedings of the 31st Intersociety Energy Conversion Engineering Conference 2 (1996) 1295–1300.

[88] C.E. Andraka, K.S. Rawlinson, N.P. Siegel, Technical Feasibility of Storage on Large Dish Stirling Systems, 2012. Sandia Report SAND2012-8352.

[89] H. Hassan, A. Accensi, Spacecraft application of low temperature heat pipes, in: Paper No. 9-1, Proceedings of the 1st International Heat Pipe Conference, Stuttgart, Federal Republic of Germany, 1973.

[90] H.R. Holmes, A.R. Field, The gas-tolerant high-capacity tapered artery heat pipe, in: AIAA-86-2343, Proceedings of AIAA/ASME 4th Joint Thermophysics and Heat Transfer Conference, Boston, MA, 1986.

[91] A. Faghri, Centrifugal Heat Pipe Vapor Absorption Heat Pump, U.S. Patent No. 5201196, 1993.

[92] A. Faghri, Centrifugal Heat Pipe System, U.S. Patent No. 5297619, 1994.

[93] Y. Cao, Rotating Micro/Miniature Heat Pipes for Turbine Blade Cooling Applications, AFOSR Contractor and Grantee Meeting on Turbulence and Internal Flows, Atlanta, GA, 1996.

[94] Y. Cao, A Feasibility Study of Turbine Disk Cooling by Employing Radially Rotating Heat Pipes, Final Report for Summer Faculty Research Program, Air Force Research Lab, Turbine Engine Division, 1997.

[95] Y. Cao, Miniature high-temperature rotating heat pipes and their applications in gas turbine cooling, Frontiers in Heat Pipes (FHP) 1 (2010) 023002.

[96] Z.J. Zuo, A. Faghri, L. Langston, Numerical analysis of heat pipe turbine vane cooling, J. Eng. Gas Turbines Power 120 (4) (1998) 735–743.

[97] T. Alleau, A. Bricard, J. Chabanne, H. Kermorgant, J. de Lallee, Sodium Heat Pipes for Thermal Energy Supply of a Stirling Engine, in: Proceedings of the 5th International Heat Pipe Conference, Tsukuba, Japan, 1984, pp. 54–58.

[98] T. Alleau, A. Bricard, A. Thouvenin, Stirling Engine Coupled with a Sodium Boiler, in: Proceedings of the 6th International Heat Pipe Conference, Grenoble, France, 1987, pp. 748–752.

[99] R. Meijer, K. Khalili, Design and Testing of a Heat Pipe Gas Combustion System for the STM4-120 Stirling Engine, in: Proceedings of the 7th International Heat Pipe Conference, Minsk, USSR, 1990.

[100] T. Nguyen, P. Johnson, A. Akbarzadeh, K. Gibson, M. Mochizuki, Design, Manufacture and Testing of a Closed Cycle Thermosyphon Rankine Engine, in: Proceedings of the 8th International Heat Pipe Conference, Beijing, China, 1992.

[101] S. Maezawa, Y. Suzuki, A. Tsuchida, Heat Transfer Characteristics of Disk-Shaped Rotating, Wickless Heat Pipe, in: Proceedings of the 4th International Heat Pipe Conference, London, UK, 1981, pp. 725–734.

[102] X. Tang, C. Park, Vibration/Shock-Tolerant Capillary Two-phase Loop Technology for Vehicle Thermal Control, in: Proceedings of the ASME Summer Heat Transfer Conference, Jacksonville, FL, 2008, pp. 531–537.

[103] J. Winship, Diecasting sharpens its edge, Am. Mach. 118 (1974) 77–78.

[104] D. Reay, Heat pipes – a new diecasting aid, in: Proceedings of the 1st National Diecasters Conference, Birmingham Exhibition Centre, UK, 1977.

[105] A. Basiulis, Follow-up on heat pipe applications, in: Proceedings of the 2nd International Heat Pipe Conference, Bologna, Italy, 1976, pp. 473–480.

[106] L.S. Fletcher, G.P. Peterson, U.S. Patent No. 5591162, 1997.

[107] A. Faghri, D. Reynolds, P. Faghri, Heat pipes for hands, Mech. Eng. 111 (6) (1989b) 72–75.

[108] A. Faghri, Temperature Regulation System for the Human Body Using Heat Pipes, U.S. Patent No. 5269369, 1993.

[109] W.J. Bilski, J. Broadbent, Feel the heat: thermal design trends in medical devices, MD+DI 32 (6) (2010).

[110] Anonymous, The Paris Exhibition – Perkins' Portable Oven, The Engineer, 1867, p. 519.

[111] T. Lamp, Development of a Heat Pipe Pan Assembly for a Gas-Fired Commercial Tilting Braiser, in: Paper No. 78-399, Proceedings of the 3rd International Heat Pipe Conference, Palo Alto, CA, 1978.

[112] I. Finlay, D. Cree, D. Blundell, Performance of a prototype isothermal oven for use with an 'X' band microwave noise standard, in: Proceedings of the 2nd International Heat Pipe Conference, Bologna, Italy, 1976, pp. 545–554.

[113] O. Brost, M. Groll, H. Mack, High-temperature lithium heat pipe furnace for space applications, in: Proceedings of the 7th International Heat Pipe Conference, Minsk, USSR, 1990.

[114] H. Chengsheng, L. Tinghan, W. Yaopu, H. Zengqi, Design and performance of low temperature heat pipe blackbody, in: Proceedings of the 5th International Heat Pipe Conference, Tsukuba, Japan, 1984, pp. 74–81.

[115] C. Bassani, J. Lighthart, G. Sciamanna, P. Marcarino, V. Fernicola, Gas controlled heat pipes for primary temperature measurements, in: Proceedings of the 7th International Heat Pipe Conference, Minsk, USSR, 1990.

[116] E.D. Waters, Heat pipes for the trans-alaskan pipeline, in: Proceedings of the 2nd International Heat Pipe Conference, Bolonga, Italy, 1976, pp. 803–814.

[117] K. Mashiko, R. Okiai, M. Mochizuki, K. Ryokai, F. Tsuchiya, M. Fukuda, Development of an artificial permafrost storage using heat pipes, in: Paper No. 89-HT-15, ASME/AIChE National Heat Transfer Conference, Philadelphia, PA, 1989.

[118] H. Suelau, E. Kroliczek, C. Brinkman, Application of heat pipes to deicing systems, in: Proceedings of the 2nd International Heat Pipe Conference, Bologna, Italy, 1976, pp. 515–528.

[119] G. Bartsch, E. Butow, D. Schroeder-Richter, Deicing of road surfaces employing heat pipes installed in the ground-water layer, in: Proceedings of the 6th International Heat Pipe Conference, Grenoble, France, 1987, pp. 715–720.

[120] M. Shiraishi, M. Mochizuki, K. Mashiko, M. Ito, S. Sughiharu, Y. Yamagishi, F. Watanabe, Snow removal and deicing using flexible long heat pipes, in: Symposium on Snow and Snow Related Problems, International Glaciological Society, 1992.

[121] S. Matsuda, G. Miskolczy, M. Okihara, T. Okihara, M. Kanamori, N. Hamada, Test of a horizontal heat pipe deicing panel for use on marine vessels, in: Proceedings of the 4th International Heat Pipe Conference, London, UK, 1981, pp. 3–10.
[122] B. Larkin, S. Dubuc, Self-De-icing navigation buoys using heat pipes, in: Proceedings of the 2nd International Heat Pipe Conference, Bologna, Italy, 1976, pp. 529–536.
[123] A.T. Marriott, T. Fujita, AIP Conf. Proc. 301, 157, Review of Scientific Instruments 65 (1994) 3449, https://doi.org/10.1063/1.1144521.

Chapter 3

Different types of heat pipes

Chapter outline

3.1 Introduction 183
3.2 Compatible fluids and materials 189
3.2.1 Freeze – thaw and thermal cycling 192
3.3 Other types of heat pipes 195
3.4 Thermosyphon 195
3.5 Loop heat pipes/capillary pumped loop 196
3.5.1 Loop heat pipe advantages 196
3.6 Pulsating heat pipes 196
3.7 Micro heat pipes (MHP) 197
3.8 Constant-condenser heat pipes (CCHP) 202
3.9 Constant-condenser heat pipes (CCHP) 203
3.9.1 Variable conductance with gas-loaded heat pipes 206
3.10 Rotating and revolving heat pipes 208
3.11 High-temperature heat pipes (liquid metal heat pipes) 211
3.12 Cryogenic heat pipes 213
3.13 Wrap-around heat pipe (WAHP) in air conditioning systems 214
3.14 Oscillating Heat Pipes 215
3.15 Liquid trap diode heat pipes 217
3.16 Vapor trap diode heat pipes 218
3.17 Diode heat pipes for Venus Landers concept 221
3.17.1 Function 1 – collecting heat 222
3.17.2 Function 2 – rejecting heat 222
3.17.3 The role of the diode heat pipe 223
3.17.4 Background - diode heat pipe 224
3.18 Annular heat pipes concept 225
3.19 HiK™ heat pipe plates 227
3.19.1 HiK™ plates CFD analysis 230
3.19.2 Cooling embedded VME and VPX systems 231
3.20 Pressure controlled heat pipes (PCHPs) 233
References 236

3.1 Introduction

Heat Pipes are one of the most efficient ways to move heat, or thermal energy, from one point to another. These two-phase systems are typically used to cool areas or materials, even in outer space. Heat pipes were first developed for use by Los Alamos National Laboratory to supply heat to and remove waste heat from energy conversion systems.

Today, heat pipes are used in a variety of applications from space to handheld devices that fit in your pocket. According to our market experts, heat

Functionality, Advancements and Industrial Applications of Heat Pipes.
https://doi.org/10.1016/B978-0-12-819819-3.00003-1

pipes are present in the cooling and heat transfer systems found in computers, cell phones, and satellite systems.

These devices are sealed vessels that are evacuated and backfilled with a working fluid, typically in a small amount. The pipes use a combination of evaporation and condensation of this working fluid to transfer heat in an extremely efficient way.

The most common heat pipe is cylindrical in cross-section, with a wick on the inner diameter. Cool working fluid moves through the wick from the colder side (condenser) to the hotter side (evaporator) where it vaporizes. This vapor then moves to the condenser's heat sink, bringing thermal energy along with it. The working fluid condenses, releasing its latent heat in the condenser, and then repeats the cycle to continuously remove heat from part of the system as depicted in Fig. 3.1.

The temperature drop in the system is minimal due to the very high heat transfer coefficients for boiling and condensation. Effective thermal conductivities can approach 10,000–100,000 W/m K for long heat pipes, in comparison with roughly 400 W/m K for copper. The choice of material varies depending on the application and has led to pairings such as potassium with stainless steel, water with copper, and ammonia with aluminum, steel and nickel. Benefits of heat pipes include passive operation and very long life with little to no maintenances as it is illustrated here in Fig. 3.2.

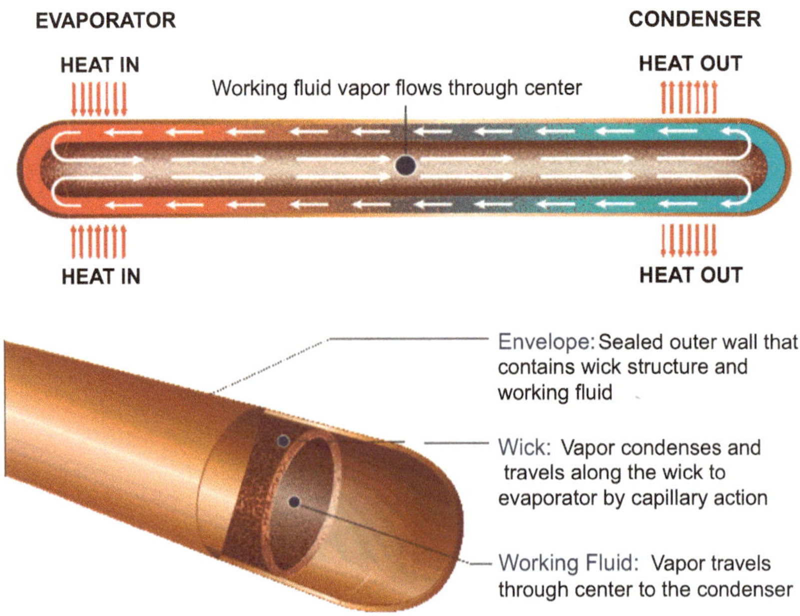

FIG. 3.1 General schematic of standard heat pipe configuration.

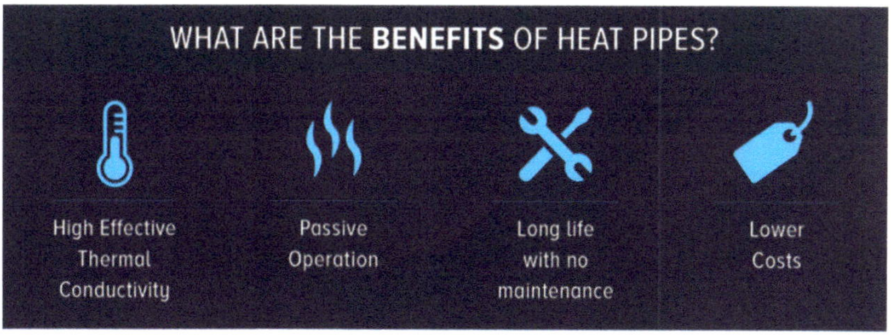

FIG. 3.2 Benefits of heat pipes.

As we have introduced at the beginning of this book, A heat pipe consists of a working fluid, a wick structure, and a vacuum-tight containment unit (envelope). The heat input vaporizes the working fluid in liquid form at the wick surface in the evaporator section.

Vapor and its associated latent heat flow toward the colder condenser section, where it condenses, giving up the latent heat. Capillary action then moves the condensed liquid back to the evaporator through the wick structure. Essentially, this operates in the same way as how a sponge soaks up water.

Phase-change processes and the two-phase flow circulation in the heat pipe will continue as long as there is a large enough temperature difference between the evaporator and condenser sections. The fluid stops moving if the overall temperature is uniform but starts back up again as soon as a temperature difference exists. No power source (other than heat) is needed.

In some cases, when the heated section is below the cooled section, gravity is used to return the liquid to the evaporator. However, a wick is required when the evaporator is above the condenser on earth. A wick is also used for liquid return if there is no gravity, such as in NASA's micro-gravity applications.

When asking what a heat pipe is, you will get a better understanding by learning about when they are used. You'll find many simple and complex systems that use these pipes in a variety of deployments based on different operating principles, thermal performance needs, conductivity requirements, spatial restrictions, overall strength, and cost.

Our thermal engineers agree that heat pipes are a smart investment if you have a device or platform that needs any of the following:

- Transfer of heat from one location to another. For example, many electronics use this to transfer heat from a chip to a remote heat sink.
- Transform heat from a high heat flux at the evaporator to a lower heat flux at the condenser, making it easier to remove overall heat with conventional methods such as liquid or air cooling. Heat fluxes of up to 1,000 W/cm^2 can be transformed with custom vapor chambers.

- Provide an isothermal surface. Examples include operating multiple laser diodes at the same temperature and providing very isothermal surfaces for temperature calibration.

A few standard examples of how heat pipes are use, the followings could be mentioned.

The most common application is a copper pipe system that uses water inside a copper envelope in order to cool electronics, operating within a temperature range of 20–150 °C.

One of the benefits for a copper/water system is that it is easy to combine with elements that are already existing in electronics. Heat sinks are present in almost every computing device and have their cooling capabilities enhanced when paired with heat pipes.

Heating, Ventilation, and Air Conditioning (HVAC) systems often turn to heat pipes for energy recovery because they require no power.

They are also used for thermal control of satellites and spacecraft. The systems provide an efficient method of heat distribution. These spacecraft systems use extremely pure fluids and are built to meet the strictest of standards, to allow operation for 0+ years. Every issue in space is mission-critical, and small failures can ruin multi-million-dollar equipment.

The common questions that what are the benefits of a heat pipe can be observed in Fig. 3.2 and associated description given below:

✔High Effective Thermal Conductivity. Transfer heat over long distances, with minimal temperature drop.

✔Passive operation. No moving parts and require no energy input other than heat to operate.

✔Isothermal operation. Very isothermal surfaces, with temperature variations as low as ±5 mK.

✔Long life with no maintenance. No moving parts that could wear out. The vacuum seal prevents liquid losses, and protective coatings can give each device a long-lasting protection against corrosion.

✔Lower costs. By lowering the operating temperature, these devices can increase the Mean Time Between Failure (MTBF) for electronic assemblies. In turn, this lowers the maintenance required, and the replacement costs. In HVAC systems, they can reduce the energy required for heating and air conditioning, with payback times of a couple of years.

There are some universal benefits of how a heat pipe works across almost all applications.

Some common questions and answers in respect to heat pipes, while some answers are specific to your needs and system requirements, these responses to standard questions will give you a better understanding as to how these devices operate:

1. What is the distance heat pipes can operate over?

   Earthbound heat pipes that work against gravity are relatively short — typically, a maximum of roughly 2 feet (60 cm) long, and a maximum elevation against gravity of roughly one foot (30 cm).

   Spacecraft heat pipes are usually under 10 feet (3 m) long, and the extra length is allowed because they operate in zero gravity.

   When a heat pipe works with gravity, called a thermosyphon, the length can be virtually unlimited, and you'll find many in lengths up to hundreds of feet (m).

2. Can a heat pipe operate against gravity?

   They can operate even when the evaporator is located above the condenser, going against gravity. This means the capillary action must return liquid against the fluid pressure drops, as well as the gravitational head. This setup will reduce the overall maximum power available to move the working fluid. Use ACT's Heat Pipe Calculator to see exact requirements and capabilities. See the link here and using configuration given in Fig. 3.3: https://www.1-act.com/resources/heat-pipe-calculator/.

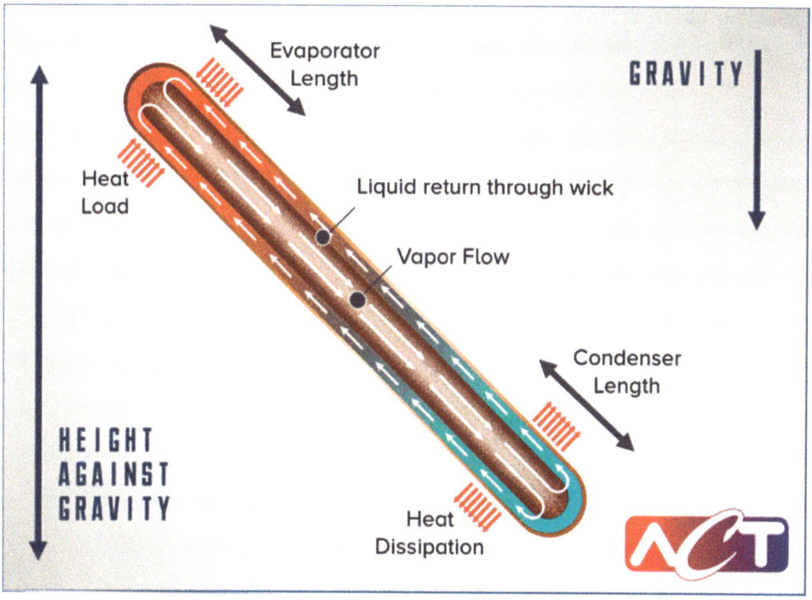

FIG. 3.3 Typical standard heat pipe operating with inclined angle.

The program in above link on Act Site, will give a performance curve of a copper/water heat pipe with the given input values. This curve is a guide for ACT's standard heat pipes; custom solutions are readily available to meet a large variety of design specific requirements. If this guide does not provide results that meet your requirements please contact our engineering experts, as they have years of design experience and can tell you if heat pipes are appropriate for your application and will work with you to assure an optimum solution. Please input the following information referenced in the picture above.

3. What is the temperature range for a heat pipe?

Heat pipes have been built to operate at a variety of temperatures ranging from −271 °C with Helium as the working fluid, and or 2000 °C or higher with Lithium or Silver as the working fluid. Fig. 3.4.

Individual two phase systems can carry at least some heat between the triple point and the critical point of the working fluid, but the power transferred near both the triple point and the critical point is very low. There is a smaller practical temperature range that shows individual capabilities and limits, e.g., copper/water heat pipes normally operate between 25 and 150 °C.

4. What materials are used for envelopes, wicks, and working fluids?

We often get asked what envelopes and wicks are made of, and what can be used for working fluids. There are a significant number of materials that can be used for each, but the important requirement is that the fluid and materials must be compatible. See next Section 3.2.

Proper selection of envelope, wick, and working fluids allow ACT Corporation to build you a system that operates maintenance-free. We've put together this list of compatible materials, but the most common envelope/wick and working fluid combinations are copper/water for electronics cooling, aluminum/ammonia for spacecraft thermal control, copper/Freon and steel/Freon for energy-recovery applications, and superalloy/alkali metal working fluids for high-temperature applications.

5. Can a Water Heat Pipe Operate After Freezing?

FIG. 3.4 Operating heat pipe temperatures.

Water heat pipes carry very little power at temperatures below ~25 °C, due to the very low vapor densities limiting the amount of power that can be transferred. At temperatures below freezing, heat transfer only occurs by conduction through the wall and wick.

Note that properly designed copper/water heat pipes can be designed to withstand thousands of freeze/thaw cycles without damage carrying power after the water is liquid. This is achieved by tightly controlling the liquid inventory, so that all of the liquid is contained in the wick. This prevents a liquid bridge from forming and damaging the device by expansion when it freezes.

3.2 Compatible fluids and materials

Since heat pipes were rediscovered by George Grover in 1963, extensive life tests have been conducted to determine compatible envelope/fluid pairs, and a large number have been found. Some of these life tests have been conducted for decades. As discussed in Most Commonly Used Envelope/Fluid Pairs, most heat pipes are fabricated for electronics cooling, and are either copper/water or copper/methanol. Most spacecraft heat pipes are aluminum/ammonia, most heat pipes for HVAC applications are copper/R134a or steel/R134a, and most high temperature heat pipes are superalloy/alkali metal. These envelope/fluid pairs cover the vast majority of heat pipes used today.

This webpage discusses the compatibility for these envelope/fluid pairs, and other pairs used in special circumstances. Table 3.1 lists most envelope/fluid pairs used today, as well as some envelopes that are known to be incompatible. Research is still ongoing in the temperature range from 150 to 400 °C, see High Temperature Water Life Tests and Intermediate Temperature Fluids Life Tests for more details.

An estimated upper and lower fluid temperature range is also shown. In most cases, the lower limit is set by the sonic limit, while the upper limit is set by the maximum vapor pressure that can be contained with a reasonable envelope wall thickness. Heat pipes can be built that operate at lower temperatures with a large diameter to maximize the sonic limit, theoretically down to the triple point. Heat pipes can also be built that operate at higher temperatures, theoretically up to the critical point, but will require a thicker envelope to withstand the vapor pressure and will have a reduced capillary limit.

The practical upper temperature limit for copper/water heat pipes is set by the vapor pressure at around 150 °C; Monel or titanium are used at higher temperatures. It is very important to note that Table 3.1 lists the generic type of material, such as Monel or Superalloy. In many cases, only some alloys are known to be compatible, while others have not been tested. Only some stainless steels are suitable for cryogenic heat pipes, since other steels become very brittle at low temperature. For additional help in selecting fluids and materials.

TABLE 3.1 Working fluid and envelope compatibility, with practical temperature limits.

Operating min temp., °C	Operating max temp., °C	Working fluid	Envelope materials	Comments
−271	−269	Helium	Stainless Steel, Titanium	
−258	−243	Hydrogen	Stainless Steel	
−246	−234	Neon	Stainless Steel	
−214	−160	Oxygen	Aluminum, Stainless Steel	
−203	−170	Nitrogen	Aluminum, Stainless Steel	
−170	0	Ethane	Aluminum, Stainless Steel	CCHPs below Ammonia Freezing point
−150	40	Propylene	Aluminum, Stainless Steel, Nickel	LHPs below Ammonia Freezing point
−100	120	Pentane	Aluminum, Stainless Steel	
−80	50	R134a	Stainless Steel	Used in Energy Recovery
−65	100	Ammonia	Aluminum, Steel, Stainless Steel, Nickel	Copper, titanium are not compatible
−60	~25 to 100	Methanol	Copper, Stainless Steel	Gas observed with Ni at 125 °C, Cu at 140 °C. Aluminum and titanium are not compatible
−50	~100	Acetone	Aluminum, Stainless Steel	Decomposes at higher temperatures
−50	280	Toluene	Al at 140 °C, Steel, Stainless Steel, Titanium, Cu-NI	Gas generation at higher temperatures (ACT life test)

20	280, short term to 300	Water	Copper, Monel, Nickel, Titanium	Short term operation to 300 °C. Aluminum, steels, stainless steels and nickel are not compatible
100	350	Naphthalene	Al, Steel, Stainless Steel, Titanium, Cu-Ni	380 °C for short term. Freezes at 80 °C
200	300, short term to 350	Dowtherm A/Therminol VP	Al, Steel, Stainless Steel, Titanium	Gas generation increases with temperature. Incompatible with Copper and Cu-Ni
200	400	$AlBr_3$	Hastelloys	Aluminum is not compatible. Freezes at 100 °C
400	600	Cesium	Stainless Steel, Inconel, Haynes, Titanium	Upper limit set by where K is the better working fluid. Monel, Copper, and Copper-Nickel are not compatible
500	700	Potassium	Stainless Steel, Inconel, Haynes	Upper limit set where Na is the better fluid. Monel and Copper are not compatible
500	800	NaK	Stainless Steel, Inconel, Haynes	Upper limit set where Na is the better working fluid. Monel and Copper are not compatible
600	1100	Sodium	Stainless Steel, Inconel, Haynes	Upper limit set by Haynes 230 creep strength
1100	1825	Lithium	Tungsten, Niobium. Molybdenum, TZM	Lithium not compatible with superalloys. Refractory metals react with air

The upper temperature limits for cesium, potassium, and NaK are set by ranking the properties of suitable alkali metals at a given temperature. For example, cesium is not normally used at higher temperatures than 600°C, since potassium is a superior working fluid. This can be seen graphically in Fig. 3.5, which compares heat pipe power versus temperature for identical heat pipes using either cesium or potassium as the working fluid. On the left side of the graph, the maximum heat pipe power is set by the sonic limit (the roughly parabolic part of the curve), while on the right side of the graph, the maximum power is set by the capillary limit (the roughly flat part of the curve). At lower temperatures, more power can be carried with cesium, since it has a higher vapor density (and higher sonic limit) at any given temperature. Once the temperature is increased above roughly about 500 °C, the potassium heat pipe carries more power (for this particular design). This is the reason that Table 3.1 states that cesium is not normally used above 600 °C.

3.2.1 Freeze – thaw and thermal cycling

Heat pipes utilize a wick structure to transport the liquid working fluid from the condenser to the evaporator. When properly made, the working fluid fully saturates the wick without making a puddle of excess fluid. With the fluid completely contained within the wick, it is not able to bridge the gap across the inside diameter of the heat pipe. This allows multiple freeze thaw cycles to occur without heat pipe deformation. A variety of working fluids may be used which directly affects the freezing temperature of the heat pipe. Fig. 3.6.

ACT Corporation routinely subjects heat pipes to thermal cycling to meet customer requirements. Typical freeze thaw tests are conducted from

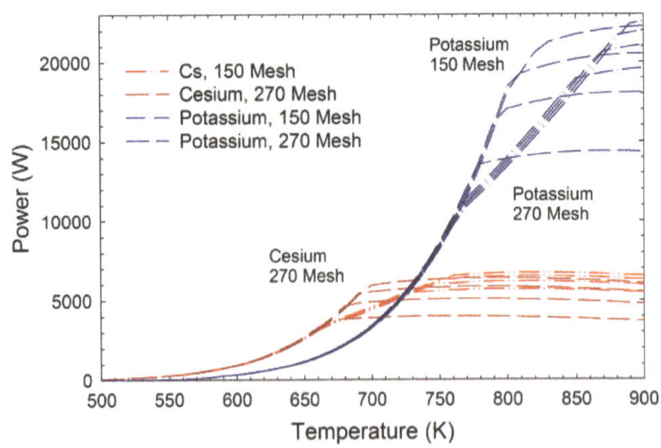

FIG. 3.5 Sonic and wicking limits for cesium and potassium heat pipes. For these specific designs, the sonic limit controls the power below 400 °C for cesium, and below 500 °C for potassium.

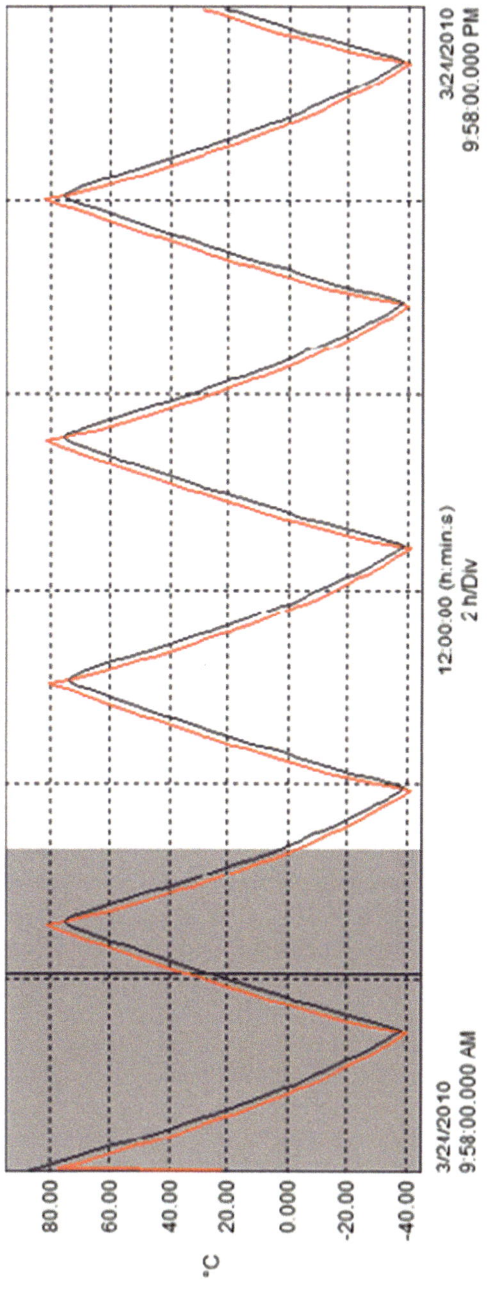

FIG. 3.6 Typical heat pipe freeze/thaw cycle data.

temperatures ranging from −20 to +20 °C and −45 to +125 °C. ACT Corporation has tested heat pipes up to 1,200 cycles, but the number of cycles is typically customer driven. When freeze/thaw screen is typically conducted with as little as 10−20 cycles. Heat pipes may be thermally cycled prior to installation into assemblies. Heat pipe assemblies are also thermally cycled in assembled units to assure system level performance. Below are three examples:

- Heat Pipes − ACT conducted tests to collect data on heat pipe thermal cycling survivability. The data set for these experiments used both fabricated flattened and bent 4 mm heat pipes as well as 0.25″ diameter copper water heat pipes. Heat pipes were exposed to as many as 1200 freeze thaw cycles without deformation or performance degradation.
- AlSiC HiK™ Plates − This project developed an innovative low Coefficient of Thermal Expansion (CTE) heat spreader by embedding heat pipes into Aluminum Silicon Carbide (AlSiC) plates (See Fig. 3.7). These plates showed similar effective thermal conductivity before and after 100 freeze/thaw cycles from −55 to 125 °C.
- Aluminum HiK™ Plates − In this project, copper water heat pipes are soldered into aluminum plates. Prior to fabrication, the heat pipes are screened by being exposed to 300 cycles from −20 to +20 °C. Once the assemblies were fabricated, the plates were exposed to an additional 50 cycles from −40 to +75 °C in two different orientations (100 cycles total) to assure freeze/thaw survivability. Fig. 3.5 shows the temperature profile of plates exposed to thermal cycling. 100% of the parts were tested, which all assemblies which all assemblies must pass prior to shipping. All plates are checked for any signs of thermal or mechanical degradation.

Aluminum Silicon Carbide (AlSiC) is a metal matrix composite that is designed to have a low Coefficient of Thermal Expansion (CTE).

This is made possible by adjusting the composition of Al and SiC to create a surface that has a low enough CTE for direct attachment of electronic components. By embedding heat pipes into these plates, ACT demonstrates thermal advantages similar to traditional aluminum HiK™ plates, with the added benefit of having a low CTE mounting surface.

Semiconductor devices tend to have low CTE, when mounted to traditional metal heat spreaders there is a significant CTE mismatch which causes the need for a stress reduction material such as a gap pad. These necessary components have poor thermal properties and often lead to lower performance from the heat spreader and overall cooling solution. With an AlSiC HiK™ plate solution, you can create higher system performance by eliminating the need for a gap pad AND increasing your thermal capability due to the embedded heat pipes.

Compared to other low CTE high conductivity heat spreaders, ACT's AlSiC HiK™ plate is a cost effective solution that can provide the necessary thermal capability to meet your system requirements.

FIG. 3.7 HiK™ AlSiC plates for electronics cooling.

3.3 Other types of heat pipes

Some of the different types of heat pipe that can be discussed in this chapter are summarized as follow;

- Variable condenser heat pipes
- Thermal diodes
- Pulsating (oscillating) heat pipes
- Loop Heat Pipes (LHPs) and Capillary Pumped Loops (CPLs)
- Micro-heat pipes
- Use of electro kinetic forces
- Rotating and Revolving heat pipes
- Miscellaneous types – Sorption Heat Pipe (SHP); magnetic fluid heat pipes

3.4 Thermosyphon

Another relatively simple, passive system, and the most popular solar water heater worldwide is the thermosyphon. Common in Japan, Australia, India, and Israel, they are easily recognizable because the tank must be located directly above the collector.

Thermosyphon systems work on the principal of heat rising. In an open-loop system (for nonfreezing climates only), potable water enters the bottom of the collector and rises to the tank as it warms. In colder climates, an antifreeze solution, such as propylene glycol, is used in the closed solar loop, and freeze-tolerant piping, such as cross-linked Polyethylene (PEX), is used for the potable water lines in the attic and on the roof.

Several international manufacturers make thermosyphon systems. The advantage of this system over the batch heater is that solar heat is stored in a well-insulated tank, so hot water can be used any time, without the penalty of overnight losses.

The following Fig. 3.8 is illustration includes the primary components of any thermosyphon system.

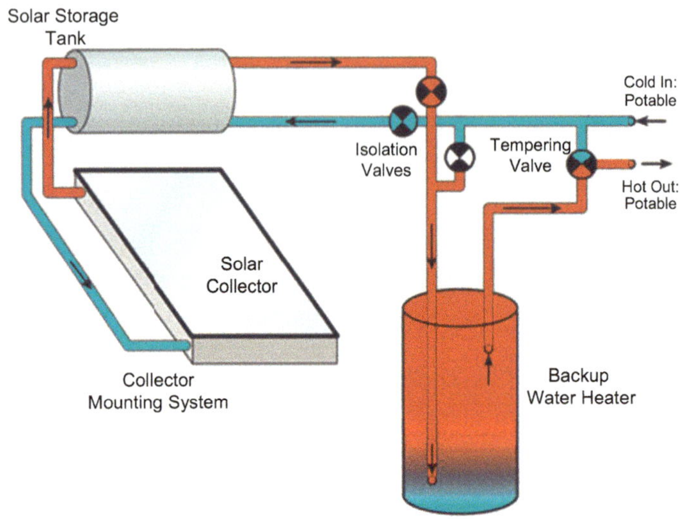

FIG. 3.8 Thermosyphon system.

3.5 Loop heat pipes/capillary pumped loop

Loop heat pipe thermal solutions are completely passive (minimal moving parts), two-phase heat transfer devices that are bendable, flexible and routable. They can even operate as thermal diodes to prevent backward heat leak. Ideal for cooling the dispersed control systems found throughout today's military aircraft, Thermacore loop heat pipes can incorporate multiple evaporators and passive/active thermal regulations.

3.5.1 Loop heat pipe advantages

- Totally passive (no external energy required)
- Transports heat up to 75 feet (23 m)
- Broad operating temperature range — for cryogenic to high-temperature applications
- Flexible and flex fatigue resistance (tested to more than 7.5 million flex cycles)
- Resists gravity loads (9 g-capable), shock, vibration, freeze and thaw
- Versatile heat load capabilities (for dissipating a few watts or many kW)

3.6 Pulsating heat pipes

Pulsating, or oscillating, heat pipes comprise a tube of capillary diameter, evacuated and partially filled with the working fluid. The implementation of pulsating heat pipe is shown in Fig. 3.9. Typically, a pulsating heat pipe comprises a serpentine channel of capillary dimension, which has been

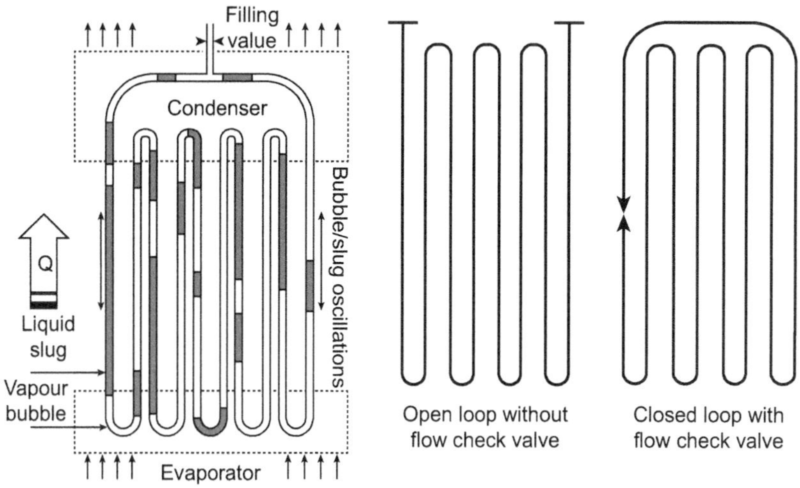

FIG. 3.9 Schematic representation of pulsating heat pipe.

evacuated, and partially filled with the working fluid. Surface tension effects result in the formation of slugs of liquid interspersed with bubbles of vapor. The operation of pulsating heat pipes was outlined in Ref. [20]. When one end of the capillary or evaporator tube is heated, the working fluid evaporates and increases the vapor pressure, thus causing the bubbles in the evaporator zone to grow. This pushes the liquid toward the low-temperature end or condenser. Cooling of the condenser results in a reduction of vapor pressure and condensation of bubbles in that section of the heat pipe. The rise and collapse of bubbles in the evaporator and condenser section results in and oscillation motion within the tube respectively [1].

Closed Loop Pulsating Heat Pipes (CLPHPs) performs better than open loop devices because of the fluid circulation that is superposed upon the oscillations within the loop. It has been suggested that further performance improvements may result from the use of check valves within the loop; however, due to the inherently small nature of the device it is difficult and costly to install such valves [19–24]. Therefore, a closed loop device without a check valve is the most practicable implementation of the pulsating heat pipe [1]. See Fig. 3.10.

3.7 Micro heat pipes (MHP)

The theory of micro heat pipe was introduced by Cotter in 1984. He defined a micro heat pipe as "so small that the mean curvature of the liquid-vapor interface is comparable in magnitude to the reciprocal of the hydraulic radius of the total flow channel."

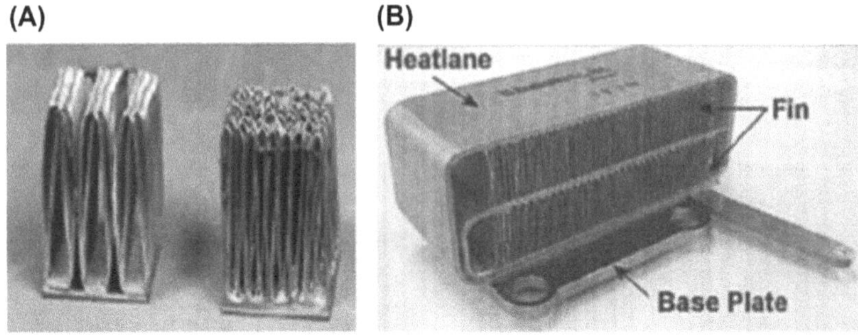

FIG. 3.10 Practical implementation of pulsating heat pipe (A) Ref. [22], (B) Ref. [23].

A Micro Heat Pipe (MHP) is a small-scale device with a hydraulic diameter on the order of 100 μ and a length of several centimeters. It differs from a conventional heat pipe in that it is much smaller, 5−500 μ - in hydraulic diameter. In general, it does not contain a wick structure to assist the return of the condensate to the evaporator section. It rather uses capillary forces generated in the sharp edges of the pipe's cross section. Fig. 3.11.

As integrated circuits become faster and more densely packed with transistors. As result the power density increases, and the heat generated as a by-product becomes more severe. Conventional methods of cooling are not an ideal way to overcome the heat problem so that small scale and high performance cooling devices are needed. A simple solution would be using micro heat pipe as integrated part in the silicon substrate of the processors.

Currently, micro heat pipe are being fabricated using micro machining technology (MEMS) and tested to verify the operation of the micro heat pipe as a thermal heat spreader.

Computers today have many components as to how fast it can run. One of the components is the processor chip and it ability to cool itself during operations. The technology of computer chips has advanced since the first computer chip dated back in 1978, with 7.14 MHz to the most recent computer chips (for home user) Duo 2 Core with maximum speed of 5.32 GHz (2.66 × 2) an increase of more than 745%. The increase is not only from the Chip speed but also in power density and power dissipation. From Fig. 3.11, there is a clear increasing trend of the speed of the chip over the last 10 year. There is also an increasing trend in Fig. 3.12 for the power density and in Figs. 3.13 and 3.14, for power dissipation of computer processor chip.

Note that it was almost impossible to find any reference for power density of any chip. Power density graph above was calculated using the power dissipation found divided by the die size of the chip, assuming the die size of the chip is the actual size of the Pentium chip.

Fig. 3.13 shows the thermal design power which is also known as the power dissipation. Through much research the power consumption could not

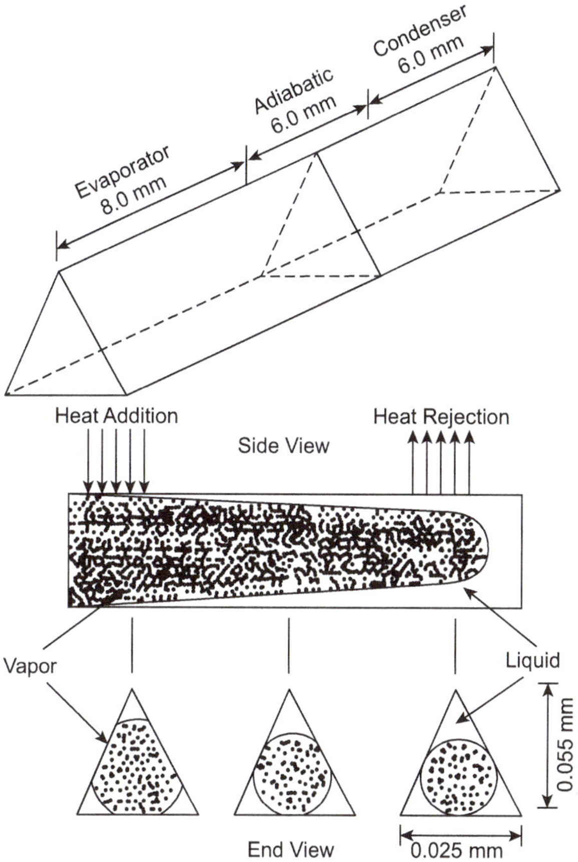

FIG. 3.11 Schematic of micro heat pipe.

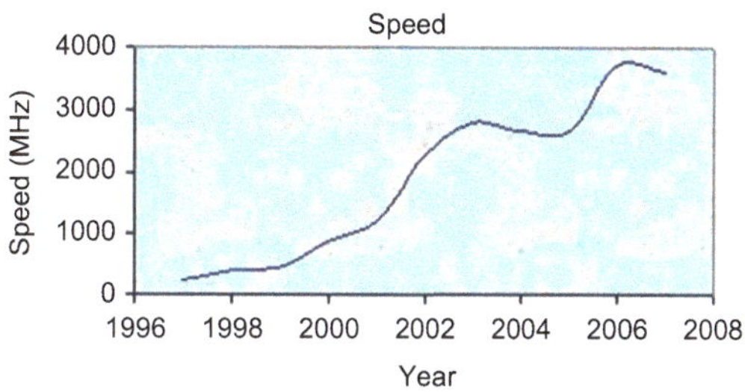

FIG. 3.12 Chip speed.

be found. Since the power needed to operate is almost always entirely dissipated in the form of heat and it is assumed that there is no power generated or stored in the chip, the power consumption is therefore be close (or the same) as the power dissipation.

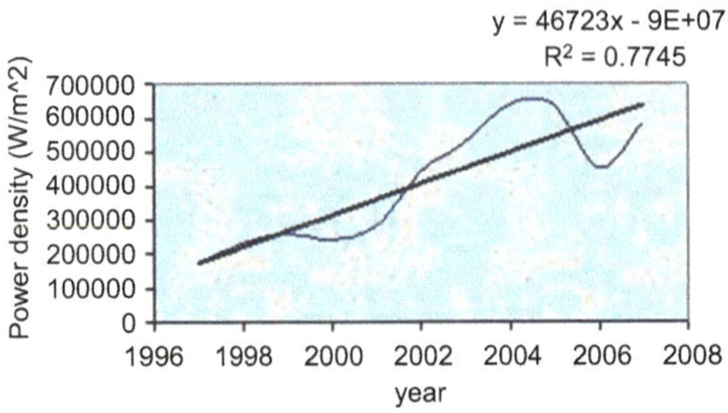

FIG. 3.13 Power density.

There are many ways to remove heat from laptops. Since laptops are design to sit on flat surfaces, and that the heat are produces from the lower end on the laptop, heat is usually removed by fans. The laptop depended on the cooled air pulled from the bottom of the laptop to cool excess heat. It is very risky to place laptop on sofas, and bed while using them, this will cause overheating and might damage the internal electronic components. Another way of cooling processor is to use heat pipe. The heat pipe contains heat transfer fluid in its center. "As the liquid evaporates, it carries heat to the cool end, where it condenses to the hot end." (Wikipedia). This method is very expensive but at the same Time very useful when space is limited.

Fig. 3.15 above shows an example of a micro-heat pipe. From this Figure, it is shown that the heat pipe is a closed cylindrical pipe (in a vacuum) which contains some sort of liquid. As explain above, the heat would be carried from the 'heat in end' to the other end of the pipe with cooler temperature. At the Cooled end, the liquid (in steam form) would then condenses and release all the heat it carried. This fluid then flows back to the hot end and the whole process repeats. Micro-heat pipe have fast thermal response and is very small. It is very useful in smaller machinery such as a laptop computer, and cameras. It also every dependable and can have a very long lifetime. Fig. 3.16 is the newly developed super-thin Micro-Heat Pipe.

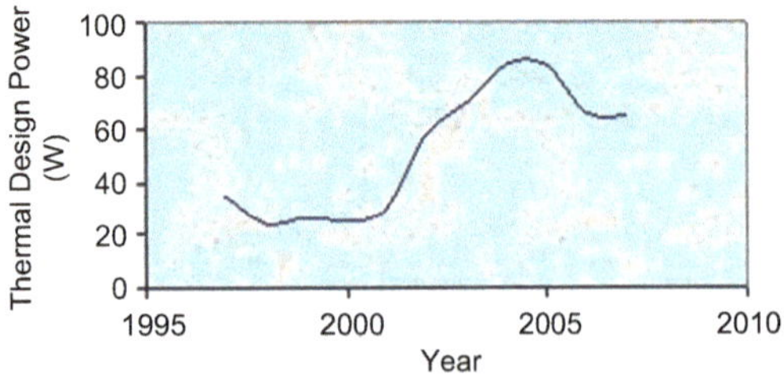

FIG. 3.14 Power dissipation.

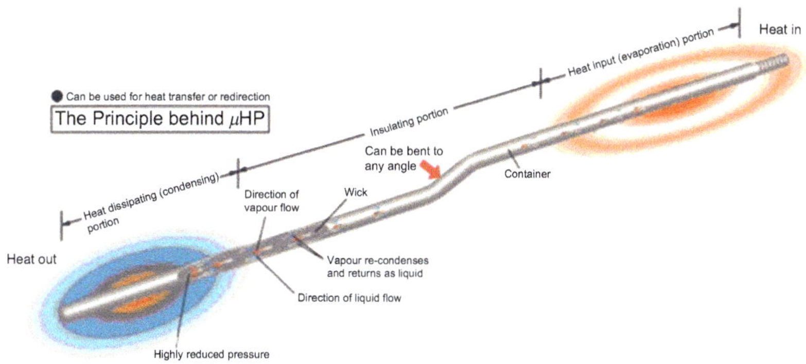

FIG. 3.15 Micro-heat pipe.

Micro-heat pipe differs from the common fins technology. Although, the two uses the same idea, which is to cool off the processor, the fins method is less effective because even though the surface area increases, it cannot transfer heat fast enough to cool the processor. It also requires air flow free area, and fin space. Micro-heat pipe doesn't take much room and it cools much faster due to the instant phase change in the cooling liquid.

Specific example of micro-heat pipe usage is the ThinkPad T60p notebook by IBM. As technology progress, there is more and more demand for smaller and thinner laptop; there for many newer the laptops from HP, Dell and Fujitsu all use micro heat pipe technology.

In conclusion, computer processor chip is a very complex subject. It was clear that the speed, power density will increase linearly as technology progress. It would be interesting to look further into the future to see where technology will take the computer chip in size, shape and its heat transfer aspect.

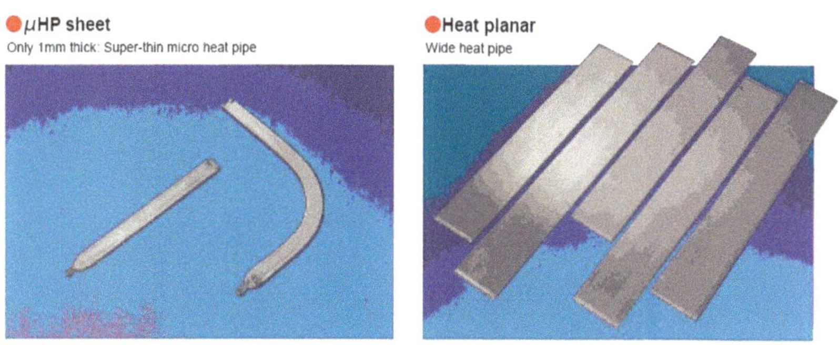

FIG. 3.16 Super-thin micro heat pipe and wide heat pipe.

3.8 Constant-condenser heat pipes (CCHP)

Constant Condenser Heat Pipes (CCHPs) also called Fixed Conductance Heat Pipes or (FCHP). Constant conductance heat pipes transport heat from a heat source to a heat sink with a very small temperature difference. Axial groove capillary wick structures are utilized because of the relative ease of manufacturing (aluminum extrusions) and their demonstrated heritage in spacecraft and instrument thermal control applications (Fig. 3.17). CCHP can transport heat in either direction and are typically used to transfer heat from specific thermal loads to a radiator panel or as part of an integrated heat pipe radiator panel. Common working fluids include: ammonia, propylene, ethane, and water. The optimum fluid charge is determined for the specific application and the effect of excess fluid charge is determined for both 0-G and 1-G operations.

The following chart below compares the model predictions and test data on a specific CCHP. ACT is Advanced Cooling Technologies Inc.

The requirement to manufacture CCHPs to achieve aerospace quality by ACT Corporation includes:

- Dedicated cleaning baths for chemically cleaning raw extrusions,
- Triple distillation apparatus for working fluid purification,
- State-of-the-art helium mass spectrometer leak detector, dedicated charging, processing, and
- Non-condensable gas venting stations, and specialized test setups for testing and characterization at various temperatures.

After the rigorous physical/chemical cleaning process, the heat pipes are further cleaned in-situ by operation at elevated temperatures with a high purity working fluid charge that is subsequently vented. A second high purity charge is loaded, and the pipe is operated over a number of days with several checks for non-condensable gases at extremely low temperatures, where the slightest amount of non-condensable gas can be detected and removed before final hermetic sealing.

FIG. 3.17 Axial groove extrusion CCHP with integral flange. *Courtesy of ACT.*

3.9 Constant-condenser heat pipes (CCHP)

The early workers in the field of cold-reservoir VCHP's were troubled by vapor diffusion into the reservoir, followed by condensation, even if liquid flow into the gas area had been arrested. It is necessary to wick the reservoir of a cold-reservoir unit in order to enable the condensate to be removed. A good report provided by Berennan and Kroliczek [18] describes various VCHP and most of this section is copied from their report. The partial pressure of the vapor in the reservoir will then be at the vapor pressure corresponding to its temperature. In principle, a variable heat pipe conductance can be achieved by modulating any one or several of the individual conductances that make up the overall conductance. A number of techniques exist to achieve variable conductance, and they can be grouped into the following four categories:

1. Gas-Loaded Heat Pipe

   This technique consists of introducing a fixed amount of non-condensable gas into the heat pipe which during operation will form a "plug" which blocks the vapor flow. A schematic of a gas-loaded VCHP is presented in Fig. 3.18. Typically, a reservoir is added to accommodate the gas when "full-on" heat pipe operation is required. As vapor flows from the evaporator to the condenser, it sweeps the non-condensing gas which accumulates in the cold end of the heat pipe. The gas therein forms a barrier to the vapor flow and effectively "shuts off" that portion of the

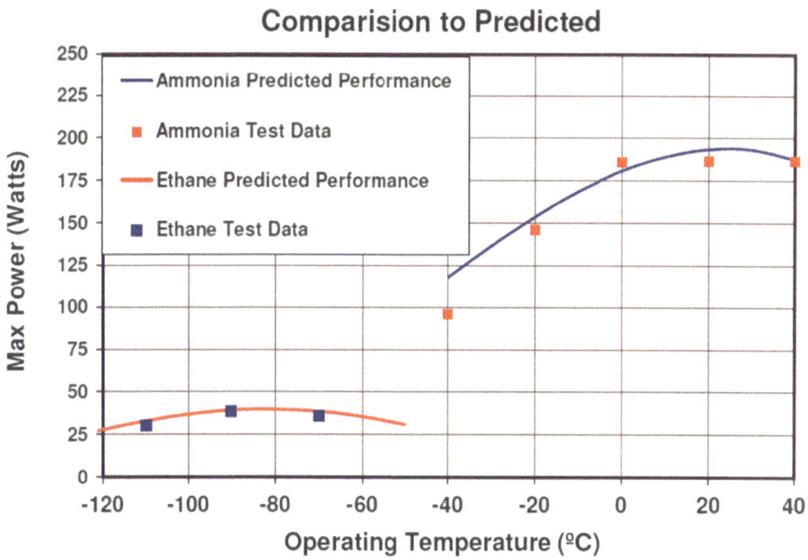

FIG. 3.18 Predicted versus actual performance for an aluminum/ammonia & aluminum/ethane CCHP. Tested at 0.1″ adverse elevation - pperated with one sided heat input and removal – 6″ evaporator, 28″ adiabatic, and 6″ condenser – nominal 0.5″ diameter – fixed fluid charge.

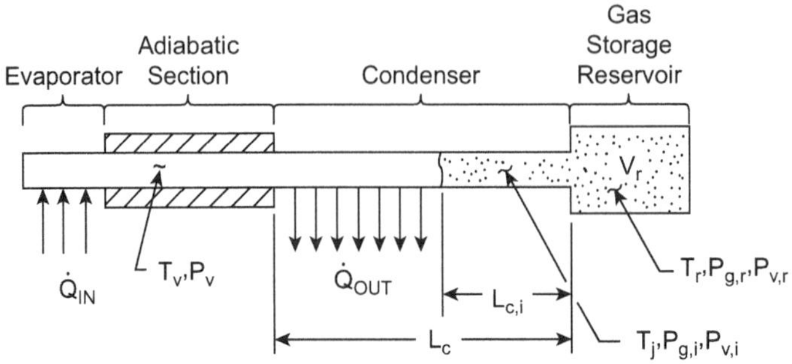

FIG. 3.19 Gas-loaded variable conductance heat pipe [18].

condenser which it fills. The length of the plug and therefore the condenser conductance depends on such factors as the system's operating temperature, heat source and sink conditions, reservoir size and reservoir temperature, etc. The influence of these parameters as well as the various methods for obtaining gas-loaded VCHP control is discussed in the next section. It should also be noted that gas blockage can also be used to affect diode and switching operations: however, the transients associated with the "shutdown" or "switching" operations can be prohibitive with a gas-loaded system [18] (Fig. 3.17).

2. Excess-Liquid Heat Pipe

This approach is analogous to the "gas-loaded" heat pipe except that excess liquid accumulates as a slug in the condenser end rather than a non-condensable gas. Control with this technique tends to be less sensitive to variations in sink conditions (however, the actual designs can be more difficult to implement). Fig. 3.20A shows one method for obtaining variable conductance with excess liquid.

Again, a reservoir is utilized, and it is located inside the heat pipe envelope. The effective volume of the reservoir is varied by means of a bellows which contains an auxiliary fluid in equilibrium with its vapor. Adjustment of the bellows to changes in system temperature changes the reservoir volume therein allowing the excess liquid to move into or out of the condenser. Fig. 3.20B illustrates a thermal diode heat pipe which utilizes liquid blockage to "shut off" the heat piping action in the reverse direction. In the normal forward mode operation, the excess liquid is swept into the reservoir at the condenser end. When conditions arise (e.g., an increase in sink temperature due to orbital conditions, etc.) which cause the condenser temperature to rise above the evaporator, the direction of vapor flow is reversed. The excess liquid is then driven from the reservoir into the normal evaporator section thus blocking the vapor flow and inactivating that section for heat rejection. Thus, the heat source is insulated from the hot condenser end with the result that the heat piping action is only effective in the forward mode [18].

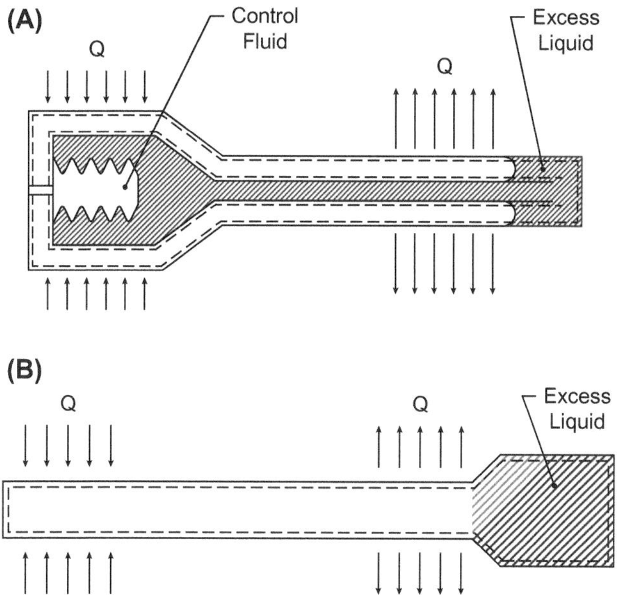

FIG. 3.20 (A) Variable conductance heat pipe [18]. (B) Variable thermal diode heat pipe [18].

3. Liquid Flow Control

Liquid flow control involves either interrupting or impeding the condensate return in the wick in order to "dry-out" part or all of the evaporator. This technique achieves control of the evaporator conductance by affecting the circulation of the working fluid and therein creating a hydrodynamic failure in the evaporator section.

Liquid flow control is limited generally to providing "on - off" control for diodes and thermal switches when the heat source is a dissipative one since the hydrodynamic failure will result in a non-uniform temperature distribution at the heat source. However, for fixed temperature sources, continuous modulation of the heat pipe conductance by varying the wick flow resistance is acceptable since partial evaporator dry-out simply results in reduced heat transfer into the pipe.

Fig. 3.21A shows a liquid trap diode heat pipe for aerospace application. In this case, a wicked reservoir is located at the evaporator end. This reservoir does not communicate with the main wick; therefore, when the temperature gradient is reversed, liquid evaporates at the hot side of the pipe and then condenses and is trapped within the reservoir. As a result, the wick becomes partially saturated and ultimately the condensate cannot return to the heat input section and the heat piping action is effectively shut "off."

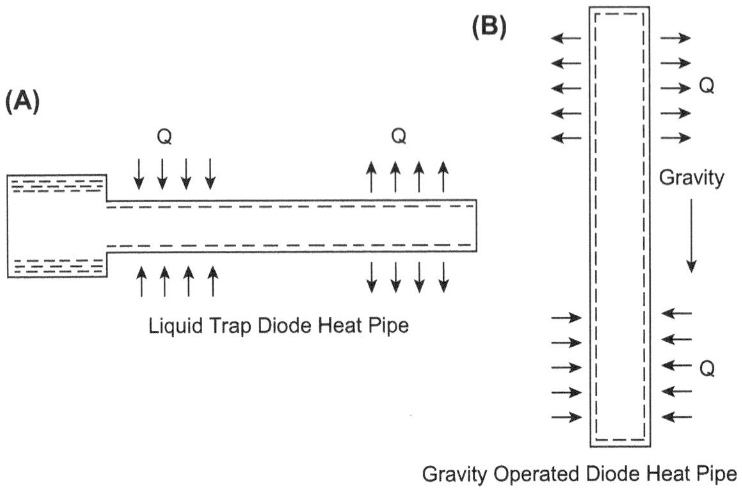

FIG. 3.21 Schematics of liquid-flow modulated heat pipes [18]. (A) horizontal variable heat pipe. (B) vertical constant heat pipe.

A gravity operated diode heat pipe is shown in Fig. 3.21B. Here a reversal of the temperature gradient causes the liquid to collect at the bottom of the pipe where it cannot be pumped back up against the gravitational force.

4. Vapor Flow Control

Vapor flow control involves throttling or interrupting the vapor as it proceeds from the evaporator to the condenser. This creates a pressure drop between the two sections, and hence a corresponding temperature drop.

A schematic of a vapor modulated variable conductance heat pipe is given in Fig. 3.22A. A bellows and auxiliary fluid are used to affect the throttling action. An increase in heat load or source temperature causes a rise in the vapor temperature which in turn causes the control fluid to expand and partially close the throttling valve therein creating a pressure differential. This method of control is substantially limited by the fact that the evaporator to condenser pressure differential must not exceed the capillary pressure developed by the fluid/wick combination.

If the valve arrangement is reversed to that shown in Fig. 3.22B, a diode action is achieved when conditions arise which reverse the normal temperature gradient [18] (Fig. 3.19).

3.9.1 Variable conductance with gas-loaded heat pipes

The principle of this technique is the formation of a gas plug at the condenser end of the pipe which prevents vapor from condensing in the part blocked by the gas. This plug is the result of introducing a fixed amount of a non-condensable gas into the heat pipe.

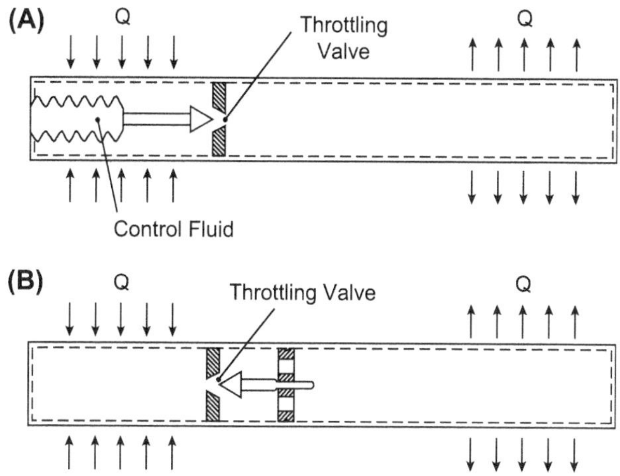

FIG. 3.22 (A) Schematics of vapor-flow modulated heat pipes [18]. Vapor modulated thermal conductance. (B) Schematics of vapor-flow modulated heat pipes [18]. Vapor modulated thermal diode.

In the absence of circulation of the working fluid (i.e., without heat transport) the gas is uniformly distributed within the vapor space except for a small amount which is dissolved in the liquid phase of the working fluid. During operation a steady flow of vapor exists from the evaporator to the condenser. The gas is swept by the vapor to the condenser. Unlike the vapor, it does not condense but forms a "plug" at the condenser end of the heat pipe.

Variable conductance variation through the addition of a non-condensable gas is particularly attractive because it accomplishes passive control of the vapor temperature. In a conventional (fixed conductance) heat pipe, the vapor temperature adjusts itself in order to meet the heat rejection requirements for a given sink condition. Thus, if the heat load and/or sink temperature increases, the vapor temperature will also rise.

In a gas-loaded heat pipe, the fixed amount of gas occupies part of the condenser; the length of the gas plug being dependent on the vapor (and sink) temperature. If the heat load is increased, the vapor temperature tends to rise as in the fixed conductance heat pipe.

However, the corresponding increase in vapor pressure of the working fluid compresses the gas plug, thereby increasing the size of the active condenser. This results in a higher conductance which effectively opposes the tendency of the vapor temperature to increase. Similarly, if the heat source and/or sink temperature decreases, the vapor temperature and pressure tend to drop which permits the gas plug to expand, the conductance of the heat pipe to decrease, and the vapor temperature decreases to be minimized. A gas-loaded heat pipe therefore reduces fluctuations of the operating temperature and behaves as a Self-controlled VCHP [18]. See Fig. 3.23.

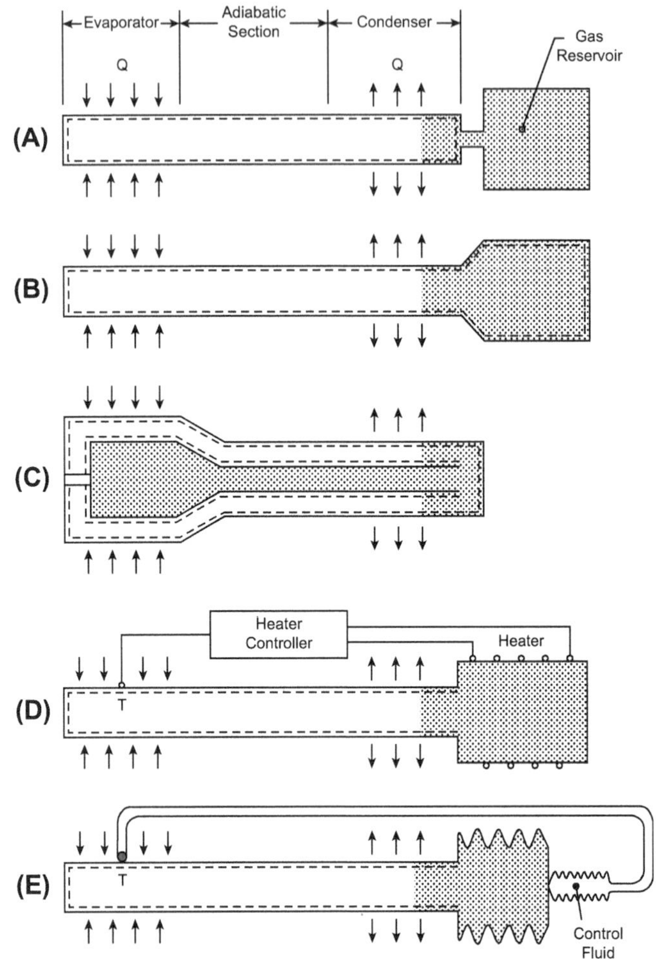

FIG. 3.23 Schematic of gas-loaded heat pipes. (A) to (E) are various configuration of Gas-loaded heat pipes.

3.10 Rotating and revolving heat pipes

An extensive review of rotating and revolving heat pipes has been conducted by Wu and Peterson [2]. In their review they first define the difference between Rotating and Revolving heat pipes, which are sometimes used interchangeably. In Rotating or Revolving heat pipe condensate returned to the evaporator through centrifugal force and there is no capillary wicks required. These types of heat pipe are used to cool turbine components and armatures for electric motors.

Fig. 3.24A is illustration of a Rotating Heat Pipe (RHP) as the shaft of electric motor, while Fig. 3.24B is illustration of a revolving heat pipe that rotates around an axis located at some distance from and parallel to the center

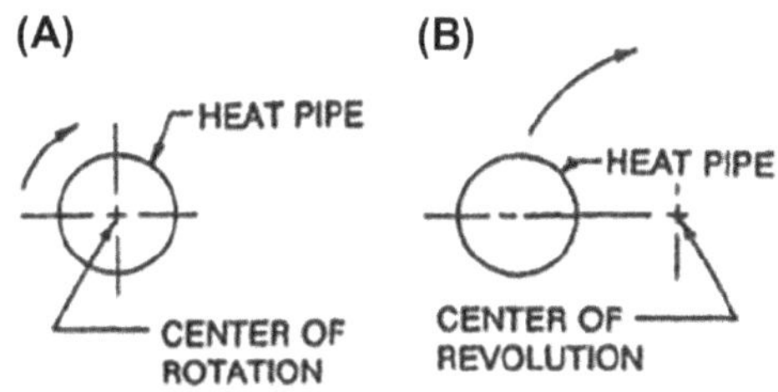

FIG. 3.24 Rotating and revolving heat pipes [3]. (A) Heat Pipe cross-section with center of rotation. (B) Heat Pipe cross-section with Center of revolution.

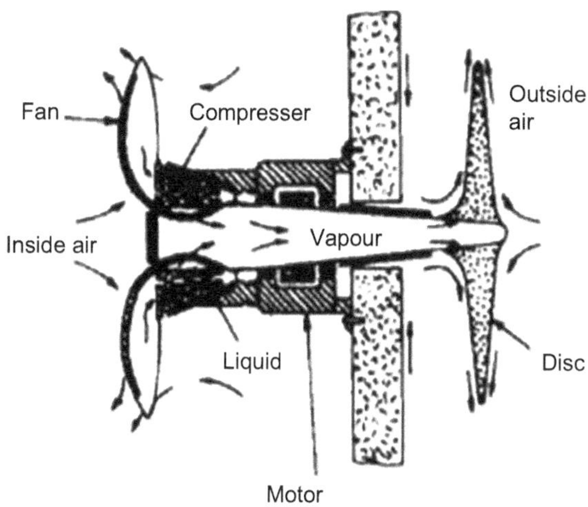

FIG. 3.25 A compact air-conditioning unit based on the wickless rotating heat pipe. *Courtesy NASA [1].*

axis of the pipe similar to the type that might be used to equalize the temperature in a rotating print head drum. The resulting performance characteristics of these two types of heat pipes are quite different and reader should refer to Peterson book for more details [3].

There have been quite bid of investigations and researches by various scientists around rotation heat pipe but in the contrast, only limited of investigations are in existence around Revolving heat pipe and some reference are work by Bontemps et al. [4], Chen and Lou [5], Chen and Tu [6,7], Keiyou and Maezawa [8], Mochizuki and Shiratori [9], and Niekawa et al. [10] and in most cases the speed of revolution was the most important parameter affecting the heat pipe transfer performance very similar to rotating heat pipes. See Fig. 3.25.

Rotating Heat Pipes can be used to remove heat generated in motors and other rotating machinery, e.g., electric motor/stator heat build-up, gear heat loads, bearing heat generation, etc. Most shaft are made of steel or stainless steel and have poor thermal conductivity. By implementing heat pipes into the shaft, the effective thermal conductivity can be increased with little strength/weight penalty.

In summary, The *Rotating Heat Pipe (RHP)* is a two phase heat transfer device that is designed to cool machinery by removing heat through a rotating shaft. As shown in Fig. 3.26 below, heat input to the evaporator vaporizes the working fluid. As in a normal heat pipe, the vapor travels down the heat pipe to the condenser, where heat is removed as the vapor condenses.

While a normal heat pipe uses a wick to return the condensate, a rotating heat pipe uses centrifugal forces. The inside of the heat pipe is a conical frustum, with the evaporator Inside Diameter (I.D.) larger than the condenser I.D. A portion of the centrifugal force is directed along the heat pipe wall, due to the slight taper. A copper-methanol rotating heat pipe is shown in Fig. 3.27.

Rotating Heat Pipes can be used to remove heat generated in motors and other rotating machinery, e.g., electric motor/stator heat build-up, gear heat loads, bearing heat generation, etc. Most shaft are made of steel or stainless steel and have poor thermal conductivity. By implementing heat pipes into the shaft, the effective thermal conductivity can be increased with little strength/weight penalty.

Figs. 3.4, 3.28 and 3.29 show a rotating heat pipe test apparatus developed by ACT. It is capable of testing at rotational speeds up to 10,000 RPM, with an Infra-Red (IR) heater capable of 4 kW supplying heat to the evaporator, and a liquid cooled condenser. Resistance Temperature Detector (RTD) temperature sensors and high speed slip ring are used with a Keithley DAQ system to provide real time data readings. See Fig. 3.29 for depiction of common components of an RTD device.

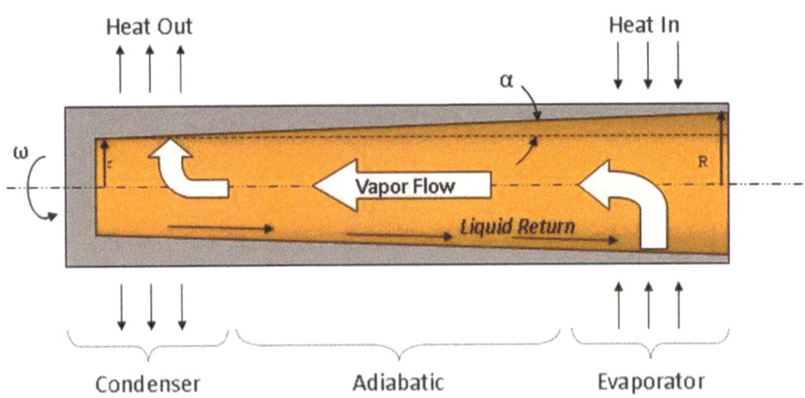

FIG. 3.26 Rotating heat pipe schematic. *Courtesy of ACT Corporation.*

As we stated in above RTD stands for Resistance Temperature Detector. RTDs are sometimes referred to generally as resistance thermometers. The American Society for Testing and Materials (ASTMs) has defined the term resistance thermometer as follows:

Resistance thermometer, n. - a temperature-measuring device composed of a resistance thermometer element, internal connecting wires, a protective shell with or without means for mounting a connection head, or connecting wire or other fittings, or both.

An RTD is a temperature sensor which measures temperature using the principle that the resistance of a metal changes with temperature. In practice, an electric current is transmitted through a piece of metal (the RTD element or resistor) located in proximity to the area where temperature is to be measured. The resistance value of the RTD element is then measured by an instrument.

This resistance value is then correlated to temperature based upon the known resistance characteristics of the RTD element.

Resistance to the flow of electricity increases. Similarly, as the temperature of the RTD resistance element increases, the electrical resistance, measured in ohms (Ω), increases. RTD elements are commonly specified according to their resistance in ohms at zero degrees Celsius (0 °C). The most common RTD specification is 100 Ω, which means that at 0 °C the RTD element should demonstrate 100 Ω of resistance.

3.11 High-temperature heat pipes (liquid metal heat pipes)

Initially Grover was interested in the development of high-temperature heat pipes, employing liquid metal working fluids, suitable for supplying heat to the emitters of thermionic electrical generators and removing heat from the collectors of these devices.

FIG. 3.27 Copper-methanol rotating heat pipe. *Courtesy of ACT Corporation.*

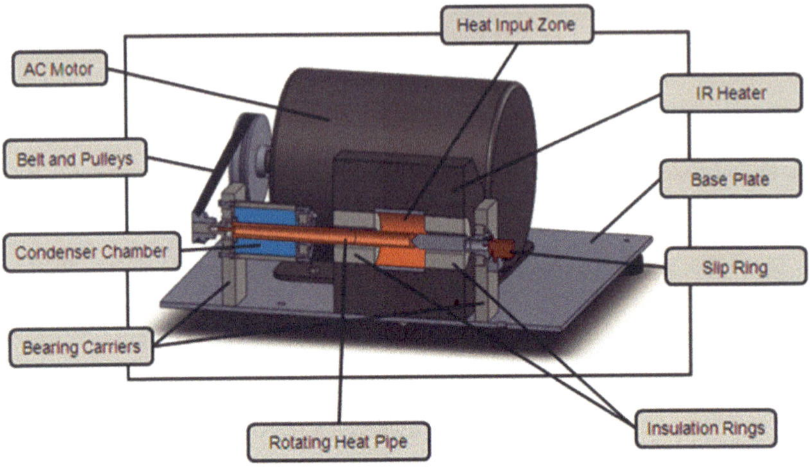

FIG. 3.28 Cross-section of the rotating heat pipe test apparatus.

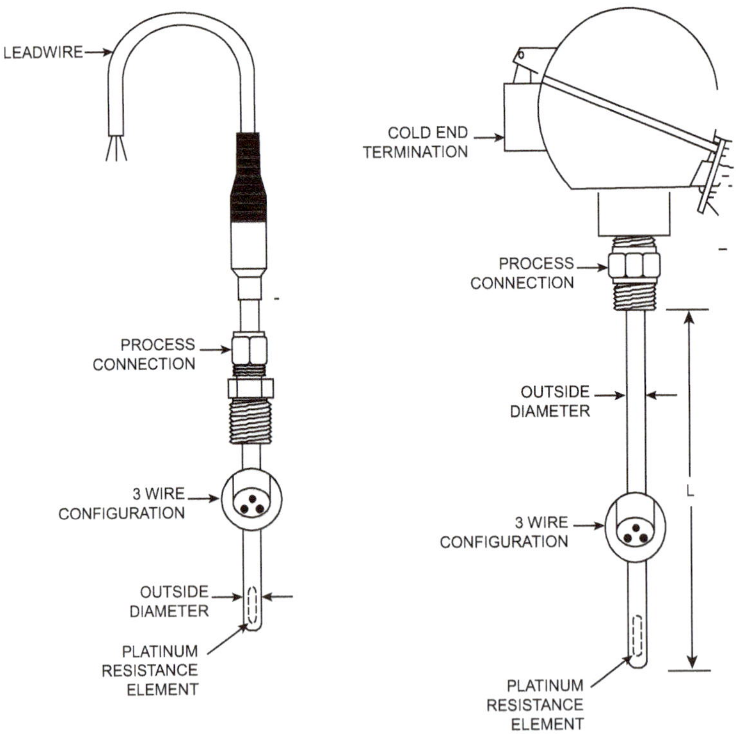

FIG. 3.29 Common components of an RTD device.

High temperature heat pipes are used in a wide variety of applications and service conditions from the ocean floor to geosynchronous orbit. These high temperature heat pipes have improved processes as mundane as glass making, as industrial as oil-shale extraction, and as high-tech as epitaxial deposition.

It is found that the heat pipe unit potentially offers better heat transfer, lower pressure drop, lower maintenance cost, and possibly lower installation cost for different applications in industry in particular for Nuclear Power plant and high heat conversion.

3.12 Cryogenic heat pipes

Most of the work on heat pipes described so far has been associated with liquid metal working fluids, and for lower temperatures water, acetone, alcohols, etc. With the need for cooling detectors in satellite infra-red scanning systems, to mention but one application, cryogenic heat pipes began to receive particular attention [11,12]. The most common working fluid in these heat pipes was nitrogen, which was acceptable for temperature ranges between 77 and 100K. Liquid oxygen was also used for this temperature range. The Rutherford High Energy Laboratory (RHEL) was the first organization in the United Kingdom to operate cryogenic heat pipes [13], liquid hydrogen units being developed for cooling targets at the RHEL.

Long-term life tests on cryogenic heat pipes started a little later than those for higher temperature units. However, there are comprehensive data from European Space Agency sources [14] on stainless steel (container was type 304L and wick type 316) heat pipes using as working fluids methane, ethane, nitrogen or oxygen, arising out of tests extending over a period of up to 13 years. The test units were 1 m in length and either 3.2 or 6.35 mm outside diameter. Heat transport capability was up to 5 Wm (meaning that the pipe transport 5 W over 1 m, or, for example, 10 W over 0.5 m), and vapor temperatures 70–270K. Tests were completed in the mid-1990s [2].

The main outcomes were as follows:

- All pipes retained maximum heat transport capability.
- All pipes maintained maximum tilt capability (capillary pumping demonstration).
- The evaporator heat transfer coefficient remained constant.
- No incompatibility or corrosion was evident in the oxygen and nitrogen pipes.
- Slight incompatibility, resulting in non-condensable gas extending over 1% of the heat pipe length and therefore affecting condenser efficiency, was noted in the ethane and methane units.

Unlike the oxygen and nitrogen TIG-welded pipes, the ethane and methane units were hard-brazed, and the implication is that the gas generation was attributed to this. Cryogenic heat pipes employing fluids such as liquid air should have special provision for pressure release or be of sufficient strength since they are frequently allowed to rise to room temperature when not in use. The critical pressure of nitrogen is 34 bar [2].

Cryogenic heat pipes should be tested in a vacuum chamber. This prevents convective heat exchange and a cold wall may be used to keep the environment at the required temperature. As a protection against radiation heat input, the heat pipe, fluid lines and cold wall should all be covered with super insulation. If the heat pipe is mounted such that the mounting points are all at the same temperature (cold wall and heat sink) it can be assumed that all heat put into the evaporator will be transported by the heat pipe as there will be no heat path to the environment. Further data on cryogenic heat pipe testing can be obtaining from Refs. [15,16].

The operating range of cryogenic heat pipes is limited by the comparatively small temperature range between the critical and triple points. Therefore, the wick of a cryogenic heat pipe must have highly effective thermal conductivity and good thermal contact with the internal wall of the pipe. An analysis of the known capillary structures such as metal mesh, felt, cermet, longitudinal grooves, and spiral threads with channels showed that the most preferred capillary structures for cryogenic heat pipes are copper wool, ceramics made of sintered metal particles, and a spiral thread combined with a channel [17].

3.13 Wrap-around heat pipe (WAHP) in air conditioning systems

Wrap around heat pipes (WAHPs) belong to a special class of recuperative heat exchangers that transfer heat from inlet to outlet locations via thermal gradient, without using any energy. In the present work, effects of various mechanical parameters on the performance of a WAHP dehumidifier system that are based on the underlying principles of heat and mass conservation are presented primarily from a theoretical point of view. A simplified methodology pertaining to wet cooling coils is applied here for defining the case of moisture condensation during precooling process at WAHP evaporator. Inlet air temperature, inlet humidity ratio, air mass flow rate, dehumidifier outlet temperature and effectiveness are the main operational parameters considered in this study. On the other hand, comparative performance study of the WAHP system is done through other set of thermodynamic parameters like the supply air temperature, supply humidity ratio, specific coil load and recovered enthalpy. The subtle variations in these factors against the operational parameters not only help in stipulating functional characteristics of the WAHP, but also allow HVAC designers to make informed decisions for system design and performance without relying entirely on manufacturer's equipment data. In this area Mridul Sarkar [24] in his study of theoretical parametric study of WAHP application in air conditioning system presents an excellent result.

3.14 Oscillating Heat Pipes

The science and technology of Oscillating Heat Pipe (OHP) has made tremendous advances in the past 20 years. Heat transfer processes in an OHP are very complex and involve liquid–vapor interfacial phenomenon, surface forces, thermally excited mechanical vibration, evaporation and condensation heat transfer, oscillated forced convection, and heat conduction.

The most outstanding feature is that an OHP can effectively integrate the state-of the-art of heat transfer processes such as thin film evaporation, oscillating flow, thermally excited mechanical vibration, nanoparticles, high heat transfer coefficient of entrance regions, and vortexes induced by the oscillating flow of liquid plugs and vapor bubbles. Therefore, the OHP can achieve an extra high effective thermal conductivity.

The OHP is a heat transfer device that functions via thermally excited oscillating motion induced by the cyclic phase change of an encapsulated working fluid. A typical OHP consists of a train of liquid plugs and vapor bubbles which exist in serpentine-arranged, interconnected capillary tubes or channels, as shown in Fig. 3.30.

The OHP is partially filled with a working fluid. The internal diameter of the OHP must be small enough so that liquid plugs can be separated by vapor bubbles. Unlike conventional heat pipes [26], the OHP can function with no wicking structure and has high manufacturability.

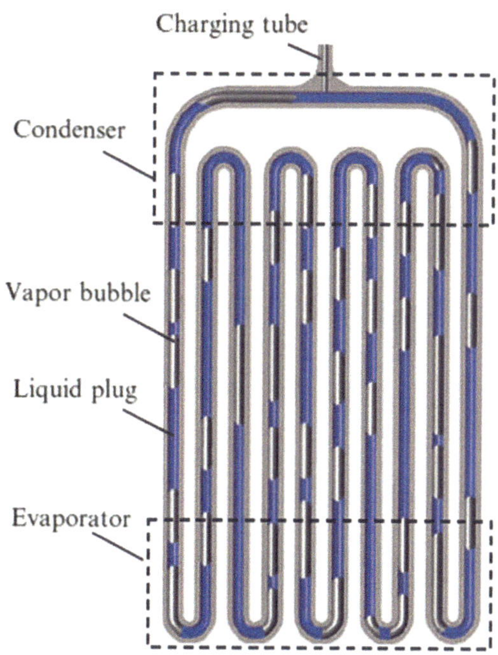

FIG. 3.30 Schematic of an oscillating heat pipe [25].

A typical OHP has three sections: an evaporating section, an adiabatic section, and a condensing section. The OHP functions by meandering its tube/channel to and through a heat reception area (evaporator) and a heat rejection area (condenser). During operation, the continual condensation (in the condenser) and evaporation (in the evaporator) of the working fluid creates a pulsating/nonequilibrium vapor pressure field that drives fluid motion between adjacent tube/channel sections.

This results in a complex flow pattern characterized by oscillatory and circulatory liquid/vapor volumes providing for both sensible (convection) and latent (phase change) heat transfer. Although the phase change heat transfer in an OHP helps to generate the oscillating motion, most of the heat is transported by sensible heat transfer. The oscillating flow and heat transfer of single phases play an important role in OHPs.

Oscillating single-phase fluids significantly enhance heat and mass transfer in a channel and have been employed in a number of heat transfer devices [27,28]. The oscillating motions generated by a variable frequency shaker [29,30] can result in a thermal diffusivity up to 17,900 times higher than those without oscillations in the capillary tubes; however, the use of mechanically driven shakers may limit its application. In 1990, Akachi [31] invented the OHP as shown in Fig. 3.31. In his invention patent (US4921041A), he defined his OHP stating: "A structure of the loop-type heat pipe includes the elongate pipe, both ends thereof being air-tightly interconnected to form a loop type container, the heat carrying fluid, at least one heat receiving portion and at least one heat radiating portion and at least one check valve for limiting a stream direction of the heat carrying fluid" [25].

This invention is responsible for the first literature, which describes how an OHP functions. In Akachi's invention, it can be found that the OHP can have different types: closed loop, open loop, check valve, tubular, and flat plate. The

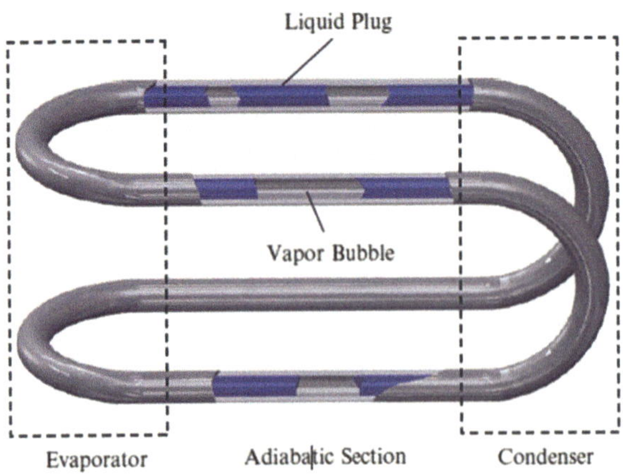

FIG. 3.31 Schematic of an oscillating heat pipe. *Courtesy of Akachi's patent [31].*

heat added in the evaporating section produces vaporization causing vapor volume expansion and the heat removed on the condensation section generates vapor condensation causing vapor volume contraction. The expansion and contraction of vapor volume produces the oscillating motion of liquid plugs and vapor bubbles in the system. In addition to the oscillating motion in the system, the pulsating motions of liquid plugs and vapor bubbles coexist at the same time. For this reason, the OHP is sometimes called a Pulsating Heat Pipe (PHP). The phase change heat transfer in the evaporator and the condenser is the primary driving force for the oscillating/pulsating motion in the system. The thermally excited oscillating/pulsating motion is the primary means used to transport heat from the evaporator to the condenser. The oscillating/pulsating motions in the OHP depend on the surface conditions, dimensions, working fluid, operating temperature, heat flux and total heat load, orientation, number of meandering turns, and, most importantly, the filling ratio [25].

For an OHP, the evaporator section has a high temperature which is in contact with a high temperature heat source and the condenser section has a low temperature which is in contact with a low temperature heat sink. During the heat transfer process from the high temperature heat source to the low temperature heat sink, the work output is generated mainly to overcome the work done by the frictional force due to the viscous fluid flow. From the thermodynamic point of view, an OHP is an engine, but the work output is directly used to generate the oscillating motion and phase change heat transfer [25].

The oscillating motion in the OHP will stop, and the OHP reaches the maximum heat transport capability, which is very different from the operating limits existing in conventional heat pipes.

3.15 Liquid trap diode heat pipes

A *Liquid Trap Diode* has a wicked reservoir located at the evaporator end of the diode heat pipe. The wicks in the heat pipe and reservoir are designed so that they can't communicate with each other. During normal operation, the heat pipe behaves like a standard heat pipe. Heat applied to the evaporator and reservoir causes liquid to evaporate. The vapor travels to the condenser, and capillary action in the heat pipe wick returns the condensate to the evaporator.

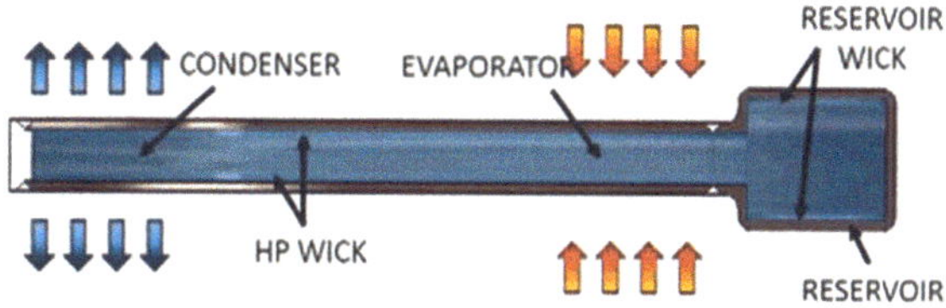

FIG. 3.32 Liquid trap diode heat pipe.

Since the reservoir wick is not connected to the main wick, the reservoir quickly dries out, and becomes inactive. See Fig. 3.32.

When the condenser becomes hotter than the evaporator/reservoir, the role of the evaporator and condenser are switched. Vapor evaporates from the hotter nominal condenser, and travels to the nominal evaporator and the reservoir, where it condenses. Since the reservoir wick does not communicate with the heat pipe wick, any liquid that condenses in the reservoir can't return to the nominal condenser. In a short time, all of the liquid is trapped in the reservoir. The main part of the pipe contains only vapor, so the only heat transfer from the condenser to the evaporator is by conduction through the heat pipe wall and wick, which has a much, much higher thermal resistance than the resistance during normal operation.

As soon as the evaporator and reservoir become hotter than the condenser, the liquid evaporates from the reservoir, and the heat pipe resumes normal operation.

3.16 Vapor trap diode heat pipes

A *Vapor Trap Diode* is fabricated in a similar fashion to a Variable Conductance Heat Pipe (VCHP). In contrast to a Liquid Trap Diode with a reservoir wick at the end of the evaporator, the vapor trap diode has a gas reservoir at the end of the condenser. In addition, wick in the vapor trap diode is connected to the wick in the liquid trap diode.

During fabrication, the heat pipe is charged with the working fluid and a controlled amount of a Non-Condensable Gas (NCG). During normal operation, the flow of the working fluid vapor from the evaporator to the condenser sweeps the NCG into the reservoir, where it does not interfere with the normal heat pipe operation. See Fig. 3.33.

When the condenser becomes hotter than the evaporator, the vapor flow is from the nominal condenser to the nominal evaporator. The NCG is dragged along with the flowing vapor. In a few minutes, it completely blocks the evaporator, greatly increasing the thermal resistivity of the heat pipe. In general, there is some heat transfer to the nominal adiabatic section. Heat is then conducted through the heat pipe walls to the evaporator.

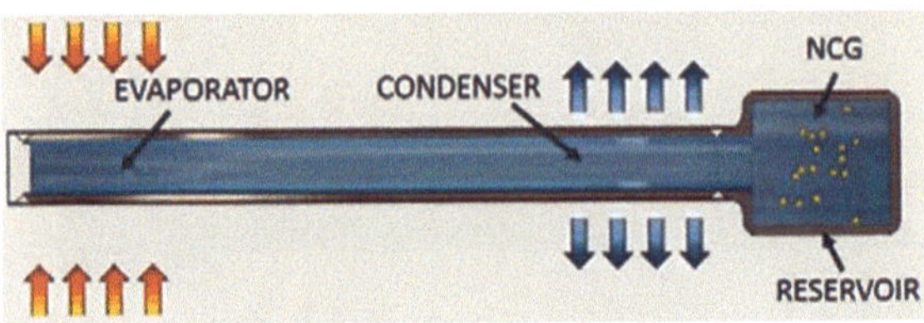

FIG. 3.33 Vapor trap diode heat pipe.

Recently, Advanced Cooling Technologies (ACTs) fabricated an alkali metal diode heat pipe for a Venus Lander application (add link to the technical paper, "Diode Heat Pipes for Venus Landers,"), as it is presented in Fig. 3.34A and B. Diode mode and heat pipe (normal) mode are shown in Fig. 3.34. The system starts in heat pipe mode, with heat transferred from the evaporator to the condenser. The evaporator, adiabatic section, and the active part of the condenser are all at a nearly constant temperature of 525 °C. When the condenser is heated, the entire condenser reaches a uniform temperature. The Non-Condensable Gas (NCG) travels to the evaporator and blocks it, dropping the evaporator temperature to ~200 °C. Finally, when heat is again applied to the evaporator, the entire system recovers, and starts transferring heat in the forward direction.

Fig. 3.34 shows two steady state temperature profiles that correspond to the initial Heat Pipe Mode, where heat was applied to the evaporator and rejected by the condenser, and to the Diode Mode, where heat was both applied to and rejected by the condenser. To a lesser extent, some heat is also rejected at the adiabatic zone. As seen in Fig. 3.35, the Diode Mode shows a separation front between Non-Condensable Gas (on the low temperature side) and potassium vapor (on the high temperature side) just before the evaporator. This separation

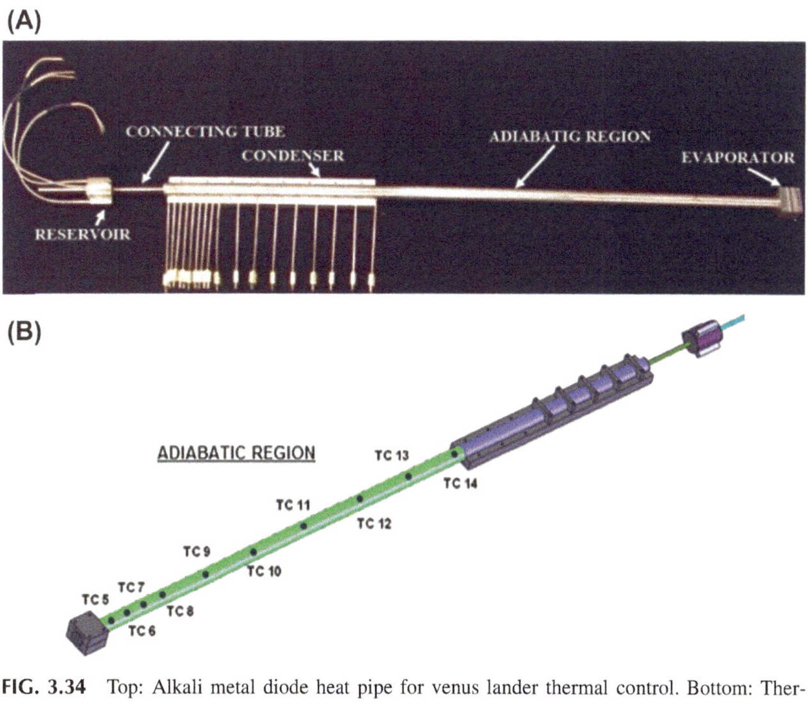

FIG. 3.34 Top: Alkali metal diode heat pipe for venus lander thermal control. Bottom: Thermocouple locations for the diode heat pipe. (A): Top (B): Bottom.

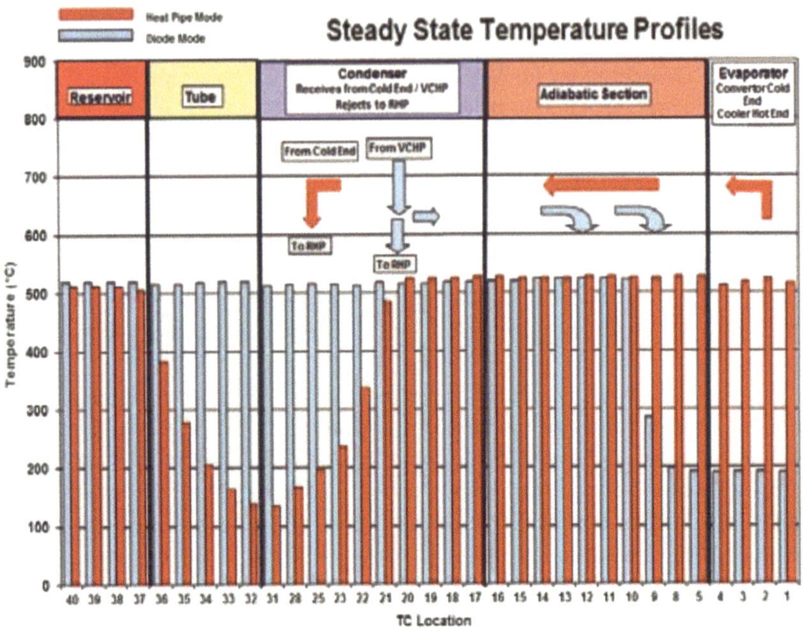

FIG. 3.35 Steady state temperature profiles along the diode heat pipe corresponding to the heat pipe mode and diode mode.

proves that the diode effect is indeed enforced by the relocated Non-Condensable Gas that now blocks the evaporator. Note that, slow transients were observed, and we believe that the cause is the small size (5 mm) of the reservoir to condenser connecting tube.

The second Diode Heat Pipe (DHP) that has a larger reservoir to condenser connecting tube (16 mm inner diameter) will also could be tested.

The difficult operating environment on the Venus surface, 460 °C and 9.3 MPa pressure, presents significant thermal design and implementation challenges for any mission. None of the previous missions operated more than 2 h on the Venus surface. For a greater science return, missions with longer operating duration on the Venus surface are needed. Cooling during normal operation of the Long-lived Venus Lander can be provided by a Stirling Duplex System [1] that uses a radioisotope Stirling power converter to energize the Stirling coolers. High temperature heat from roughly 10.

General Purpose Heat Source (GPHS) modules must be delivered to the Stirling convertor with minimal temperature drop (ΔT). In addition, the cooling system must be able to provide the following features:

1. Allows the Stirling convertor to shut off during transit to Venus without overheating the GPHS modules.

2. Pre-cool the system before the entry of the Venus atmosphere.
3. Work at nominal temperature on Venus surface.
4. Briefly shut-off on Venus surface to allow scientific measurements.
5. Reject the excess heat during the entire mission when short-lived energy sources are used.

This high temperature (~1150 °C) heat from the GPHSs is managed by a passive High Temperature Thermal Management System (HTTMS) [2] that is capable of working within the above mentioned features. During normal operation, waste heat is produced at both the cold end of the main Stirling converter and the hot end of the highest rank Stirling cooler. Rejecting this waste heat to the environment, also with a minimal temperature drop, is critical to maintaining a high efficiency cooling system. A passive Intermediate Temperature Thermal Management System (ITTMS) approximately about (~520 °C) that will reject this waste heat is under development and consists of the following two major components:

1. Heat Transport System – alkali metal Diode Heat Pipes (DHPs).
2. Heat Rejection System – alkali metal Radiator Heat Pipes (RHPs).

In terms of functionality, the ITTMS will collect (function 1) and reject (function 2) this waste heat. These two main components and two functions of the ITTMS are strongly related and their description is given below. Note that only the first component of the ITTMS, the alkali metal Diode Heat Pipe (DHP), is the object of this section.

3.17 Diode heat pipes for Venus Landers concept

As already mentioned, the Intermediate Temperature Thermal Management System (ITTMS) works with the High Temperature Thermal Management through all its features. Fig. 3.14 shows one of the five identical segments of the entire thermal management system where both the ITTMS and the HTTMS with all the inputs, outputs and thermal interactions are shown. The HTTMS consists of Variable Conductance Heat Pipes (VCHPs) that collect and transport the heat from the GPHS modules to the Stirling converter and also provide backup cooling for the GPHS modules [32–34].

As seen, a VCHP evaporator is attached to the GPHS stack, the first condenser is attached to the Stirling converter's heater head and the second condenser is attached to the backup cooling radiator. During normal operation, heat is transmitted from the GPHSs to the first condenser attached to the heater head. When the Stirling is shut off, the heat is transmitted to the second condenser, and then radiated to the Diode Heat Pipes.

A segment of the ITTMS consists of a Diode Heat Pipe (DHP) and a Rotating Heat Pipe (RHP) that are physically and thermally connected. A Diode Heat Pipe is designed so that heat is readily transmitted in the forward direction but blocked in the reverse direction. Two groups of DHPs are used within the entire ITTMS, one for removing the Stirling converter's waste heat and the other for removing the highest rank Stirling cooler's waste heat. The only difference between the two DHP groups is the location of the evaporator. The condensers are identical, and they are attached to the inner side of the shell. Any segment of the ITTMS has just one DHP of either category. Three out of the five segments have DHPs of the first category while the other two have DHPs of the second category.

The DHPs are entirely (evaporator, adiabatic zone and condensers) inside the shell. To minimize the thermal resistance of the waste heat path, the condensing vapor comes in direct contact with the shell's material. The Radiator Heat Pipes are located entirely on the outside of the shell, and also have the working fluid in direct contact with the shell. As described below, the ITTMS is used for collecting and rejecting the waste heat from the Venus Lander.

3.17.1 Function 1 – collecting heat

Three categories of heat are collected from three sources:

- Waste heat from the cold end of the Stirling convertor and from the hot end of the highest rank Stirling cooler at ~520 °C. This heat is available only when the Stirling convertor is operating, and it will be transported by a set of alkali metal gas charged Diode Heat Pipes (DHPs).
- Bypass heat from the VCHP radiators of the HTTMS (~800–900 °C). This heat is available when Stirling convertor is not operating and Pu238 based GPHS modules are used.
- Excess heat from the VCHP radiators of the HTTMS (~800–900 °C) if alternative isotopes are used. This decaying excess heat must be continuously removed. This heat is added to the bypass heat at all the times, regardless that Stirling convertor is operating or not. Both excess and bypass heat are transported by the HTTMS through its VCHPs.

3.17.2 Function 2 – rejecting heat

The entire amount of heat that is collected in *Function 1* is further transferred and rejected to the ambient (space/Venus atmosphere) by the Radiator Heat Pipes (RHPs). As seen, the RHPs form the ultimate heat rejection device. Although they will reject heat by both radiation in space and convection on Venus, they are referred as Radiator Heat Pipes (RHPs) for simplicity.

As seen in Fig. 3.36 each DHP condenser rejects heat to two RHPs, a vertical one and a horizontal one. Unlike the vertical RHP, the horizontal RHP will also receive excess and/or bypass heat from the VCHP Radiator. The reason for splitting the RHP into a vertical one and a horizontal one is that the

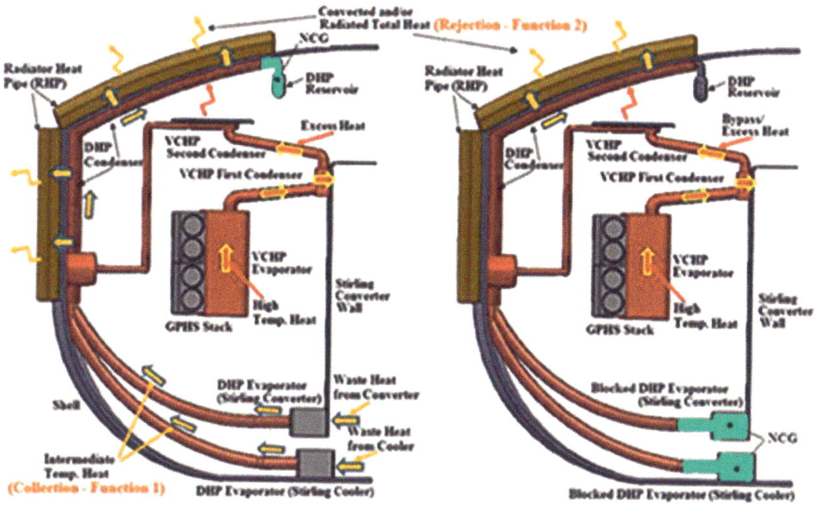

FIG. 3.36 Segment of the intermediate temperature thermal management system (ITTMS).

cylindrical portion of the pressure vessel may be heated during transit, when excess and/or bypass heat is rejected at a relatively high temperature (500–700 °C). This heating is not desired since the VCHP reservoir is attached to the inner side of the shell and an excessively warm reservoir would determine an increase in temperature of the Stirling converter's heater head.

In Fig. 3.36, the Segment of the Intermediate Temperature Thermal Management System (ITTMS). (a) Waste heat is generated (Stirling is working) and the Diode Heat Pipe is working. (b) Waste heat is not generated (Stirling is not working) and the Diode Heat Pipe is not working. This prevents heating of the Stirling cold end from the excess/bypass heat.

3.17.3 The role of the diode heat pipe

During transit, the cooling system is inactive, and no waste heat is generated. When the Stirling systems are shut off, the HTTMS must rejects the high temperature heat that bypasses the Stirling converter's heater head. This heat is rejected to the Radiator Heat Pipes (RHPs) that in turn will further reject it into the environment. However, during this process, heating the condensers of the heat collecting part of the ITTMS, the Diode Heat Pipes (DHPs), is unavoidable and the direction of heat flow reverses. Since this heat is at a higher temperature than 520 °C, reversing the heat flow is not desirable as this may overheat both the cold end of the Stirling converter and the hot end of the highest rank cooler. The role of the Diode Heat Pipe within the Thermal Management System of the Long-lived Venus Lander is to allow the radiator to work higher than 500 °C during transit when no waste heat is generated. This is also beneficial for the case when alternative isotopes are used (e.g., polonium) and the excess heat is significant at the Beginning of Mission (BOM) when higher heat rejection temperatures will allow reasonable size for the heat pipe radiator. This gas charged alkali metal Diode Heat Pipe (DHP) is the focus of this section.

3.17.4 Background – diode heat pipe

Diode heat pipes are designed to allow the heat to flow from the evaporator to condenser, while preventing flow in the reverse direction. In Fig. 3.37, a *gas charged* Diode Heat Pipe is presented in principle. The alternate diode heat pipe, with a liquid trap, is not suitable for the current application. Fig. 3.37A shows the pipe under normal operation when the evaporator temperature (T_{evap}) is higher than the condenser temperature (T_{cond}). In this case, the NCG is kept beyond the condenser in a reservoir. If the condenser temperature becomes higher than the evaporator temperature, $T_{cond} > T_{evap}$, (See Fig. 3.37B), the NCG is swept to the evaporator and prevents vapor from condensing in the evaporator. However, as seen in Fig. 3.37B, a small amount of heat is allowed to be transferred from the condenser to the adiabatic zone or the evaporator, which is necessary to continuously sweep and maintain the NCG in the evaporator.

Fig. 3.32 presents the charged Diode Heat Pipe (DHP) schematic of principle. (a) Normal operation (Heat Pipe Mode), $T_{evap} > T_{cond}$ while the NCG is kept in reservoir. (b) Non-conducting state (Diode Mode) when $T_{evap} < T_{cond}$ and the NCG is swept to the evaporator blocking it.

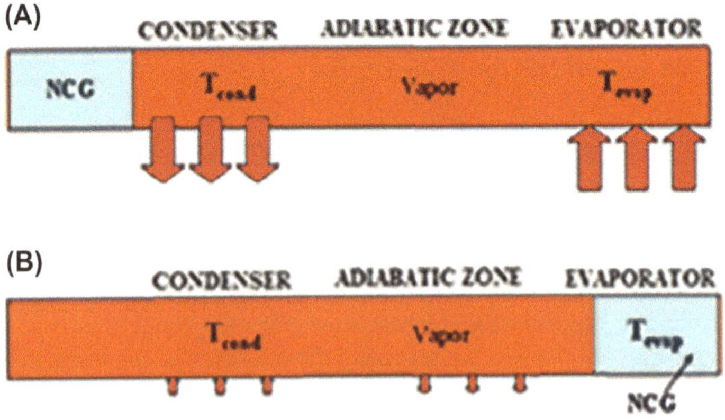

FIG. 3.37 Gas charged diode heat pipe (DHP) schematic of principle. (A) Heat pipe mode. (B) Diode mode.

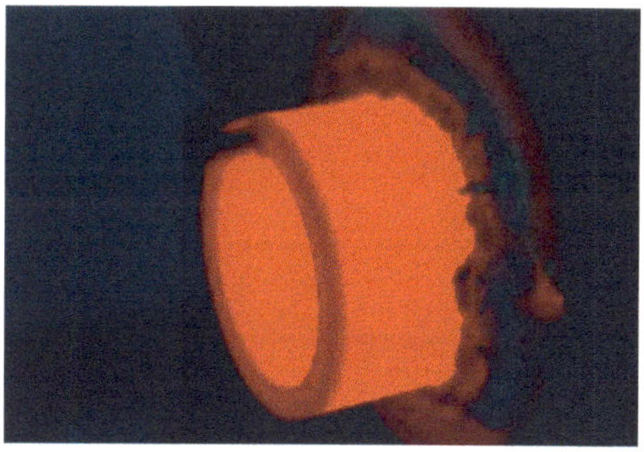

FIG. 3.38 Annular heat pipe schematic.

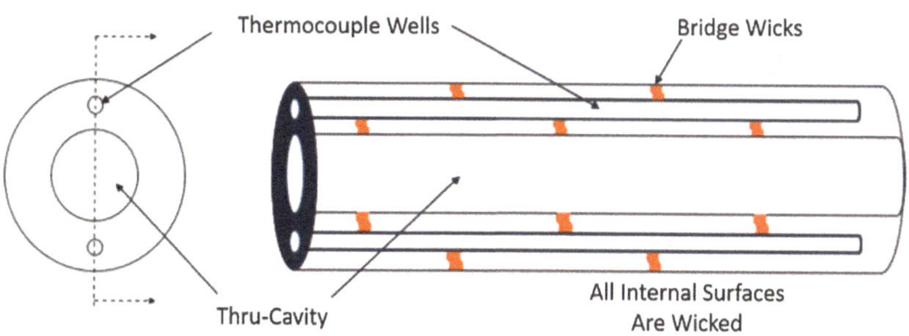

FIG. 3.39 Annular heat pipe cross-section schematic.

3.18 Annular heat pipes concept

Most standard heat pipes are used to transmit heat from one location to another, with very high effective thermal conductivity. In contrast, annular heat pipes, with an internal cavity, are most commonly used to provide a very high degree of isothermality. An annular heat pipe, also known as an Isothermal Furnace Liner (IFL) is shown in Figs. 3.38 and 3.39. All of the interior surfaces are wicked. Bridge wicks connect the inner and outer cylinders of the heat pipe to allow the fluid to return from the inner to the outer cylinder.

Note that ACT's Isothermal Furnace Liners provide Precise and Repeatable Temperature Environments.

In most applications, temperature uniformity is within 0.1 °C over the liner length. Energy can be saved, and productivity increased because useable reaction zone length in a given furnace becomes larger than the active heater length. Two or more liners may be used in series to create multiple, individually controlled zones for special effects such as step changes in temperature profile.

Annular heat pipes are most often used as Isothermal Furnace Liners.

Annular heat pipes are most commonly used as Isothermal Furnace Liners. When used for temperature calibration, thermocouple wells are installed in the annular heat pipe. With a closed end, the interior cavity can be used for pyrometer calibration.

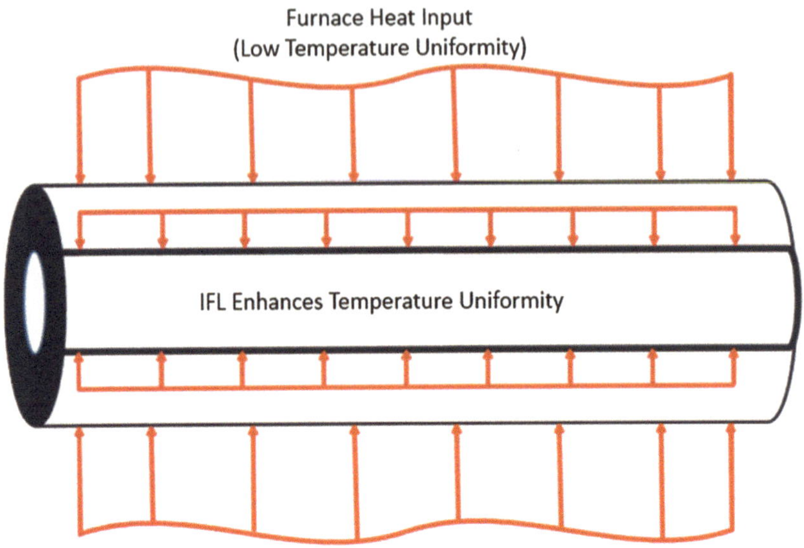

FIG. 3.40 An isothermal furnace liner annular heat pipe schematic.

During operation, the annular heat pipe is placed inside an electrically heated oven, in either a horizontal or vertical orientation. Such ovens provide non-uniform heat, both axially and radially. As shown in Fig. 3.40, the heat vaporizes the working fluid in the wick against the outside wall. The vapor travels radially and condenses on the wick against the outside wall (as well as the wick around the thermal wells). The liquid then travels back by capillary action to the outside wick through a series of bridge wicks, and then the cycle repeats.

An Isothermal Furnace Liner is an annular heat pipe, designed so that the interior temperature is very uniform.

Most heat pipes are designed to transport large amounts of power with a minimal temperature drop. In contrast, temperature uniformity is more important in IFL applications. Most of the power in a high temperature Isothermal Furnace Liner (IFL) is radiated from the exterior surfaces, with just enough heat transfer to the interior cylinder to replace heat losses. The IFL temperature in the heat pipe, and on the inner cylinder, is very uniform due to the very high evaporation and condensation heat transfer coefficients. Additional vapor evaporates from the outer cylinder wick where the heat flux is higher, however, the evaporation heat transfer coefficient is so high that the temperature difference is minimized. Similarly, a slightly colder patch on the inside cylinder will receive a higher heat flux until it is at the overall temperature. The expected uniformity inside the heat pipe is on the order of mK.

Annular heat pipes are most commonly used in the temperature calibration industry to calibrate primary temperature standards at temperatures up to

1100 °C. Thermal wells can be inserted inside the annular heat pipe to directly calibrate equipment, typically with a freeze-point cell; as it can be seen in Fig. 3.40. For calibrating pyrometers, the interior cavity forms an isothermal black body cavity.

When very precise temperature control is also required, the annular heat pipe is incorporated into a Pressure Controlled Heat Pipe. Advanced Cooling Technologies (ACTs) Corporation has fabricated a PCHP furnace with dual heat pipes that can operate from 400 to 1100 °C, while maintaining better than 3 mK stability. For more information, read our technical paper, A Novel Closed System, Pressure Controlled Heat Pipe Design for High Stability Isothermal Furnace Liner Applications. See Section 3.20 of this book for further details on Pressure Controlled Heat Pipe (PCHP).

Another important application for annular heat pipes is to isothermalize materials processing reactors for applications such as sintering, annealing, and crystal growing.

3.19 HiK™ heat pipe plates

As we know by now, Standard heat pipes only transfer heat along the axis of the heat pipe, so they are best suited to cooling discrete heat sources. High Conductivity Plates (HiK™ plates) or Vapor Chambers are used to collect heat from larger area sources, and either spread the heat, or conduct it to a cold rail for cooling. Vapor Chambers are generally used for high heat flux applications, or when genuine two-dimensional spreading is required. The lower cost HiK™ plates are used when only high conductivity in a tailored direction is required.

Aluminum and aluminum alloys have thermal conductivities around 180–200 W/m K. Copper, with a thermal conductivity of around 400 W/m K can be used when higher thermal conductivities are required, however, it is more expensive than aluminum, and weighs more than three times more than copper. Materials with higher thermal conductivity than copper are significantly more expensive.

When high conductivity structures are required in thermal management, heat pipes can be embedded in aluminum to create a HiK™ plate, achieving effective thermal conductivities that can be as high as 1200 W/m K (2400 W/m K for large HiK™ plates), which is higher than any material other than high quality diamond heat sinks.

The first step in fabricating a HiK™ plate is to determine the location of the high power components on the aluminum board, as well as the location of the cooling areas (typically water-cooled cold rails at the sides of the circuit board). Slots are then milled in the board from the high power components to the heat sink, and flattened copper/water heat pipes are inserted into the slots; see Fig. 3.41. The heat pipes are soldered in place, and then the surface is machined to leave a smooth surface, as shown in Fig. 3.42. In this figure, two of the high power locations are one-quarter and three-quarters of the way up

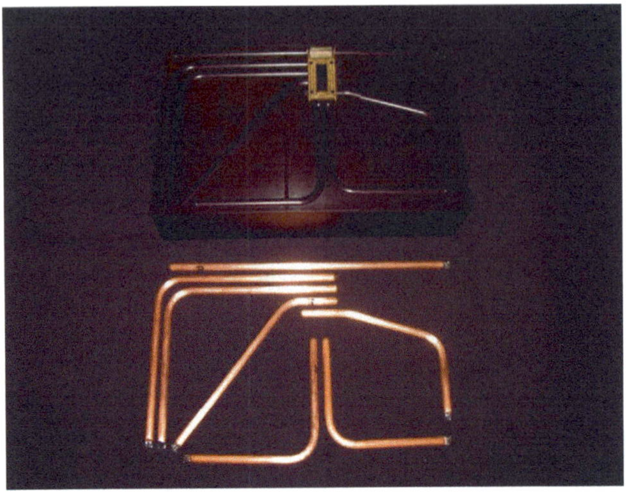

FIG. 3.41 HiK™ heat pipe Configuration. *Courtesy of ACT Corporation.*

FIG. 3.42 Heat pipes position in HiK™. *Courtesy of ACT Corporation.*

near the right side. Note the three sets of heat pipes that spread the heat over the right hand side of the cooling rails on the top and the bottom.

Note that: A HiK™ plate is fabricated by inserting flattened heat pipes into slots milled in aluminum (or other metals). Heat pipes are used to spread heat from the gold-colored region to the rest of the box.

Note that: Heat pipes are positioned to remove heat from the 3 high heat flux areas: left center, and two areas one-quarter and three-quarter of the way up on the right.

A thermal analysis was conducted on the HiK™ plate in Fig. 3.37 to help determine the heat pipe locations in the HiK™ plate. As shown in the top half of Fig. 3.38, there were three hot spots in the aluminum plate design, one on

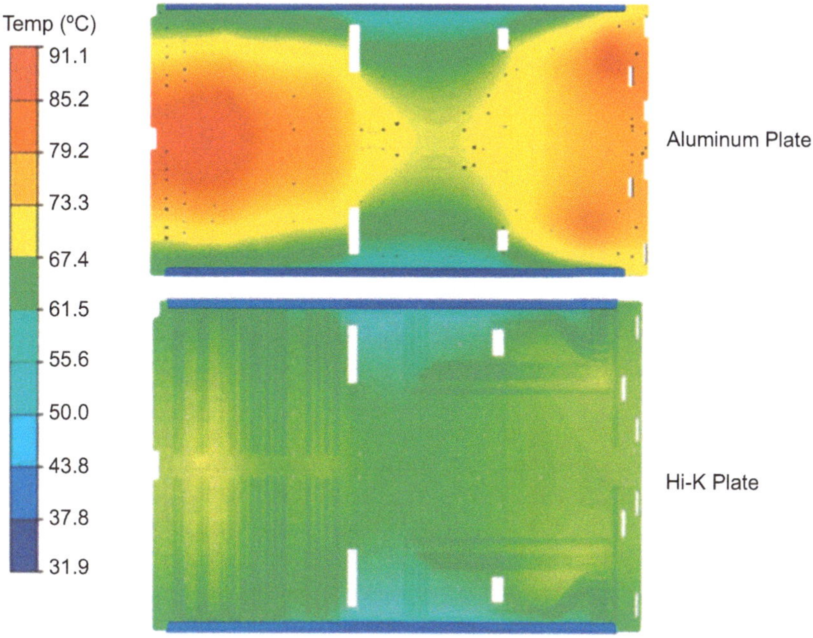

FIG. 3.43 HiK™ plate reduced the temperature. *Courtesy of ACT Corporation.*

the left, and two smaller areas on the right. The bottom half of what is shown in Fig. 3.43 shows the benefits of the embedded heat pipes. The addition of the heat pipes reduced the peak temperature by 22.1 °C, as verified by experimental testing.

Note that in Fig. 3.43, the HiK™ plate reduced the temperature by 22.1 °C when compared to an aluminum plate of identical thickness.

Analyses such as that shown in Fig. 3.43 are used to calculate the effective thermal conductivity of the HiK™ plate. The thermal conductivity of the plate is increased in the Computational Fluid Dynamics (CFDs) model until the temperature profile measures the experimentally measured temperature profile. The effective conductivity is dependent on distance (it is higher over longer distances, since the internal heat pipe ΔT is very low). Typically, the effective thermal conductivity of a HiK™ plate ranges from 500 to 1200 W/m K, depending on the specific application.

While most HiK™ plates are flat, ACT also has the ability to embed heat pipes so that the condenser is oriented at an angle from the evaporator; see Fig. 3.44. In this case, the heat pipes are bent into an L-shape, so that heat can be removed from the flange in the front of the picture.

Note that is presenting, a 3-Dimensional HiK™ plate, with the condenser oriented 90 degree from the evaporator.

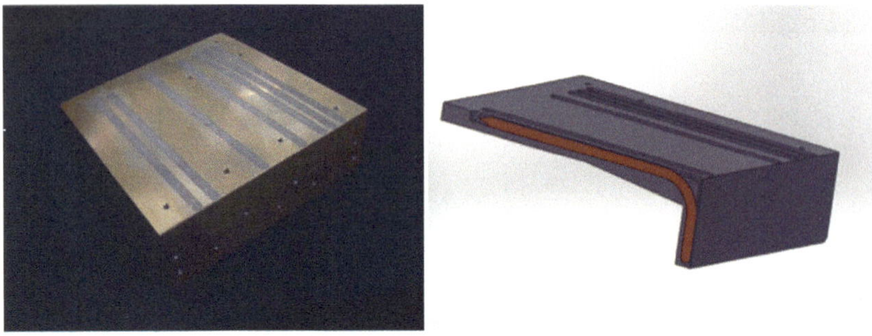

FIG. 3.44 3-Dimensional HiK™ plate. *Courtesy of ACT Corporation.*

3.19.1 HiK™ plates CFD analysis

HiK™ or high conductivity plates are heat spreaders with embedded heat pipes to transport heat as desired in your system. These plates are particularly useful for cooling of multiple high power components. The HiK™ plate collects and moves the heat from these discrete heat sources to the liquid cooled edge or air cooled heat sinks with minimal temperature gradients. As electronics continue to move forward with higher power and smaller packaging, HiK™ plates are a great way to move heat to boost performance. Whether it is adding more power to the system or reducing the hot spot temperature in hot ambient environments, HiK™ plates provide a reliable, easily integrable thermal solution as illustrated in Fig. 3.45 using software application such COMSOL™ Multiphysics package.

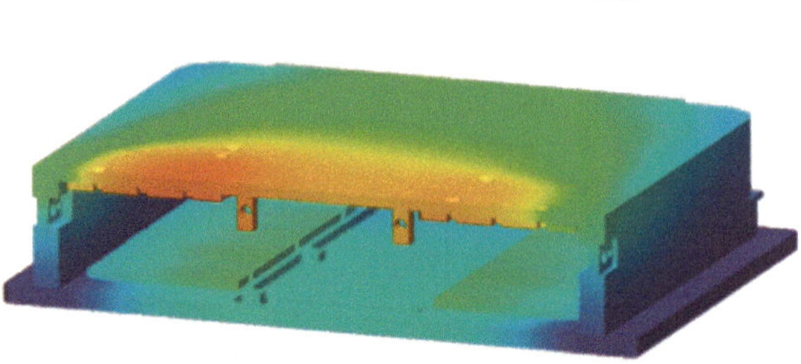

FIG. 3.45 Thermal plots showing aluminum cover and base (top) and aluminum cover with HiK™ base (bottom). *Courtesy of ACT Corporation.*

Aluminum has a thermal conductivity of 180 W/m K. As discussed in When to Use Heat Pipes, HiK™ Plates, Vapor Chambers, and Conduction Cooling, the effective conductivity of HiK™ plates ranges from 600 to 1200 W/m K or more. An additional advantage is that HiK™ plates are less expensive than vapor chambers (which have higher effective thermal conductivities), and much, much less expensive than encapsulated graphite conduction cards (which also have lower effective thermal conductivities). HiK™ plates can also use L-shaped heat pipes to extend the effective high conductivity around corners.

3.19.2 Cooling embedded VME and VPX systems

Many embedded electronics systems with VME/VPX boards have the following configuration as illustrated in Fig. 3.46:

- Metal frames under the electronics that serve as heat spreaders to move heat to the card edge
- Card retainer clamp/wedge lock to mechanically and thermally attach the card to the chassis
  - Allows for easy assembly and rigid attachment
  - Easy to service replace the cards
- A chassis that removes heat by one of two ways:
  - Liquid cooling, typically at the base, relying on conduction from chassis
  - Air cooling, using fins directly attached to the chassis side walls

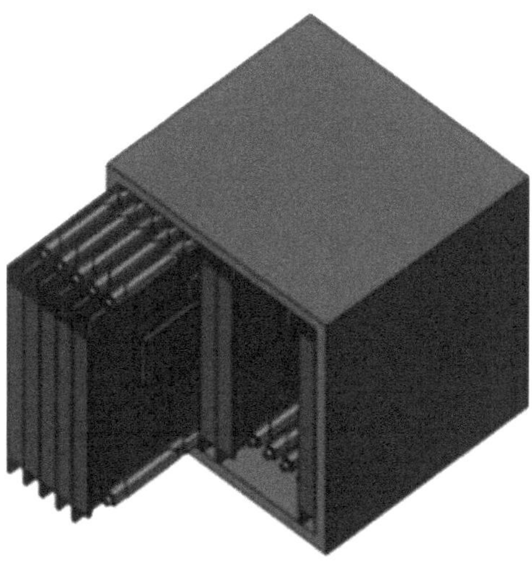

FIG. 3.46 Embedded VME or VPX systems.

In embedded Versa Module Europe (VME) or VPX systems the electronics boards attach directly to heat spreaders which move the heat outward toward the chassis. The edge of the card is attached using a card retainer usually referred to as a wedge lock.

VME (Versa Module Europe) is one of the early open-standard backplane architectures. It was created as a way to have a community of companies creating interoperable computing systems with the same form factors and framework. Typical components of the system include boards such as processors, IO boards, etc., as well as enclosures, backplanes, power supplies, and other subcomponents. Among the benefits for customers were:

- Multiple vendors to choose from (not locked to one vendor, less risk).
- A modifiable standard architecture versus a costly proprietary solution.
- A forward-looking upgradeable platform that doesn't require forklift upgrades.
- Shorter development times (not starting from scratch).
- Lower prototyping and development costs (not starting from scratch).
- More options with the latest and greatest in features.
- An open specification to do any portion of the system in-house.

The VME specification was cleverly designed with upgrade paths so that the technology would be useable for a long time. In fact, despite VME being over 30 years old, it's still used in many legacy applications today. Based on the Eurocard form factor, where boards are typically 3U or 6U high (there were also 9U and specialty versions in the earlier days), the design was quite rugged. With shrouded pins and rugged connectors, the form factor became a favorite for many mil/aero and industrial applications. The boards typically had 160-mm depths, but versions were available in other 60-mm increments, including 220, 280, and even 320 mm in some cases.

VPX (also known as VITA 46) is the next generation of ruggedized compact embedded systems. After years of VME systems dominating the military/aerospace field, users have finally reached the limit of available bandwidth on the VMEbus. VPX expands the possible bandwidth, compared to the traditional VME system, by replacing the parallel bus with high speed serial busses. Much as the desktop market is transitioning from PCI to PCIe, the VME standard has been transformed to embrace the new VPX standards. The serial busses offer higher data rates while using a fraction of routing resources. This allows the new VPX standard to focus more physical backplane resources on improving other design aspects such as supporting larger power draw and more User I/O.

As shown in Fig. 3.47, these chassis have a number of thermal challenges that can be reduced with HiK™ products:

(1) Thermal Conductivity (k) of the thin aluminum frames is not high enough to move heat efficiently. Copper causes weight issues

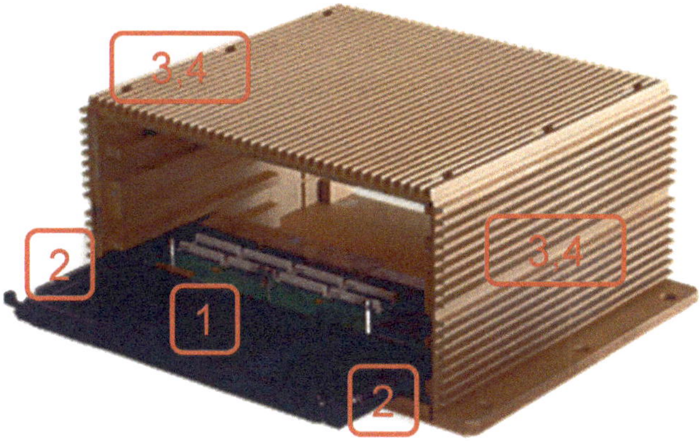

FIG. 3.47 Air-cooled chassis with a single card installed.

(2) Card clamps/wedge locks transfer heat disproportionately into the chassis.
- 80% through the card frame
- 20% through the wedge lock

(3) The thermal conductivity of the chassis is not high enough to move heat to a liquid-cooled base or spread the heat evenly across full fin stack.

(4) Fin Stacks must be optimized for volume and available air flow.

As illustrated in Fig. 3.47, air-cooled chassis with a single card installed. HiK™ technology can reduce the overall temperature drop in the following areas: (1) Improving the thermal conductivity of the frame, (2) Improving the temperature drop through the card clamps, (3) Improving the thermal conductivity of the chassis, which (4) Can help with fin stack optimization.

3.20 Pressure controlled heat pipes (PCHPs)

A *Pressure Controlled Heat Pipe (PCHP)* is a variation of a Variable Conductance Heat Pipe (VCHP), where the amount of Non-Condensable Gas (NCG) or the reservoir volume is varied. Like a VCHP, the NCG is swept toward the condenser end of the heat pipe by the flow of the working fluid

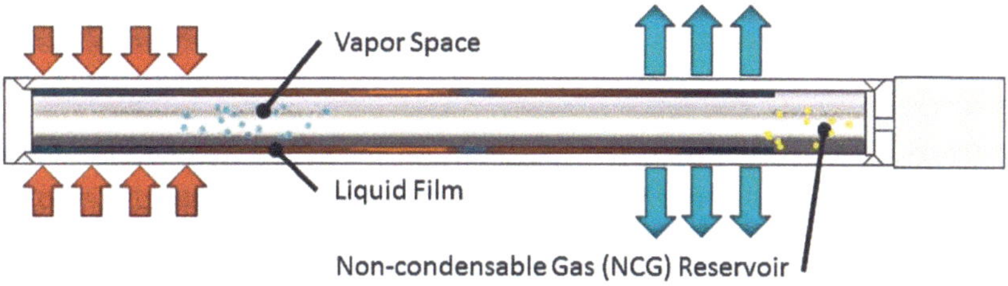

FIG. 3.48 Pressure controlled heat pipe (PCHP) Schematic.

vapor. The NCG then blocks the working fluid from reaching a portion of the condenser, inactivating a portion of the condenser.

In a VCHP, the fraction of condenser blockage is determined by the reservoir size, the non-condensable gas charge, and the operating pressure, and cannot be adjusted once the VCHP is sealed. In contrast, the condenser blockage in a PCHP is actively controlled. In some designs, an actuator drives a bellows to modulate the reservoir volume. By decreasing the reservoir volume, more of the condenser is blocked. In other designs, non-condensable gas is added and removed to the reservoir, also allowing active control of the condenser length.

The operation of a bellows/piston type of PCHP is shown in Fig. 3.48. Initially, the piston is withdrawn at higher powers, so that most of the condenser is open. When the heat load is reduced, the piston pushes additional gas into the condenser, helping to maintain the heat pipe at a constant temperature. While a VCHP will passively also increase the condenser blockage, PCHPs are able to react faster, and more precisely.

Pressure Controlled Heat Pipe (PCHP) varies the reservoir volume for precise temperature control.

The two applications for Pressure Controlled Heat Pipes are:

- PCHPs for Precise Temperature Control
- PCHPs for High Temperature Power Switching

A Pressure Controlled Heat Pipe (PCHP) is essentially an actively controlled Variable Conductance Heat Pipe (VCHP). The operating principles are similar. The vapor/Non-Condensable Gas (NCG) interface position in the condenser moves to vary the conductance of the heat pipe. When the heat load increases or the radiator sink temperature increases, the temperature (and pressure) of the heat pipe also increases. In a VCHP, the increase in vapor pressure forces more of the non-condensable gas into the reservoir, which moves the vapor/non-condensable gas interface further into the condenser. In a PCHP, the control system senses this increase (or decrease) in pressure and/or temperature and actively changes either the gas charge in the reservoir or the volume of the reservoir to maintain the operating temperature precisely at the set point [35].

There are two ways to achieve precise temperature control using a pressure controlled heat pipe (PCHP). One is to modulate the amount of NCG in the reservoir; and, the other is to modulate the volume of the reservoir. The modulation of the amount of NCG in the reservoir is the conventional means of making a terrestrial based PCHP. In these applications, primarily high-temperature precision-calibration systems, the NCG is added to or removed from the reservoir by means of a high pressure gas cylinder and a vacuum pump [35].

The challenge for adapting this type of control system for use in space is to miniaturize the NCG supply tank and vacuum pump. More likely, a

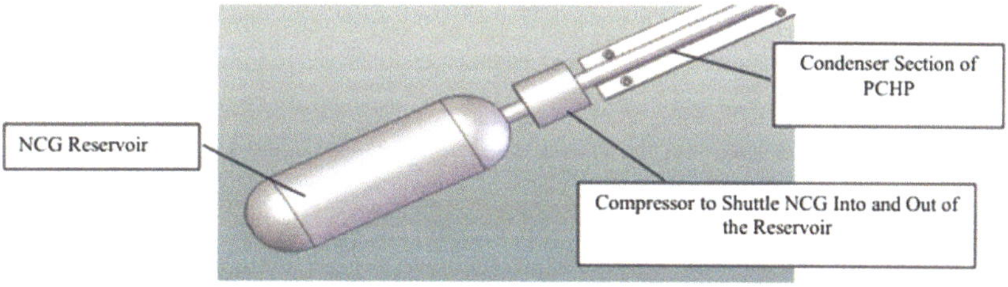

FIG. 3.49 NCG gas modulated PCHP [35].

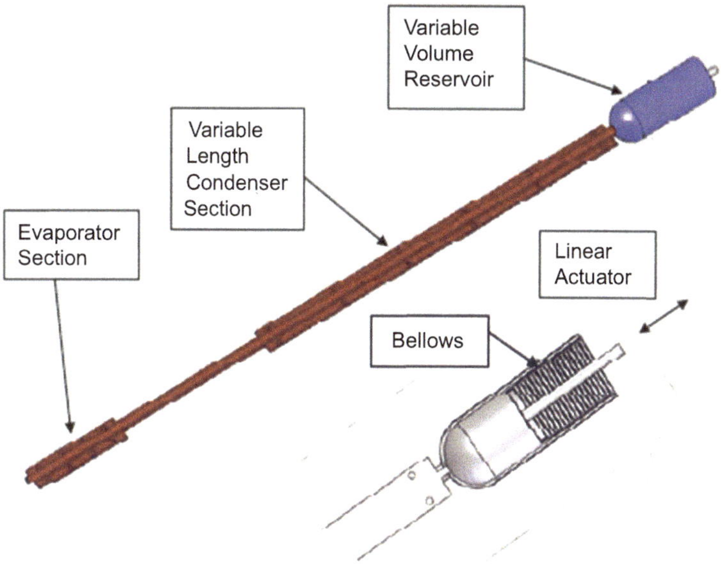

FIG. 3.50 Volume modulated PCHP schematic [35].

space-based system would incorporate a small compressor and a small reservoir. The reservoir would be high pressure biased so that when NCG must be added to the PCHP, a simple solenoid valve would be activated and NCG would flow back into the PCHP. If NCG needs to be removed, the compressor would cycle on and pump NCG from the PCHP to the small reservoir. A sketch of the concept is shown below in Fig. 3.49.

Modulation of the reservoir volume is the other method of controlling the PCHP. In this concept, the NCG reservoir includes a bellows structure. A linear actuator is used to drive the position of the reservoir, thus modulating the volume of the reservoir. This concept is relatively simple and requires only one active device. The challenge for this concept is to design and build a bellows type reservoir (mass and volume optimized), that can be varied with minimal power usage and with fine enough resolution to achieve milli-Kelvin

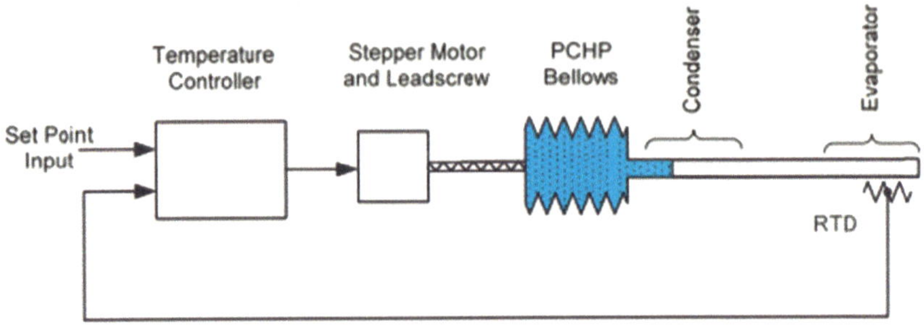

FIG. 3.51 Schematic of pressure controlled heat pipe showing feedback control of reservoir volume and condenser thermal resistance [35].

control. A sketch of the concept is shown below in Fig. 3.50. The bellows system was selected for the PCHPs, since it is simpler to implement for spacecraft applications.

Pressure Controlled Heat Pipes have three major advantages over conventional Constant Condenser Heat Pipe (CCHP), Variable Conductance Heat Pipe (VCHP)VCHP, and Loop Heat Pipe (LHP) solutions and those utilizing cold biasing and trim heaters:

- Precise temperature set point control to the milli-Kelvin level without power-wasting trim heaters and without a massive reservoir.
- Nearly instantaneous reaction to changes in the environmental conditions (low thermal mass lag).
- Ability to adjust the set point in-situ. Thermal analysis and ground testing can differ by as much as ± 10K from the results in space. The PCHP compensates for these discrepancies in real time after the satellite has been placed in orbit.

The control scheme for the PCHP for precise temperature control is illustrated schematically in Fig. 3.51. The PCHP is an enhancement to a conventional Variable Conductance Heat Pipe (VCHP) that adds the ability to actively control the reservoir volume, and subsequently the thermal resistance of the heat pipe condenser. With a suitable feedback control system, the PCHP can achieve milli-Kelvin temperature control of evaporator temperature while compensating for changes in sink temperature or input power without the need for the large VCHP reservoir.

Readers who are interested in Pressure Controlled Heat Pipe (PCHP) can refer to the paper published by William G. Anderson et al. [35] for more detailed and information on this type of heat pipe.

References

[1] D. Reay, P. Kew, Heat Pipes Theory, Design and Application, fifth ed., Elsevier, 2006. Butterworth-Heinemann is an imprint of Elsevier.

[2] D. Wu, G.P. Peterson, A review of rotating and revolving heat pipes, in: National Heat Transfer Conference Paper No. 91-HT-24, Minneapolis, MN, American Society of Mechanical Engineers, New York, 1991.
[3] P. Peterson, An Introduction to Heat Pipes – Modeling, Testing and Applications, John Wiley & Sons, Inc., 1994.
[4] Bontemps, C. Goubier, C. Marquet, J.C. Solecki, Theoretical analysis of a revolving heat pipe, in: Proc. 5th Int. Heat Pipe Conf., Tsukube Science City, Japan, May 14–18, 1984, 1984, pp. 274–279.
[5] J. Chen, Y.S. Lou, Investigation of the evaporation heat transfer in the rotating heat pipe, in: Proc. 7th Int. Heat Pipe Conf., Minsk, USSR, May 21–25, 1990, 1990.
[6] J. Chen, C. Tu, Theoretical and experimental research of condensation heat transfer in parallel rotating heat pipe, in: Int. Heat Pipe Symposium, Osaka, Japan, 1986, pp. 155–165.
[7] J. Chen, C. Tu, Condenser heat transfer in inclined rotating heat pipe, in: Proc. 6th Int. Heat Pipe Conf., Grenoble, France, May 25-29, 1987, 1987.
[8] Keiyou, S. Maezawa, Heat transfer characteristics of parallel rotating heat pipe, in: Proc. 7th Int. Heat Pipe Conf., Minsk, USSR, May 21–25, 1990, 1990.
[9] S. Mochizuki, T. Shiratori, Condensation heat transfer within a circular tube under centrifugal acceleration field, Trans. ASME 202 (Feb 01, 1980) 158–162.
[10] J. Niekawa, K. Matsumoto, T. Koizumi, K. Hasegawa, H. Kaneko, Performance of revolving heat pipes and application to a rotary heat exchanger, in: D.A. Reay (Ed.), Advances in Heat Pipe Technology, Pergamon Press, London, England, 1981, pp. 225–235.
[11] P.E. Eggers, A.W. Serkiz, Development of Cryogenic Heat Pipes, ASME 70-WA/Ener-1, American Society of Mechanical Engineers, New York, 1970.
[12] P. Joy, Optimum Cryogenic Heat Pipe Design, ASME Paper 70-HT/SpT-7, American Society of Mechanical Engineers, New York, 1970.
[13] R. Mortimer, The heat pipe, in: Engineering Note-Nimrod/NDG/70-34, Rutherford Laboratory, Nimrod Design Group, Harwell, October 1970.
[14] S. Van Oost, B. Aalders, Cryogenic heat pipe ageing, in: Paper J-6, Proceedings of the 10th International Heat Pipe Conference, Stuttgart, September 21–25, 1997, 1997.
[15] J.P. Marshburn, Heat Pipe Investigations, NASA TN-D-7219, August 1973.
[16] Rice, D. Fulford, Capillary pumping in sodium heat pipes, in: Proceedings of 7th International Heat Pipe Conference, Minsk, 1990, Hemisphere, New York, 1991.
[17] L.L. Vasil'ev, V.G. Kiselev, M.A. Litvinets, A.V. Savchenko, Experimental study of heat and mass transfer in a cryogenic heat pipe, J. Eng. Phys. Thermophys. (December 2004) 19–21.
[18] P.J. Berennan, E.J. Kroliczek, "Heat Pipe Design" from B & K Engineering Volume I and II Written by, Published June 1979 Under NASA Contract NAS5-23406.
[19] Akachi, United States Patent Office Search, US Patent No. 5490558, 1996. See the following link on the web: http://www.freepatentsonline.com/5490558.pdf.
[20] F. Polasek, L. Rossi, Thermal Control of Electronic Equipment and Two-Phase Thermosyphons, 11th IHPC, 1999.
[21] P. Charoensawan, S. Khandekar, M. Groll, P. Terdtoon, Closed loop pulsating heat pipes, part A: parametric experimental investigations, Appl. Therm. Eng. 23 (2003) 2009–2020.
[22] M. Vogel, G. Xu, Low profile heat sink cooling technologies for next generation CPU thermal designs, Electron. Cool. 11 (1) (February 2005).
[23] S. Duminy, Experimental Investigation of Pulsating Heat Pipes (Diploma thesis), Institute of Nuclear Engineering and Energy Systems (IKE), University of Stuttgart, Germany, 1998.

[24] M. Sarkar, Theoretical parametric study of Wrap-Around Heat Pipe (WAHP) in air conditioning systems, in: International Journal of Air-Conditioning and Refrigeration, World Scientific Publishing Company, 2019.
[25] H. Ma, Oscillating Heat Pipes, Springer publishing Company, New York, NY, 2015.
[26] B. Zohuri, Heat Pipe Design and Technology: Modern Applications for Practical Thermal Management, Springer Publishing Company, New York, NY, 2016.
[27] M. Kaviany, Performance of a heat exchanger based on enhanced heat diffusion in fluids by oscillation: analysis, ASME J. Heat Transf. 112 (1990) 49–55.
[28] M. Kaviany, M. Reckker, Performance of a heat exchanger based on enhanced heat diffusion in fluids by oscillation: experiment, ASME J. Heat Transf. 112 (1990) 56–63.
[29] U.H. Kurzweg, Enhanced heat conduction in fluids subjected to sinusoidal oscillations, ASME J. Heat Transf. 107 (1985) 459–462.
[30] U.H. Kurzweg, L.D. Zhao, Heat transfer by high-frequency oscillations: a new hydrodynamic technique for achieving large effective thermal conductivities, Phys. Fluids 27 (1984) 2624–2627.
[31] H. Akachi, Structure of a heat pipe, US Patent #4,921,041, 1990.
[32] C. Tarau, W.G. Anderson, W.O. Miller, R. Ramirez, Sodium VCHP With Carbon-Carbon Radiator for Radioisotope Stirling Systems, SPESIF, Washington, DC, February 2010.
[33] C. Tarau, W.G. Anderson, K. Walker, Sodium variable conductance heat pipe for radioisotope stirling systems, in: IECEC, Denver, CO, August 2–5, 2009, 2009.
[34] C. Tarau, W.G. Anderson, K. Walker, NaK variable conductance heat pipe for radioisotope stirling systems, in: IECEC, Cleveland, OH, July 25–27, 2008, 2008.
[35] W.G. Anderson, J.R. Hartenstine, C. Tarau, D.B. Sarraf, K.L. Walker, Pressure Controlled Heat Pipes, American Institute of Aeronautics and Astronautics, July 2011, https://doi.org/10.2514/6.2011-5232. https://www.researchgate.net/publication/264888950.

Chapter 4

Heat pipe manufacturing

Chapter outline

4.1 Introduction 239
4.2 Manufacturing of heat pipes 240
4.2.1 Envelope 241
4.2.2 End cap 242
4.2.3 Fill tube 243
4.2.4 Wick 244
4.2.5 Working fluid 245
4.3 Heat pipe manufacturing procedures 245
4.4 Cleaning of parts 245
4.5 Assembly of heat pipes 253
4.5.1 Wick forming and insertion 253
4.5.2 End cap installation 253
4.5.3 End closure and welding 255
4.5.4 Summary of assembly of heat pipes procedures 255
4.6 Evacuation and charging 259
4.6.1 Fluid charging 262
4.6.2 Fluid purity and inventory 262
4.6.3 Analysis of gas blockage 264
4.6.4 Effect of heat pipe design and operating conditions on gas blockage 266
4.7 Full tube closure 267
4.8 Heat pipe testing techniques 267
4.8.1 Mechanical sounders 270
4.8.2 Wick wetting 271
4.8.3 Performance versification 271
References 273

4.1 Introduction

Heat Pipes in form of Copper/Water heat pipes have a copper envelope, use water as a working fluid and typically operate in the temperature range of 20−150 °C. Heat pipes are an extremely effective method of reducing hot spot temperatures and increasing allowable power in a system by moving heat to an external sink or spreading heat along a surface.

A detailed heat pipe manufacturing process is introduced, including primary design consideration, material preparation and treatment, component assembly, evacuation and working fluid charging, and the final sealing and finishing process. Among these processes, the challenges related to the evacuation and fluid charging processes as well as corresponding impacts are analyzed.

In parallel, several heat pipe manufacturing techniques have been developed. Euratom patented a groove manufacture technique for heat pipes in 1966 (UK 1125485) and a method for mesh forming and welding for heat pipes in 1969 (UK 1228103). In 1969, Metallgesellschaft Ag. patented a method for fabricating lithium/refractory heat pipes (UK 1194530). In 1970,

Functionality, Advancements and Industrial Applications of Heat Pipes.
https://doi.org/10.1016/B978-0-12-819819-3.00004-3

the McDonnel Douglas Corporation patented a means for joining arteries in heat pipes having right angle bends (US 3620298). In the same year, Brown Boveri patented a method for porous wick manufacturing by means of vapor deposition (UK 1313525). In 1971, Q-Dot in the USA patented a technique for forming spirally grooved heat pipes using a special cutting tool (US 3753364), and in 1973 filed the same patent (UK 1433542) [5]. In 1975, the RCA Corporation obtained a Canadian patent for a method of making a heat pipe and wick (CA 960413). In 1979, Hydro-Quebec (Canada) was issued a patent regarding fabrication of ceramic heat pipes (CA 1066964).

4.2 Manufacturing of heat pipes

Key goal of this chapter is to give reader a basic idea of how to develop standardized manufacturing procedures which would assure the reliability of the final product. By evaluating and defining acceptable cleaning procedures, for example, the reliability of a properly fabricated heat pipe will be increased. Moreover, since each manufacturer is currently using essentially independent procedures and are built to different specifications and quality. There is a good deal of duplication of effort in the field; many individual company's procedures have been established iteratively, which is a costly and time consuming way to develop a technology. This effort, therefore, is only the forerunner of others which will ultimately develop basic heat pipe manufacturing specifications and details for all manufacturers.

Essentially major infrastructure of heat pipe is based on five elements or components which are as follows;

1. Envelope
2. Wick
3. End Cap
4. Fill Tube
5. Working Fluids

Working fluid is introduced into the pipe through the fill tube, which is subsequently sealed by a pinch-off. When heat is applied to a section of the heat pipe (evaporator), the working fluid evaporates, causing a local increase of pressure driving the vapor toward the other end of the pipe (condenser). Cooling this end causes the vapor to condense on the walls of the pipe. Finally, by capillary forces the working fluid flows back through the wick to the evaporator section. This cycle is repeated as long as heat is supplied to the evaporator and is removed at the condenser. However, if heat is applied to the pipe at a rate higher than the wick can carry, the pipe will fail to function (dry-out) [1]. A serious problem with heat pipes is the generation of non-condensable gases which will inhibit their performance. Basically, these gases will accumulate at the cold (condenser) end of the pipe, decreasing the effective conductance of the unit, until the condenser is completely "blocked" and the pipe fails to function. A flow chart of the basic operations involved in manufacturing a heat pipe is presented in Fig. 3.44. A brief summary of the major elements of the manufacturing cycle discussed in this report follows [1].

Details of these components from manufacturing and operation point are discussed below.

4.2.1 Envelope

The envelope may be of any cross section desired by the designer - e.g., circular, square, etc, and may contain mounting flanges to simplify installation, and bent in various shapes. The wick may be grooves extruded into the envelope, or may be assemblies - such as Grumman's spiral artery design, made from fine wire mesh (Fig. 4.1), sintered screen, or felt metal slabs, etc. The most common configuration is the round shape. The procedures for determining the tube diameter and the wall thickness has been discussed in Chapter 3 of this book and great deal of details can be found in Chapter 7 of Chi [2].

Heat pipe manufacturers are rarely concerned with tube manufacturing, because tubes of different materials and sizes are commercially available at reasonable cost. Both seamless and butt welded tubes may be used as the heat pipe envelope. However, when a tube is cut to the required length, care should be taken so as not to distort the tube ends since this will make it difficult to obtain a strong joint to the end cap. Although round tubes with smooth inside surfaces are common for heat pipe applications, tube with internal fins have also become popular. Several techniques have been developed for production of tubes with internal fins. Kemme [3] was the one of the design pioneer to develop the first axial-grooved heat pipes were formed by milling grooves on a flat plate and then rolling and butt welding a tube. Today commercially integral fin tubing is now available.

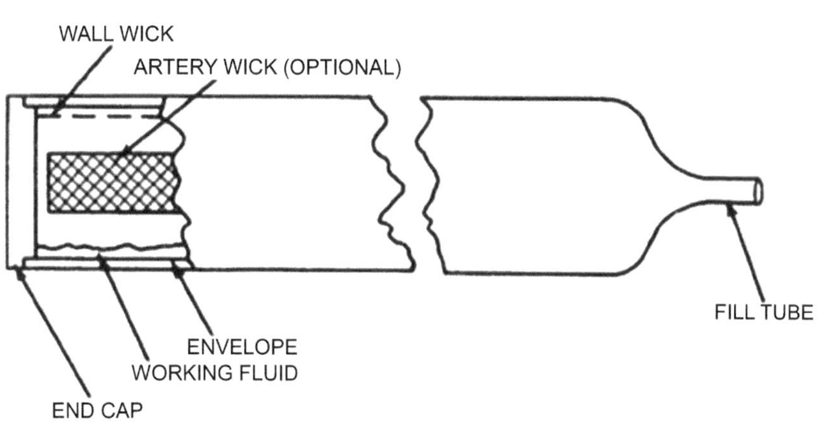

FIG. 4.1 Typical heat pipe components [1].

4.2.2 End cap

Although a seemingly simple operation, end closure and welding has been a problem for many heat pipe manufacturers. Common concern is porosity or cracks in the weld, which can lead to a loss of the working fluid. To minimize the probability of this failure, inspection should be performed to verify the adequacy of the seal.

Experience and research indicates that it is not uncommon for heat pipes to fail this inspection, requiring that the closure be repaired. A reliable process can lead, therefore, to a "hit-or-miss" manufacturing cycle, where the welding operation is repeated several times to repair a fault. Obviously, these added steps will adversely impact both time and cost budgets. In keeping with the overall objectives of this study, the "optimum" sealing process should be easy to perform, repeatable, require moderately priced equipment, and be reliable and easy to inspect.

The minimum required thickness for heat pipe end cap can be determined by stress analysis as described in Chapter 3 and available computer codes mentioned give you such analysis capabilities. Careful design of the joint for welding is important and plays a great deal to weld end caps.

Dealing with alignment of these joint is another important steps and among the four kinds the more difficulty arises with first one (a) (The butt joints) with respect to other three. In addition, it should be noted that to obtain a strong, leak-proof welded joint the end cap must be machined to provide a welded area thickness roughly speaking equal to that of the tube wall [2]. This careful step ensures that both surfaces melt equally resulting in homogeneous fusion.

Detail design of a heat pipe incorporating welded end caps should consider the following points:

- Fusion welding joining processes result in the best combination of joint strength and leak tightness.
- A fully mechanized fusion welding process should be given preference over a manual process. A mechanized process is significantly more consistent with respect to joint quality, e.g., strength, size of heat affected zone, and weld bead geometry.
- The use of the aluminum alloy 6061 or the stainless steel alloy 304L eliminates the need for post weld heat treatment, such as a stress relief anneals to prevent either corrosion susceptibility or embrittlement.
- Joint efficiency is a prime consideration. The 304L stainless steel alloy is not heat treatable and a relatively high joint efficiency in the "as welded" condition can be obtained. A conservative assumption for this alloy is a weld joint strength that is 85% (automatic process) or 70% (manual process) of the minimum strengths guaranteed for the 304L product form in the annealed condition. In the case of the 6061 aluminum alloy, the condition of

the material prior to welding and the use of post weld heat treatment must be taken into consideration in establishing weld joint strength.

- The geometry of the weld bead produced by the welding process selected should be known by the designer beforehand since the weld root bead cannot be machined after welding. An excessively large drop through the root bead can interfere with the operation of the heat pipe. Machining of the crown of the weld bead should be avoided. The removal of the skin material of the weld bead crown may expose interdendritic porosity and result in gas leakage from the heat pipe.
- The use of a square butt joint design is preferable.
- The end cap detail should be designed to be self-aligning during the welding operation. The end cap detail can also provide filler metal to the joint.
- The possible damage to interior details of the heat pipe resulting from welding heat should be considered and, if necessary, determined experimentally.

For more process of end cap welding along with details reader should refer to Edelstein and Haslett [1] report (Fig. 4.2).

4.2.3 Fill tube

An end cap and fill tube are used to complete the envelope. Working fluid is introduced into the pipe through the fill tube, which is subsequently sealed by a pinch-off. When heat is applied to a section of the heat pipe (evaporator), the working fluid evaporates, causing a local increase of pressure driving the vapor toward the other end of the pipe (condenser). Cooling this end causes the vapor to condense on the walls of the pipe. Finally, by capillary forces the working fluid flows back through the wick to the evaporator section. This cycle is repeated as long as heat is supplied to the evaporator and is removed at the condenser. However, if heat is applied to the pipe at a rate higher than the wick can carry, the pipe will fail to function (dry-out). A serious problem with heat pipes is the generation of noncondensable gases which will inhibit their performance. Basically, these gases will accumulate at the cold (condenser) end of the pipe, decreasing the effective conductance of the unit, until the condenser is completely "blocked" and the pipe fails to function [1].

The fill tube provides the only access to the heat pipe for evacuation and fluid charging, and it too must be closed and sealed. A crimp seal followed by welding has been found to be convenient. For a good crimp, the fill tube must not exceed ¼ in. (6.35×10^{-3} m) o.d. Tubes of 1/16 in. (1.59×10^{-3} m) id. are often used. Additionally, the thickness of the fill tube wall is also critical for certain materials [2].

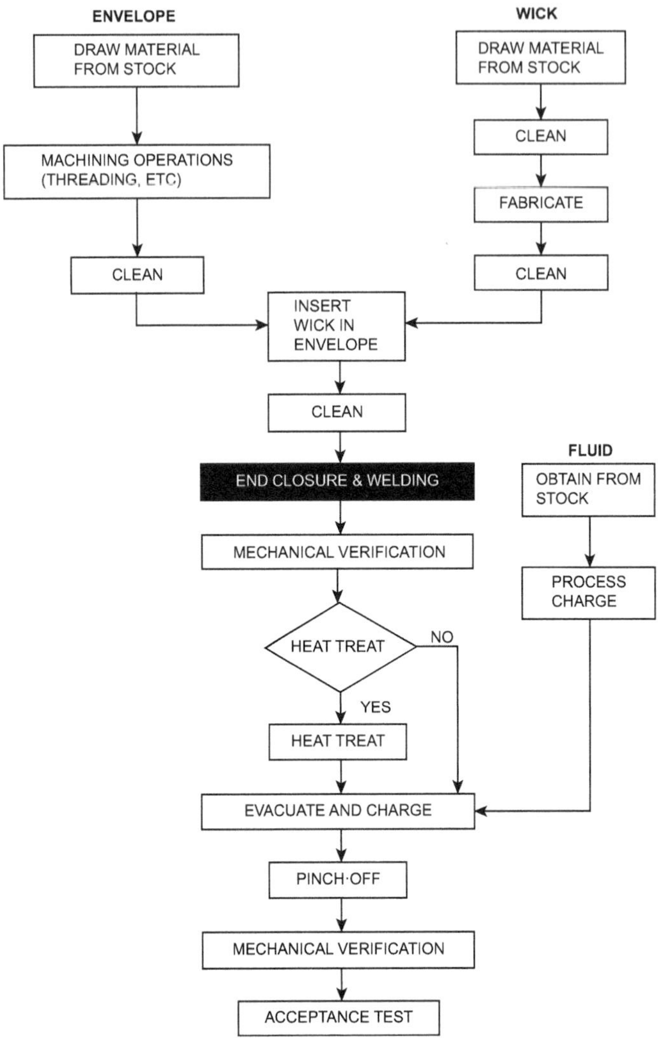

FIG. 4.2 Typical manufacturing steps for welding [1].

4.2.4 Wick

If wick structure is an integral part of a heat pipe, e.g., an axial-grooved heat pipe, the operation for the preparation of the wick is not needed. If the wick is formed from wire cloth, e.g., wrapped-screen and artery wicks, the wire cloth must first be cut to the required size. Wire cloth can be difficult to cut accurately, because it is not rigid in the plane of the material. By sandwiching a wire cloth between two flat plates a wire cloth may be cut along a straight edge [2].

4.2.5 Working fluid

A heat pipe should not be overfilled or under filled with working fluid. Heat pipe working fluid must be of high quality with minimum impurities-gaseous, dissolved solid, or liquid. If these conditions are not met, the designed heat pipe performance may not be reached. Fluid inventory and purification techniques will be discussed below in a section dealing with pipe evacuation and charging [2].

4.3 Heat pipe manufacturing procedures

A flow chart of the basic operations involved in manufacturing of a heat pipe is presented in Fig. 4.3. A bullet topic of the major elements are listed here below. Three for a brief summary of these major elements of the manufacturing cycle discussed in this report follows. These major elements are as follows;

- Envelope and Wick Cleaning.
- End Closure and Welding.
- Mechanical Verification.
- Evacuation and Charging.
- Fluid Parity.
- Charge Tube Pinch-Off.

Baseline heat pipe manufacturing procedure is summarized and illustrated in Fig. 4.3 here. Overall Restrictions and Assumptions unless otherwise specified in the individual manufacturing procedures, the following restrictions or assumptions shall apply [1]:

- Envelope materials considered are aluminum and stainless steel tubing
- Working fluids are restricted to ammonia, Freon-21 and methanol
- The pipes are intended for operation in the moderate temperature range, nominally from -100 to 200 °F (200–366K).
- Typical pipe sizes are approximately 0.5 in. (0.0127 m) diameter and up to about 12 ft (3.6 m) long
- The procedures described are generally based on small quantity runs up to approximately 10–20 units. Large quantity production runs would require more automated operations, the procedures of which might be different from those described here.
- The procedures described are based on a heat pipe manufacturing sequence typical of that shown in Fig. 4.3.

4.4 Cleaning of parts

Just as extreme care is required in selecting proper heat pipe materials to avoid compatibility problems, cleaning of the component heat pipe parts is critical to avoid similar consequences. Apparently, this is an area in which no accepted

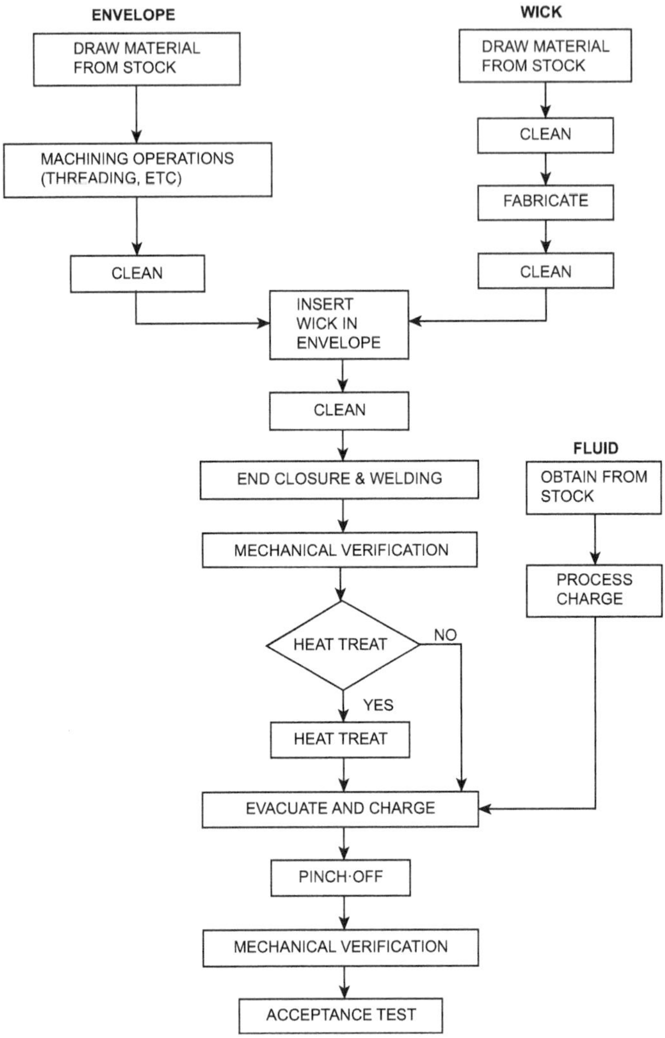

FIG. 4.3 Typical heat pipe manufacturing flow chart [1].

standard exists in the industry at the present time [1]. Although there are similarities in the procedures, each manufacturer uses a different process, essentially developed independently. The effectiveness of each process can only be evaluated on the basis of past experience (Fig. 4.4).

As seen in the heat pipe manufacturing cycle of Fig. 4.3, there are many areas where contaminants can be introduced into the heat pipe, i.e., through a dirty wick, dirty envelope, impurities in the working fluid, etc. In fact, every operation not properly performed can be a source of contamination. This

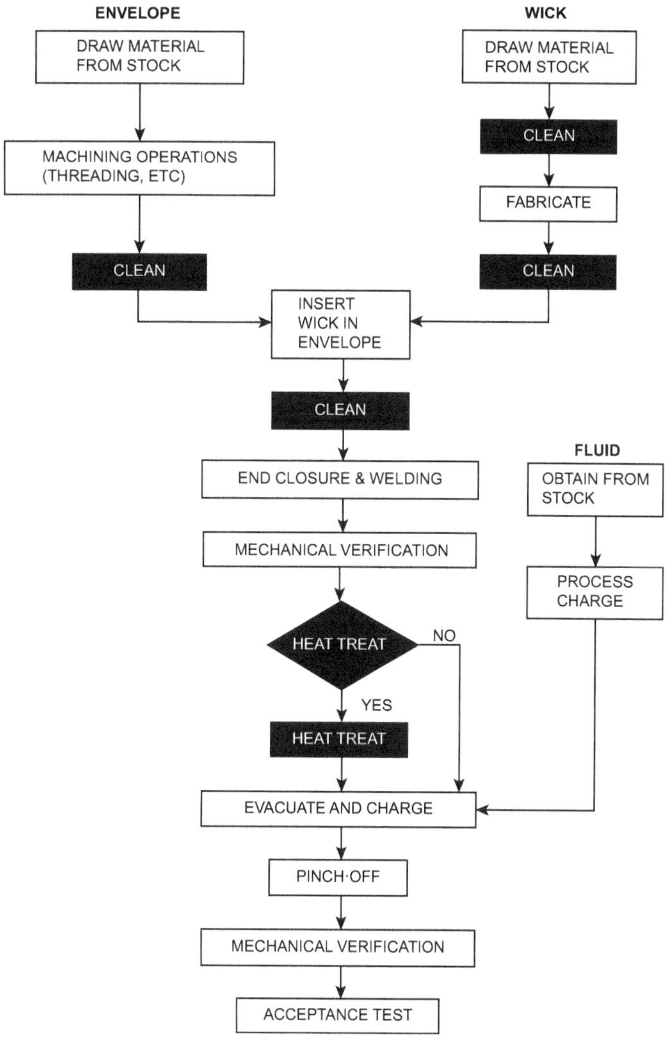

FIG. 4.4 Typical manufacturing steps for cleaning [1].

section discusses envelope and wick cleaning and pretreatment for aluminum and stainless steel envelopes and stainless steel wicks. Originally, it appeared that this area should be treated as two topics, wicks and envelopes. However, during the research effort for this study, it became clear that little wick cleaning information was available. Although most manufacturers were willing to discuss cleaning in general, few were willing to discuss wick cleaning in particular. Each manufacturer considers wick construction and fabrication details as highly proprietary, since it is this element which essentially determines heat pipe performance and separates one company from another. Of the various organizations contacted, therefore, very few were willing to supply information concerning wick cleaning.

Wick cleaning and pretreatment is at least as important as the need for envelope cleaning. Obviously, gas generation is just as likely to come from a "dirty" wick as from other improperly cleaned parts. Oil and greases imbedded in either the fine wire Mesh or sintered screening material used to construct wicks must be removed to assure proper heat pipe performance. Foreign substances conducive to gas generation might be introduced in the construction process, requiring that a post wick construction cleaning process be designed to remove this impurity. For example, if a copper electrode is used to assemble a wick with tack-welds, some copper particles may become imbedded in the stainless steel screen. To remove this material, which is incompatible with ammonia, a nitric acid rinse would be required. However, where possible, it is preferable to eliminate this potential problem by using tungsten electrodes [1].

The heat pipe tube or envelope receives its primary cleaning after dirty operations (such as machining) have been completed. Machining may involve preparing the tube ends for welding, tube bending, and in some cases, cutting fine circular threads on the inside surface to provide a circumferential wicking surface. An assortment of debris such as metal chips, cutting oil, grease, moisture, etc., can be expected after these operations. The overall cleaning operation, therefore, has a number of aims, namely to:

- Mechanically remove particulate matter, such as metal chips which may clog capillary and artery surfaces and/or damage these surfaces during subsequent artery insertion.
- Remove water that can cause corrosion, attacking both aluminum and stainless steel, as well as providing a galvanic coupling between the two. Buildup of particulate reaction products, as well as gas generation, is the principal results. Loss of container structural integrity due to crevice corrosion and porosity may also result from the presence of this contaminant.
- Remove contaminants, not necessarily corrosive, but which may impair the heat pipe wicking and fluid properties. Examples of these contaminants are the variety of oils and greases used in metal cutting and removal operations, extruding, forming, etc. These contaminants may coat the internal surfaces and increase the contact angle, or may dissolve in the working fluid, changing its transport properties.
- Chemically clean and prepare the surface so as to be nonreactive with subsequent manufacturing environments, the wick, and working fluid.
- Treat the wicking surface in a manner that enhances "wetability" with the working fluid.

Failure to achieve these objectives through either improper procedures or operator error has led to heat pipe problems ranging from performance degradation to complete failure. In one reported incident, for example, water inadvertently left on the inside surface of a threaded aluminum tube was judged to be responsible for failure of a heat pipe to achieve its performance goals. Analysis led to the discovery of huge amounts of aluminum hydroxide on the inside surface which were clogging the radial flow passages.

Contaminants can also chemically react with the wall, wick, or fluid to produce non-condensable gaseous products that block the condenser and decrease heat pipe conductance. In the case of arterial wicks, gas bubbles within the wick can, and have, severely limited the heat transport capacity [3]. On the ATS program, groove failure in a single heat pipe was attributed to embrittlement and porosity caused by the presence of water in a closed pipe during heat treatment [1].

A summary of the foregoing and other problems that can be caused by improper cleaning techniques is as follows:

- Physical clogging of wall and wick capillary surfaces, thereby impairing both heat pipe transport capacity and conductance.
- Non-condensable gas generation reducing both heat pipe conductance (loss of condensation area) and transport capacity (bubbles in arterial wicks).
- Decrease in wetability of wick.
- Adverse change in fluid properties, such as surface tension, wetting angle, and viscosity.
- Loss of structural integrity of container wall due to galvanic corrosion, crevice corrosion, and porosity.

Unfortunately, many of these problems cannot be uncovered until the pipe is charged, sealed and tested. In some cases, a long time can occur until some of these effects are noticed; by then, it is usually too late to provide corrective action. Hence, the objective must be to develop cleaning procedures which will prevent these problems from occurring, and produce, therefore, a more reliable product. Moreover, in keeping with the overall aims of this study, the cleaning procedure should also be simple, inexpensive, and as free as possible from human error [1].

Table 4.1 summarizes the procedures currently employed in the heat pipe industry. A brief general description of various cleaning techniques follows, and each manufacturer's approach will be presented and analyzed. Finally, using this evaluation, recommended cleaning procedures are presented. Note that the procedures used by various manufacturers are typical. If more information is required, it is suggested that the individual company be contacted [1].

To summarize the cleaning procedure for aluminum, copper and stainless steel we can show the following [1];

1. Envelope Cleaning and Pre-treatment
   - Assumptions:
     i. Applicable tubing: Aluminum 6061 or 6063; stainless steel 300 series

  ii. Tube condition: Threaded, wicked or otherwise internally machined prior to cleaning
  iii. Tube size: Typically, 0.5 in. (0.0127 m) dia. up to 12 ft (3.6 m) long.
- Materials:
  i. Non-etch alkaline cleaner (Refer to Table 4.2).
  ii. Chromated deoxidizer (Refer to Table 4.3).

TABLE 4.1 Summary of heat pipe cleaning procedures is currently in use [1].

Manufacturer	Envelope cleaning		Wick cleaning (stainless steel)
	Aluminum	Stainless steel	
Dynatherm	• Solvent • Acid		
NASA/GSFC	• Solvent • Acid		
Grumman	• Solvent • Alkaline/acid	• Solvent • Passivation	• Solvent • Passivation
TRW	• Solvent • Alkaline/acid • Ultrasonic	• Ultrasonic • Vacuum fire	• Ultrasonic • Vacuum fire
DWDL/MDAC	• Solvent		
ESRO/MBB	• Ultrasonic		• Ultrasonic
GE	• Alkaline/acid	• Alkaline/acid	
University of Stuttgart		• Ultrasonic • Passivation	
NASA/MSFC	• Solvent • Alkaline/acid	• Solvent • Alkaline • Passivation	

TABLE 4.2 Examples of non-etch alkaline cleaners [1].

Material	Concentration	Temperature
Ridoline No. 53 (Amchem Products Co.)	2–10 oz/gal.	140–180 °F
Oakete No. 164 (Oakite Products Co.)	2–10 oz/gal.	140–180 °F
Kelite Spray White (Kelite Corp)	40–60% by volume	Ambient
A-38 (Pennwatt Corp)	4–8 oz/gal.	160–180 °F

TABLE 4.3 Examples of chromated deoxidizer solutions (immersion type) [1].

Material	Concentration	Temperature	Immersion lime
Mixture of: Chromated deoxidizer replenisher No. 17[a] (Amchem Products Co.)	2–6 oz/gal.	Ambient to 120 °F	5–30 min
Nitric acid 42° Be	10%–20% by volume		
Mixture of: Chromated deoxidizer replenisher No. 17[a]	2–6 oz/gal.	Ambient	5–30 min
Sulfuric acid 66′ Be	4%–7% by volume		

[a]*Deoxidizer make up No 7 to be used for initial makeup.*

TABLE 4.4 Examples of passivating solutions [1].

Material	Concentration	Temperature	Immersion time
Nitric acid	35%–65% by volume	Ambient	30 min–2 h
Mixture of: Sodium dichromate or potassium dichromate	1–4 oz/gal.	Ambient	30 min–2 h
Nitric acid	15%–30% by volume		

- **iii.** Filtered air.
- **iv.** Anhydrous isopropyl alcohol.
- **v.** Dry nitrogen.
- **vi.** Passivating solution (Refer to Table 4.4).

- Procedure - aluminum tubes:
  - **i.** Clean in cold 1, 1, 1 trichloroethane with bristle brush on wire extension. Periodically clean brush between strokes.
  - **ii.** Flush internal surface with cold trichloroethane; dry with filtered air and cap pipe ends.
  - **iii.** Immerse in non-etch alkaline cleaner for 5 min minimum. Refer to Table 4.2 for materials and temperatures.
  - **iv.** Follow with a two (2) minute tap water rinse, raising and lowering tube during rinsing.

v. Immerse in chromated deoxidizer. Refer to Table 4.3 for material, time, and temperature.
vi. Follow with a two (2) minute tap water rinse, raising and lowering tube during rinsing.
vii. Thoroughly dry inside surface with forced filtered air.
viii. Rinse with anhydrous isopropyl alcohol.
ix. Force dry with clean, filtered, dry nitrogen heated to 160 °F.
x. Cap pipe ends.
xi. If applicable, insert artery, rinse with isopropyl alcohol, and dry as in step 9.
xii. If heat treat is required after welding, then

- Evacuate pipe for 4 h at 600 °F and leak check
- Seal evacuated heat pipe
- Perform heat treat operations on sealed pipe.
- Procedure - stainless steel:

i. Clean in cold 1, 1, 1 trichloroethane with bristle brush on wire extension. Periodically clean brush between strokes.
ii. Flush internal surface with cold trichloroethane; dry with filtered air and cap pipe ends.
iii. Immerse in passivating solution. Refer to Table 4.3 for materials, temperatures, and time.
iv. Follow with a two (2) minute tap water rinse, raising and lowering tube during rinsing.
v. Thoroughly dry inside surface with forced filtered air.
vi. Rinse with anhydrous isopropyl alcohol.
vii. Force dry with clean, filtered, dry nitrogen heated to 160 °F.
viii. Cap pipe ends. If applicable, insert artery, rinse with isopropyl alcohol, and dry as in step 7.

- General notes:
  - The cleaning procedure must be as free from operator error as possible, since improperly executed procedures may also lead to unwanted contamination. Training of personnel regarding the cleanliness requirements for heat pipes is mandatory. So are adequate safeguards and inspection points during the steps of the procedure; these must be consistent with good quality assurance practices.
  - Do cleaning and charging operations in fairly rapid sequence and in proximity to one another. Avoid storing the pipe for long duration, which increases the likelihood of contamination. Performing these operations near each other lessens the danger of contamination during transportation.

For more details of this operation reader should refer to Edelstein and Haslett [1] report.

4.5 Assembly of heat pipes

Assembly of heat pipe parts includes welding of end caps and fill tube, and forming and inserting of wick if applicable. Since parts have been thoroughly cleaned, they should be assembled immediately following the cleaning if practicable. Otherwise, cleaned heat pipe parts should be stored in a clean dry atmosphere to prevent contamination by vapor, smoke, and dust suspended in air. Plastic gloves should be worn while handling parts to prevent contamination by skin oils and acids.

4.5.1 Wick forming and insertion [4]

Manual forming and insertion of wrapped-screen wicks can be accomplished as follows. The assembled wick must not contain wrinkles, and to prevent this, the screen may be wrapped on a cleaned mandrel. The total diameter of mandrel and wrapped screen should be only slightly less than the heat pipe inside diameter, so that the residual stress in the coiled screen will force it against the pipe wall when it is released from the mandrel. Obviously, the screen ends must be even and the screen must be properly positioned so that the installation of end caps will not interfere with or crush the screen. To ensure physical contact between the screen layers and the pipe wall, a tapered plug or ball may be forced through the wick. A helical spring with an unstretched diameter slightly larger than the wick inside diameter is sometimes used to hold the screen layers in contact. It too can be installed with the aid of a mandrel. In this case, the spring is stretched while it is held by the mandrel. The length of the stretched spring must not greatly exceed that of the installed length or when it is released from the mandrel, axial forces may displace the screen. With the wick properly positioned in the pipe, the end caps are next welded on [2].

4.5.2 End cap installation [2]

End cap installation: If the fill tube is not an integral part of the end cap, it should first be welded to its end cap. End caps with or without a fill tube are usually welded to the pipe ends. High quality welded joints are required at all seams since porosity or cracks in the weld can lead to a loss of the working fluid. To minimize the probability of this failure, inspections should be performed to verify the adequacy of the seal. A number of welding techniques are available. However, gas torch welding, e.g., using oxyacetylene gases, are generally not recommended, because of the presence of flux. Oxygen and filler metals tend to re-contaminate the cleaned parts. Tungsten inert gas welding (TIC), either manual or automatic, and electron beam welding (EBW) have been found to be satisfactory for heat pipe welding.

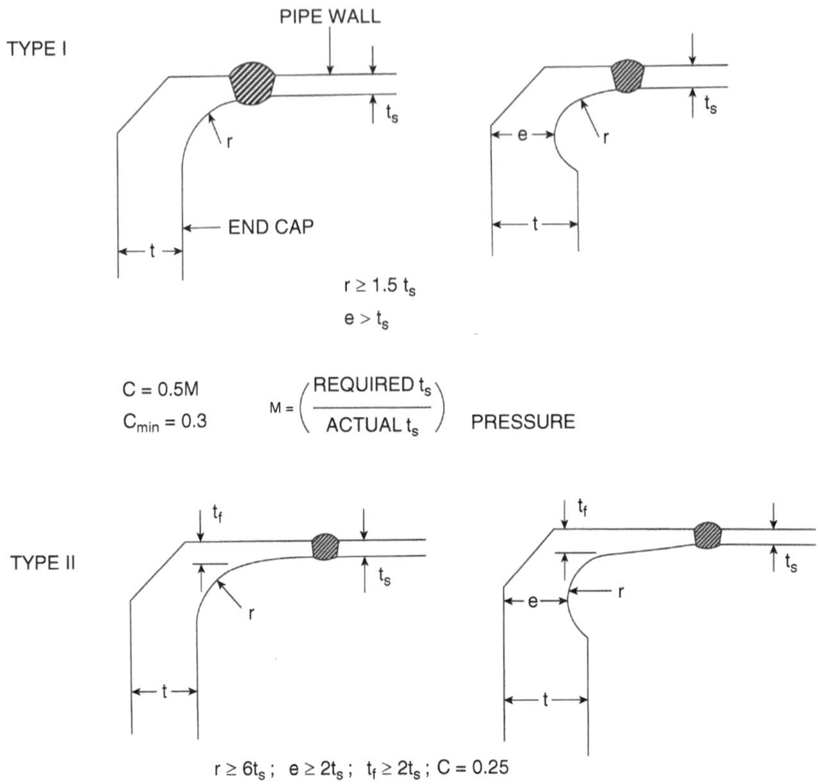

FIG. 4.5 End cap design details [1].

The TIC welding is an electric arc welding process using a sharp tipped tungsten electrode surrounded by an annular shield of inert gas flowing from a torch tip. Filler metals are not generally used for heat pipe welding, but they may be integral parts of the end caps, e.g., the lip of the lipped butt joint shown in Fig. 4.5 can serve as filler metal. In addition, this process does not employ a flux. Hence, TIC welding does not contaminate the cleaned heat pipe part. The EB weld is made in a vacuum chamber, and it eliminates the formation of the surface compounds from the metal and air. In addition, the EBW offers minimal heat input with maximum thermal density. It enables one to produce a welded joint with a minimum heat affected zone, and consequently, the joint properties may approach those of the parent metal. It is, therefore, ideal for heat pipe welding. However, the investment cost in equipment for EBW may exceed that for automatic TIC by over 100% and that for manual TIC by over 2000%. Hence, the choice of welding process is dependent upon the availability of the equipment; and the investment in equipment depends greatly upon the quantity of production and the required quality of the products. Nevertheless, TIC and EB welding have been found satisfactory for heat pipe welding [2].

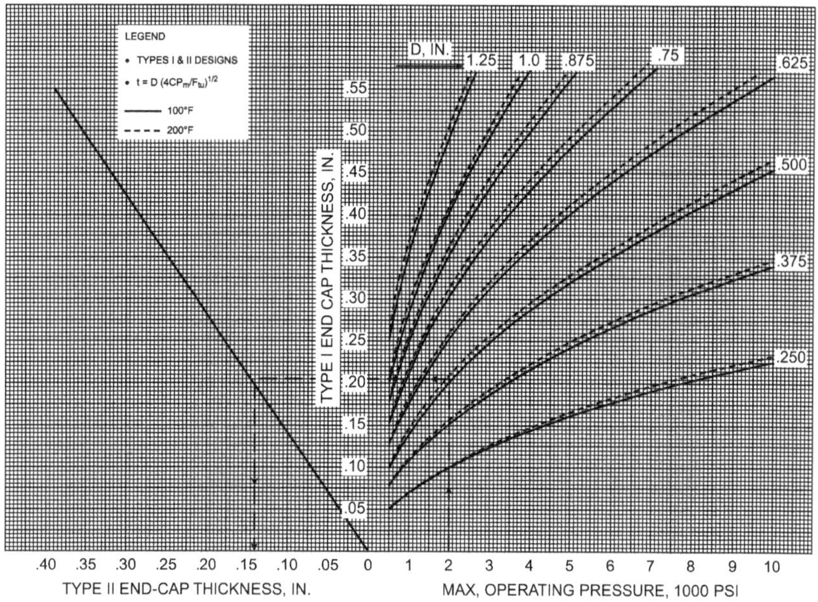

FIG. 4.6 End-cap design *curve* 6061-T6 aluminum (as welded) [1].

4.5.3 End closure and welding [1]

1. End closure design guidelines. - Observe the following general guidelines:
   - Acceptable end cap designs for flat circular heads may be Type I or II as shown in Fig. 4.5.
   - End cap thickness can be determined from Figs. 4.6 and 4.7.
   - The welded end should have a square butt joint design (see Fig. 4.8).
   - Self-alignment during welding should be provided, as per the lipped butt joint for stainless steel (Fig. 4.9), the consumable filler insert for aluminum (Fig. 4.10), or an equivalent design.
   - Full weld penetration should be achieved.
   - Machining of the crown of the weld bead should be avoided.

4.5.4 Summary of assembly of heat pipes procedures

The following is a list of the procedures described above, which should be followed during heat pipe assembly [5]:

i. Select container material
ii. Select wick material and form
iii. Fabricate wick and end caps, etc.
iv. Clean wick, container and end caps
v. Outgas metal components

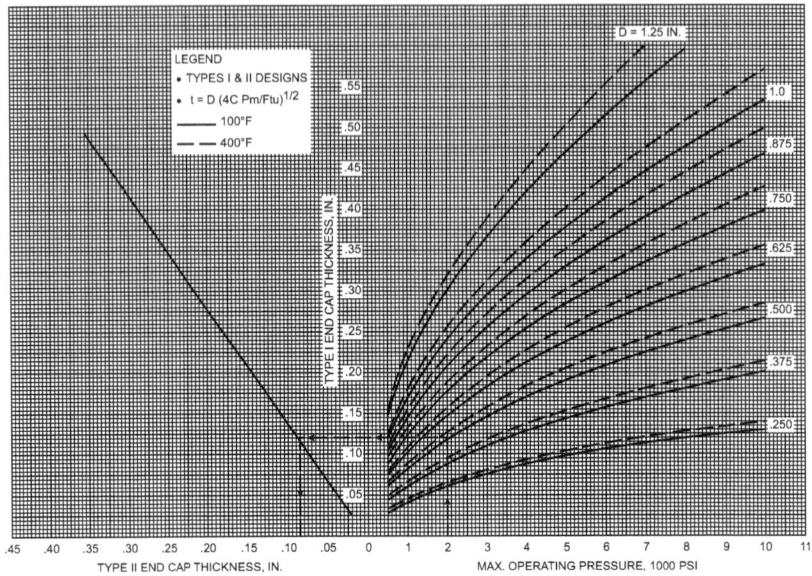

FIG. 4.7 End-cap design *curves*, 304 stainless steel (as welded) [1].

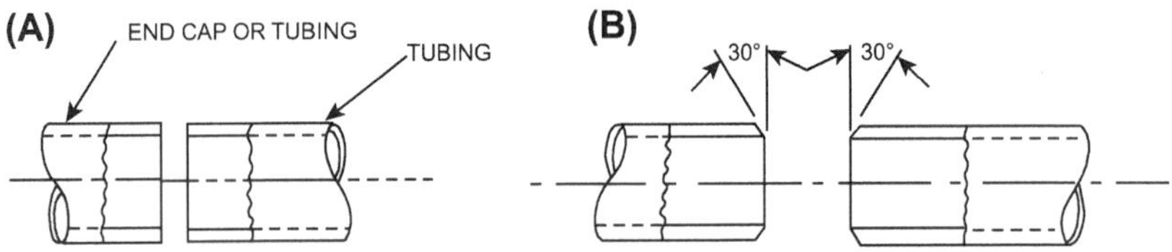

FIG. 4.8 Square butt joint designs [1]. (A) Standard design. (B) Single bevel design.

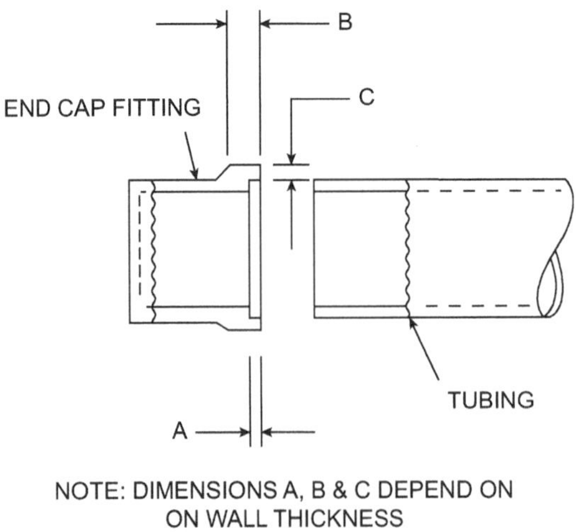

FIG. 4.9 Lipped butt joint design [1].

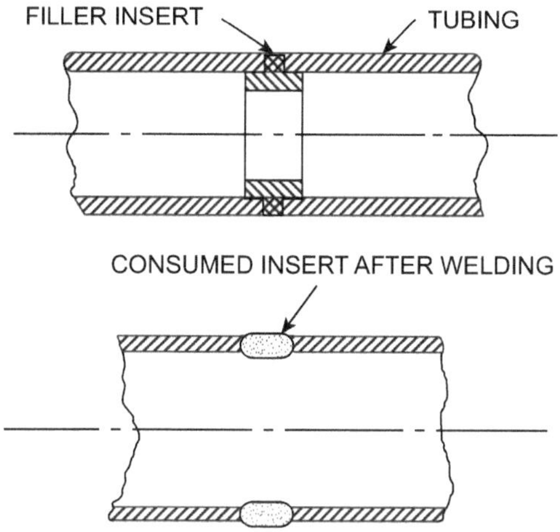

FIG. 4.10 Square butt joint with consumable filler insert [1].

vi. Insert wick and locate
vii. Weld end caps
viii. Leak check welds
ix. Select working fluid
x. Purify working fluid (if necessary)
xi. Degas working fluid
xii. Evacuate and fill heat pipe
xiii. Seal heat pipe.

It may be convenient to weld the blank end cap before wick insertion, and in cases of sintered and diffusion bonded wicks, the outgassing may be done with the wick in place in the container.

For the manufacturer considering the production of a considerable number of identical heat pipes, for example 50 or more units following prototype trials, a number of the manufacturing stages may be omitted. Outgassing of metal components may be unnecessary, and it may be found that, depending upon the filling and evacuation procedure used, the fluid degassing may be eliminated as a separate activity [5].

Prior to charging, a heat pipe must be evacuated to remove materials that may subsequently appear as unwanted non-condensable, or that or chemically react with the working fluid forming undesirable corrosion product.

The non-condensable are due not only to the free gas in the pipe but also to the molecules absorbed on the metal surface. Removal of free gases in the pipe can be done simply by pumping down with a vacuum pump. Removal of absorbed gas requires the evacuation of the pipe at elevated temperatures. The time required to absorb surface contaminants is usually reduced with the increased temperature. However, metals may lose their strength at high temperature [2].

Heat pipe assemblies design guidelines

Orientation with respect to gravity For the best performance, the application should have gravity working with the system; that is, the evaporator section (heated) should be lower, with respect to gravity, than the condenser (cooling) section. In other orientations where gravity is not aiding the condensed liquid return, the overall performance will be degraded. Performance degradation depends on a number of factors including wick structure, length and working fluid of the heat pipe along with heat flux of the application. Careful design can minimize the performance loss and allow an accurate prediction performance.

Temperature limits Most pipes use water and methanol/alcohol as the working fluids. Depending on the wick structure, pipes will operate in environments with as low as −40 °C. Upper temperature limits depend on the fluid, but 60–80 °C is the average limit.

Heat removal Heat can be removed from the condenser using air cooling in combination with either extrusion, bonded-fin heat sinks, or flat-fin stock. Enclosing the condenser in a cooling jacket allows liquid cooling.

Reliability Heat pipes have no moving parts and have demonstrated life of over 20 years. The largest contributor to heat pipe reliability comes from control of the manufacturing process. The seal of the pipe, purity of the materials used in the wick structure and cleanliness of the internal chamber have measurable effect on the long term performance of a heat pipe. Any leakage will eventually render the pipe inoperable. Contamination of the internal chamber and wick structure will contribute to the formation of non condensable gas (NCG) that will degrade performance over time. Well developed processes and rigorous testing are required to ensure reliable heat pipes.

Forming or shaping Heat pipes are easily bent or flattened to accommodate the needs of the heat sink design. Forming heat pipes may affect the power handling capability as the bends and flattening will cause a change in fluid movement inside the pipe. Therefore design rules that take into consideration heat pipe configurations and the effect on thermal performance ensure the desired solution performance.

Effects of length and pipe diameter The vapor pressure differential between the condenser end and the evaporator end controls the rate at which the vapor travels from one end to the other. Diameter and length of the heat pipe also affect the speed at which the vapor moves and must be considered when designing with heat pipes. The larger the diameter, the more cross sectional area available to allow vapor to move from the evaporator to the condenser. This allows for greater power carrying capacity. Conversely, length when in opposition to gravity has a negative effect on heat transport as the rate at which the working fluid returns from the condenser end to the evaporator end is controlled by the capillary limit of the wick which is an inverse function of the length of the pipe. Therefore, shorter heat pipes carry more power than longer pipes when used in application not assisted by gravity.

Heat pipe assemblies design guidelines—cont'd

Wick structures: Heat pipe inner walls can be lined with a variety of wick structures. The four most common wicks are:

(a) groove
(b) wire mesh
(c) sintered powder metal
(d) fiber/spring

The wick structure provides a path for liquid to travel from condenser to the evaporator using capillary action. Wick structures have performance advantages and disadvantages depending on the desired characteristics of the heat sink design. Some structures have low capillary limits making them unsuitable for applications where they must work without gravity assist.

4.6 Evacuation and charging

Prior to charging, a heat pipe must be evacuated to remove materials that may subsequently appear as unwanted non-condensable. Evacuation and charging are two processes that are closely related.

Fig. 4.11 shows a flow chart for the evacuation and charging procedure used by heat pipe manufacturers, e.g., TRW, Grumman, and McDonnell Douglas, as presented by Edelstein and Haslett [1].

Initially, the pipe is evacuated, typically on a 4-ino Veeco pumping station at a station pressure of about 10^{-6} mm Hg. Heaters wrapped around the pipe provide the elevated temperatures at which evacuation occurs. A schematic of the setup is shown in Fig. 4.12. The temperatures shown in Fig. 4.11 are representative for aluminum tubes. TRW bakes-out at about 325 °F for 16 h while Grumman bakes out at lower temperatures of 170–250 °F, but for a longer time of up to 48 h. Although higher evacuation temperatures may be beneficial in removing additional adsorbed molecules, loss of mechanical properties of aluminum can occur at temperatures in the 300 °F range. Figs. 4.11 and 4.12 show the effects of elevated temperature exposure on the room-temperature ultimate and yield strength of aluminum 6061-T6 [1]. It is seen that loss of strength begins to occur at a temperature of 380 °F for 1/2 h exposure. The close relation between evacuation and charging can be seen in this Fig. 4.11; hence, evacuation and charging are often performed in the same equipment set-up.

Typically, a 1/16 in. (0.159 cm) i.d. tube, about 4 in. (10:160 cm) long, serves as the charge (pinch-off) tube. Based on the results presented in the previous section, this size tube is mildly restrictive during pump down for 1/2 in. (1.270 cm) d pipes. A technique which avoids the use of a small diameter charge tube will be presented later.

Fig. 4.13 shows a possible set-up for combined evacuation and charging. A possible procedure is as follows: with valve B closed and valves A and C open,

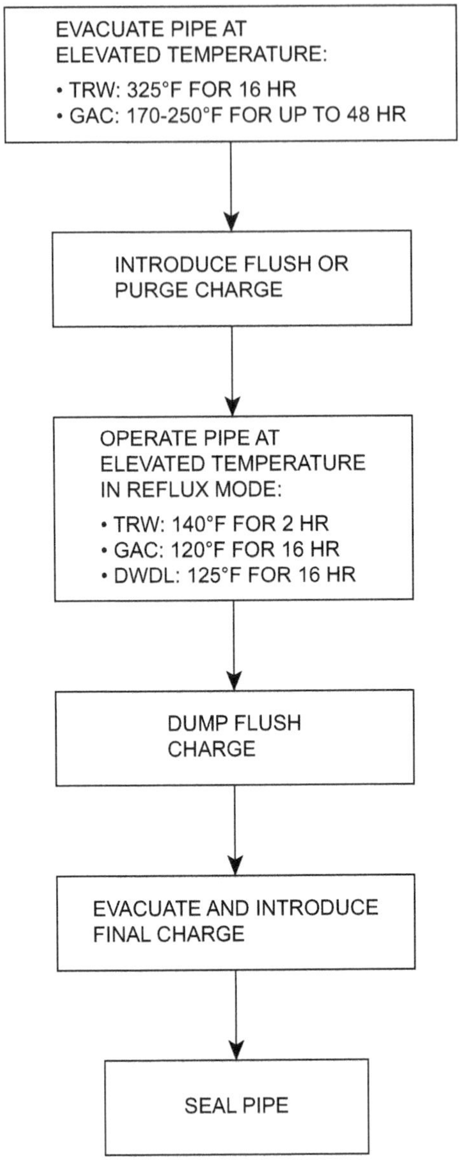

Note: Temperatures shown are for aluminum envelopes.

FIG. 4.11 Over-all evacuation and procedure used by TRW, Grumman, and DWDL/McDonnell Douglas [1].

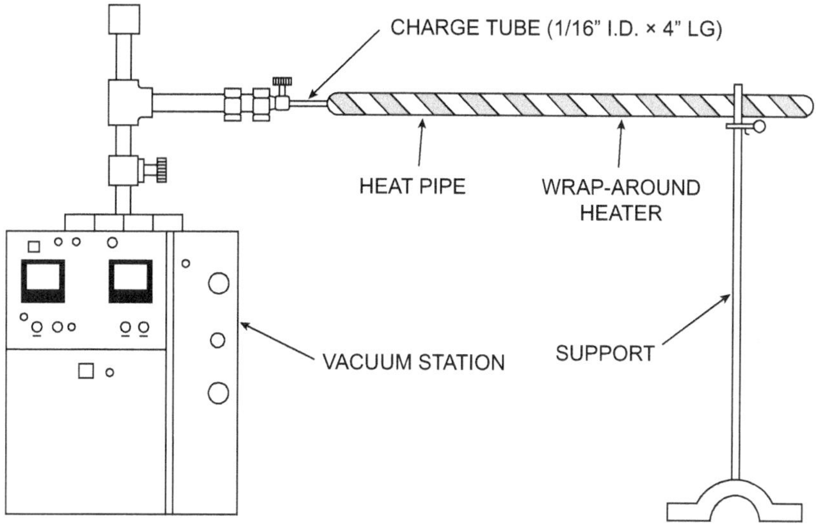

FIG. 4.12 Heat Evacuation Setup [1].

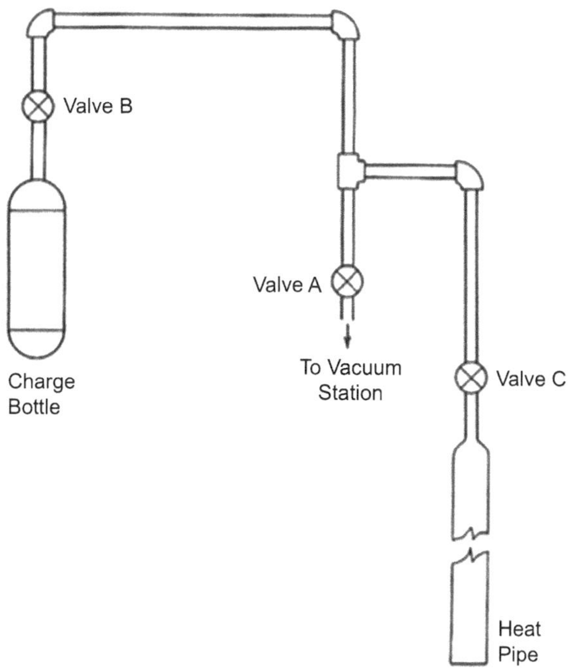

FIG. 4.13 Schematic of equipment for heat pipe evacuation and charging [2].

the pipe is first pumped down at the ambient temperature; and then, the pumping is continued while the pipe is heated. The temperature of the heat pipe and the pumping time depend, of course, on the pipe material and its eventual operating temperature as described in the previous paragraph. This process is sometimes called vacuum bake out. After completion of this vacuum bake out process, the pipe is flushed with a small amount of fluid. For this purpose, the fluid in the charge bottle is first heated above the fluid boiling temperature at the pressure of the vacuum system. Momentary opening of valve B then allows a small amount of flush charge to be dumped into the pipe. After the pipe is flushed in this manner once or twice, the pipe is ready for charging [2].

4.6.1 Fluid charging

Details of the charging process depend on the state of the working fluid at the ambient temperature. If the fluid is in the gaseous state at room temperature, such as the case for the cryogenic heat pipe fluid, the charge can be introduced from a cylinder containing gas of high quality. The amount of charge can be measured by the gas pressure in the pipe at room temperature. Hence, the charging process consists of closing valve A and opening valves B and C. When the required amount of fluid has been charged into the pipe, valves B and C are closed. The pipe is then ready for pinching-off and final sealing.

4.6.2 Fluid purity and inventory

Should a manufacturer use almost 99.999% pure ammonia costing $40 per pound, or will 99.99% pure ammonia costing $2.50 per pound be satisfactory? What impurities are present in the ammonia? How will they affect heat pipe performance? These are some typical questions asked by both manufacturers and users of heat pipes concerning the requirements for the purity of heat pipe working fluids. These considerations can strongly affect both product reliability and cost [1].

The most prominent manifestation of impurities in a heat pipe is the accumulation of non-condensable gas in the condenser zone, with consequent loss of heat pipe conductance. Depending on the design and operating conditions, the presence of gas may not be serious and may go completely unnoticed. In other cases, significant blockage can occur; for some artery pipes, loss of pumping capacity can result [1].

Impurities may be in the pipe in the form of adsorbed gas molecules even before the working fluid is introduced. It may be brought in during the fluid transfer operation, or it may be present in the fluid itself. Often, the fluid purchased from the supplier may have significantly higher impurity levels than what was nominally specified [1].

Certain techniques are available to purify the fluid to a higher state than it is "as received" from the supplier. Additional cleaning may be necessary to satisfy functional requirements, or as a safeguard against uncertain impurity levels in the suppliers fluid [1].

We can now discuss effects of impurities on heat pipe performance. In general, the impurities found in working fluids, such as ammonia, may consist of:

- Gases, such as nitrogen, oxygen, argon, carbon dioxide, carbon monoxide and methane.
- Water.
- Miscellaneous materials such as oils, hydrocarbons and non-volatile solids.

Of the miscellaneous materials, the most detrimental effect on heat pipe performance can be loss of wetability of the wick due to oily residues. As previously mentioned, a significant amount (130 ppm) of an oily contaminant was found in the liquid phase of ammonia and identified to be dioctyl phthalate. These materials, some of which may be soluble in the working fluid, may also adversely affect fluid properties, such as surface tension, wetting angle, and viscosity [1].

The presence of water in an aluminum or stainless steel pipe can cause corrosion, resulting in the loss of structural integrity. However, the quantities of water generally found in working fluids can be minimized by various purification techniques, some of which will be discussed later. The resultant water quantities, in terms of parts per million, are usually small enough so as not to present a serious corrosion loss-of-strength problem, since the corrosive reaction will generally cease when the water is consumed. The reaction products, however, may be far more serious in terms of non-condensable gas generated [1].

Fluid used for the heat pipe must obviously be of high purity. However, non-condensable may be dissolved in the supposedly pure liquids and solids. For example, non-condensable that are transferred to the charge bottle during vapor transfer may be removed by subjecting the fluid to repeated freeze-thaw cycles [1]. A schematic of this processing technique is shown in Fig. 4.14. The fluid is frozen in the charge bottle using, for example, liquid nitrogen for ammonia and ambient air for sodium.

It is also necessary that the heat pipe is neither under-filled nor over-filled with its working fluid. An under-fill may result in degradation of its

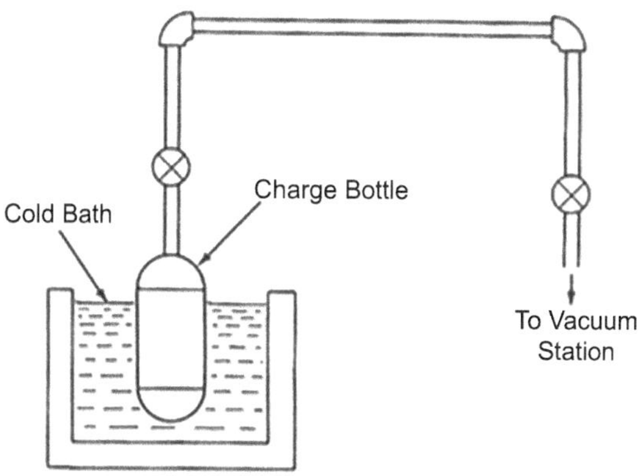

FIG. 4.14 Schematic of fluid purification process [2].

performance. On the other hand, an over-fill can result in condenser blockage. It slight over-fill can be tolerated; the fluid inventory can be calculated by Eqs. (4.1). The fluid inventory calculated by this equation is larger than what is required, because it discounts the meniscus recession. Nevertheless, the depletion of liquid due to this recession is usually negligibly small.

$$m = A_v L_t \rho_v + A_w L_t \rho_l \tag{4.1}$$

where,

m = Fluid inventory.
A_v = Vapor cross-sectional area.
A_w = Wick cross-sectional area.
L_t = Total heat pipe length.
ρ_v = Vapor density at pipe operating temperature.
ρ_l = Liquid density at pipe operating temperature.
ε = Wick porosity.

4.6.3 Analysis of gas blockage

In Fig. 4.15, the volume occupied by the non-condensable is, from the ideal gas law:

$$N = \frac{n_G R T_G}{P_G} \tag{4.2}$$

where,

n_G = Moles of non-condensable gas in pipe.
R = Gas constant.

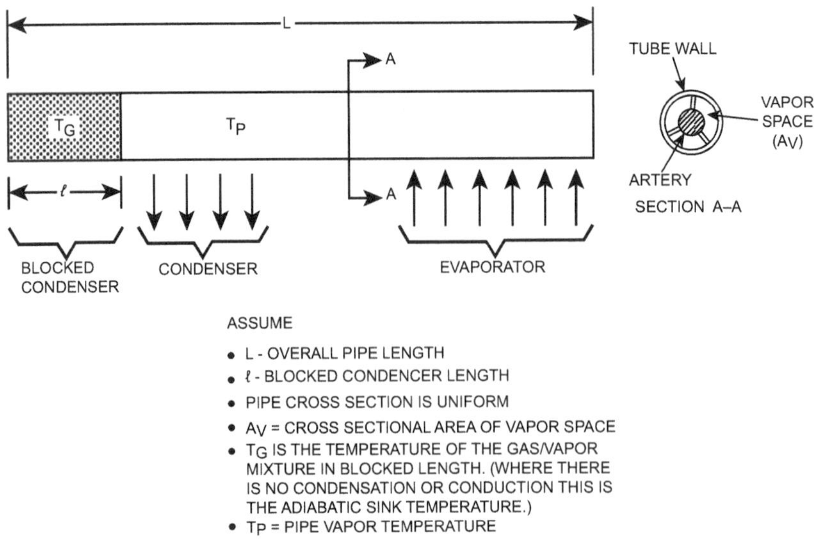

FIG. 4.15 Schematic for analysis of gas blockage [1].

T_G = Temperature of the Gas/Vapor mixture in blocked length.
P_G = Partial pressure of inert gas in length, ℓ.
P_G furthermore, is the difference between the working fluid pressure in the active and inactive portions of the pipe, i.e.,

$$P_G = P(T_p) - P(T_G) \tag{4.3}$$

where,

$P(T_p)$ = Working fluid pressure at temperature T.
$P(T_G)$ = Gas pressure at temperature T.
ℓ = Blocked condenser length.
T_p = Pipe vapor temperature.

From Fig. 4.15, the gas volume can be related to the blocked length by

$$V = A_v \cdot \ell \tag{4.4}$$

Substituting Eqs. (4.3) and (4.4) in Eqs. (4.2) and solving for ℓ, yields

$$\ell = \frac{n_G R T_G}{A_v\{P(T_P) - P(T_G)\}} \tag{4.5}$$

The moles of non-condensable can now be defined in terms of a working fluid impurity level, f,

$$f = \frac{n_G}{n_p} \tag{4.6}$$

where,

n_p = Moles of working fluid in pipe
n_G = Moles of non-condensable gas in pipe.
f = Mole fraction of non-condensable in working fluid.

Substituting Eqs. (4.6) into Eq. (4.5) and dividing by the overall pipe length, L, yields the blockage as a fraction of the total pipe length,

$$\frac{\ell}{L} = \frac{n_p R T_G f}{L A_v\{P(T_p) - P(T_G)\}} \tag{4.7}$$

If n_p is defined as the heat pipe charge per unit length, i. e. $n_p = n_p/n_L$ then Eqs. (4.7) becomes

$$\frac{\ell}{L} = \frac{n_p R T_G f}{A_v\{P(T_p) - P(T_G)\}} \tag{4.8}$$

This expression now relates the blockage to the operating conditions, T_p and T_G; to pipe design parameters n, p, and A_v as well working fluid; and to an impurity level factor, f.

4.6.4 Effect of heat pipe design and operating conditions on gas blockage

In Fig. 4.16, the pipe blockage is presented as a function of the temperature difference between the operating and non-operating sections for 0.5 in. (1.270 cm) axial groove ammonia heat pipe. An impurity content, ΔH_{vap}, of 0.0001 was assumed which is equivalent to 100 ppm (molar). The gases can be

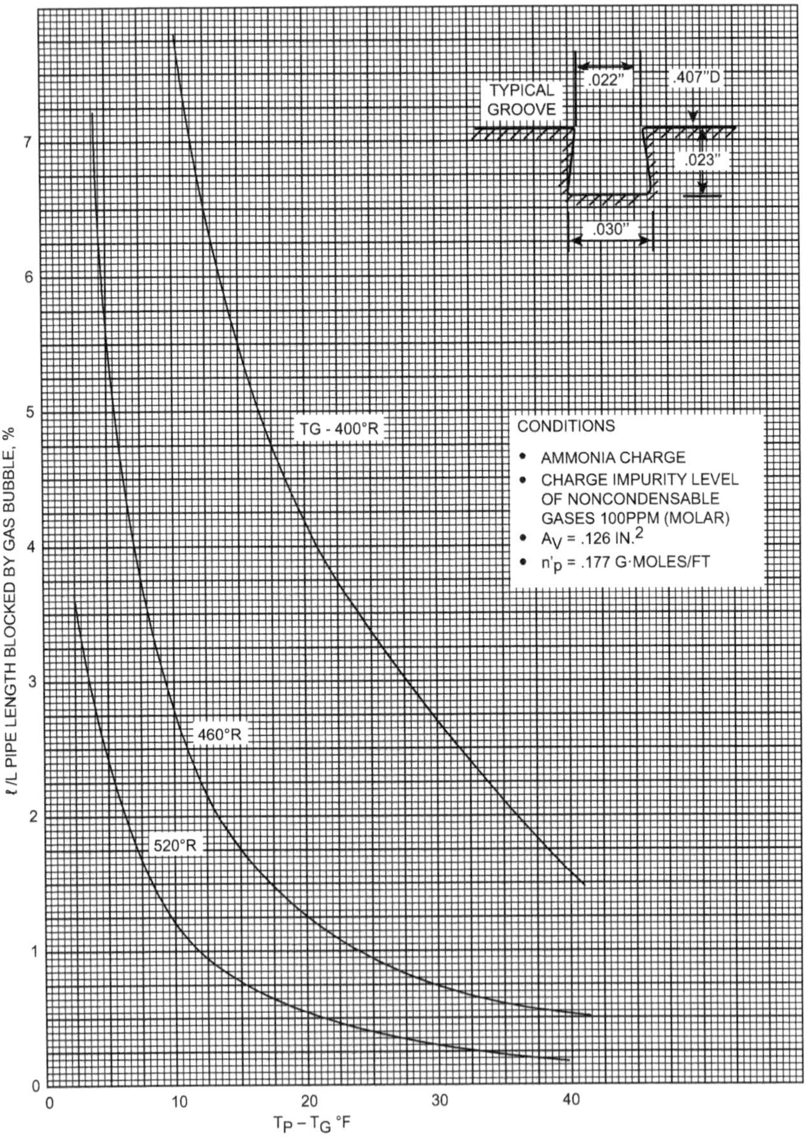

FIG. 4.16 Gas blockage in typical grooved heat pipe [1].

considered as being present in the charge fluid or as gas remaining in the pipe after evacuation.

The curve shows that for a fixed temperature difference (T_p–T_G), the blockage decreases at higher temperatures. Also, the blocked length increases for small differences in temperature between the pipe and gas temperature (thermal sink), such as, in the case of isothermalizing heat pipes. As an example, consider a 10-ft (3 m) long axial groove/ammonia heat pipe whose temperature T_p is equal to 420 ^{0}R, attached to a sink T_G at 400 ^{0}R, i.e., $T_p - T_G = 20\ ^0$R. The blockage produced by an impurity level of 100 ppm would be 4.2% or 0.42 ft (5.0 in.) (12.7 cm). For an impurity level of 10 ppm (not shown) the blockage, which is proportional to f, would be 0.50 in. (1.27 cm).

The effect of a different pipe configuration is shown in Fig. 4.17, where a 0.5-in., (1.270 cm) spiral artery design is compared to the 0.5-in. groove design. Because of its smaller vapor space, the spiral design produces a greater blockage then the groove under the same conditions.

The effect of different working fluids is shown in Fig. 4.18. Ammonia, acetone and Freon-21 are plotted for the 0.5 in. axial groove design. It is seen that ammonia results in a smaller blocked length than acetone, or Freon-21. This is because the pressure of ammonia changes more rapidly with temperature than the other fluids. The value of these curves is that it allows the designer, for his particular application, to make a judgment regarding the allowable quantity of gaseous impurities, or conversely, to design the condenser length so as to accommodate some estimated amount of gas that may be present (Fig. 4.19).

4.7 Full tube closure

The manufacture of the heat pipe is near its completion with the exception that the pipe is still attached to a closed valve at the fill tube. It is necessary to sever the pipe from the valve and to form a permanent seal at the fill tube. In this process, no gas must enter the pipe and no fluid should be lost from the pipe. A closure technique, which has proved both economical and reliable, consists of:

1. Crimp seal (flatten and pinch) the fill tube to form a temporary leak-proof closure.
2. Sever the valve from the pipe by making a cut in the flattened area of the fill tube on the valve side of the crimp seal.
3. Weld the cut end of the fill tube by TIG or EB welding and remove the crimping tool.

4.8 Heat pipe testing techniques

Per Chi [2], testing of the heat pipe may answer many questions. A simple test on the wetting of the wick and on the absence of leakage will insure that the pipe

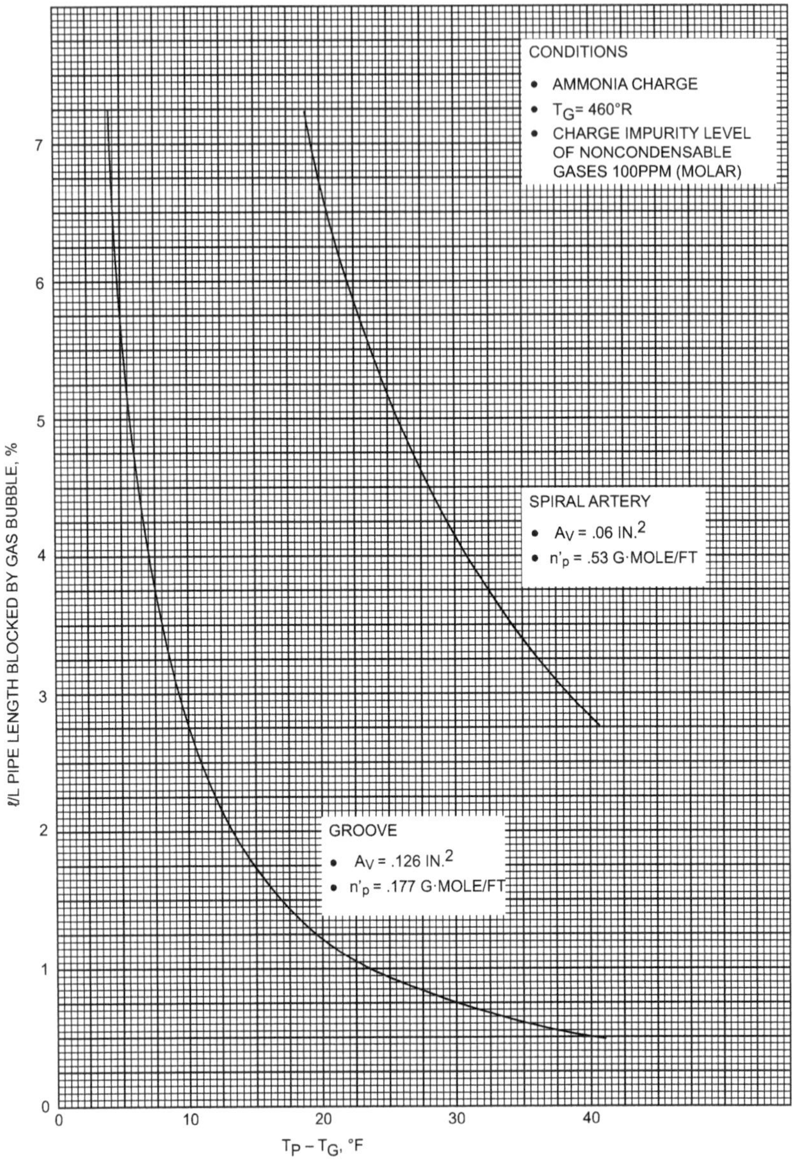

FIG. 4.17 Effect of heat pipe configuration on gas blockage [1].

will operate as a heat pipe. A heat conduction test with an appropriately designed heat source and sink may verify the heat transfer capability and characteristics of the heat pipe. A test on the life of a heat pipe may include recording the performance of sample heat pipes over a long period of time, or an accelerated life test may be performed with metallurgical examination of

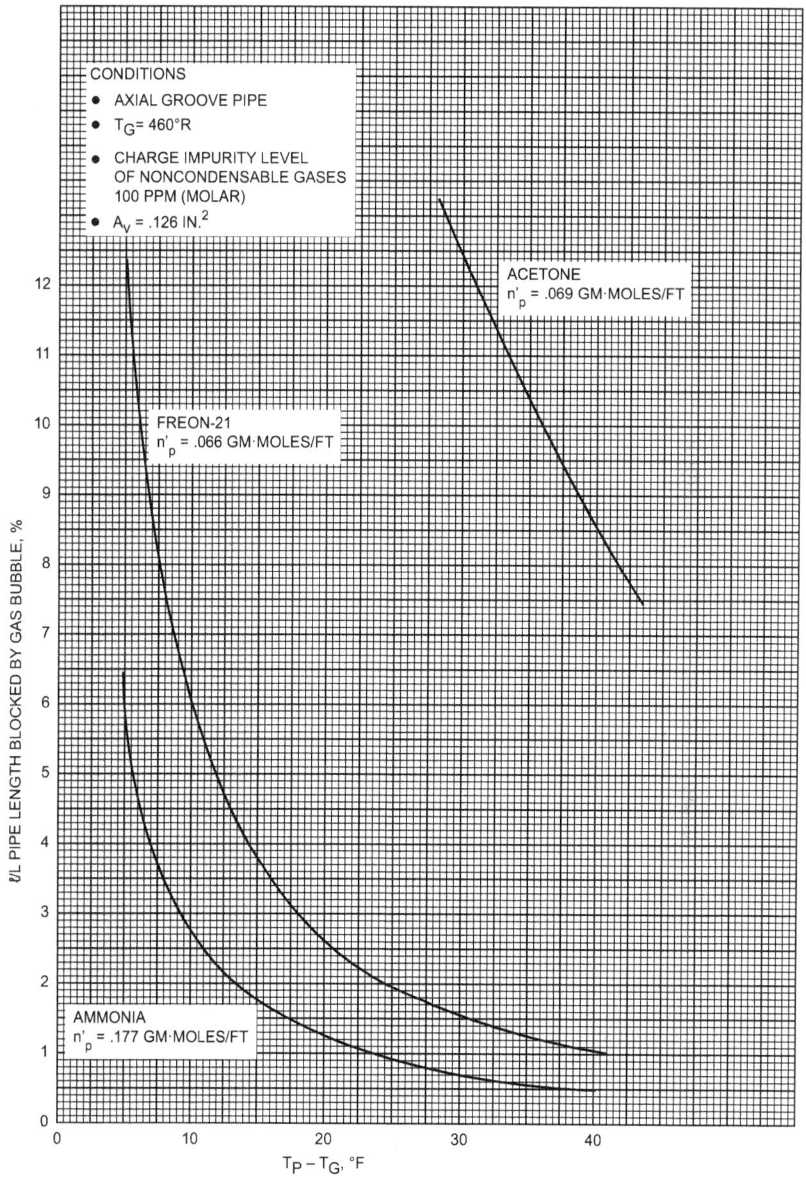

FIG. 4.18 Effect of working fluid on gas blockage [1].

materials over predetermined intervals of time. In addition, tests may be required to answer questions on transient characteristics of the pipe. We will concentrate our discussion of the pipe testing's on verifying the mechanical soundness of the pipe, proper wetting of the wick, and operating characteristics of the pipe.

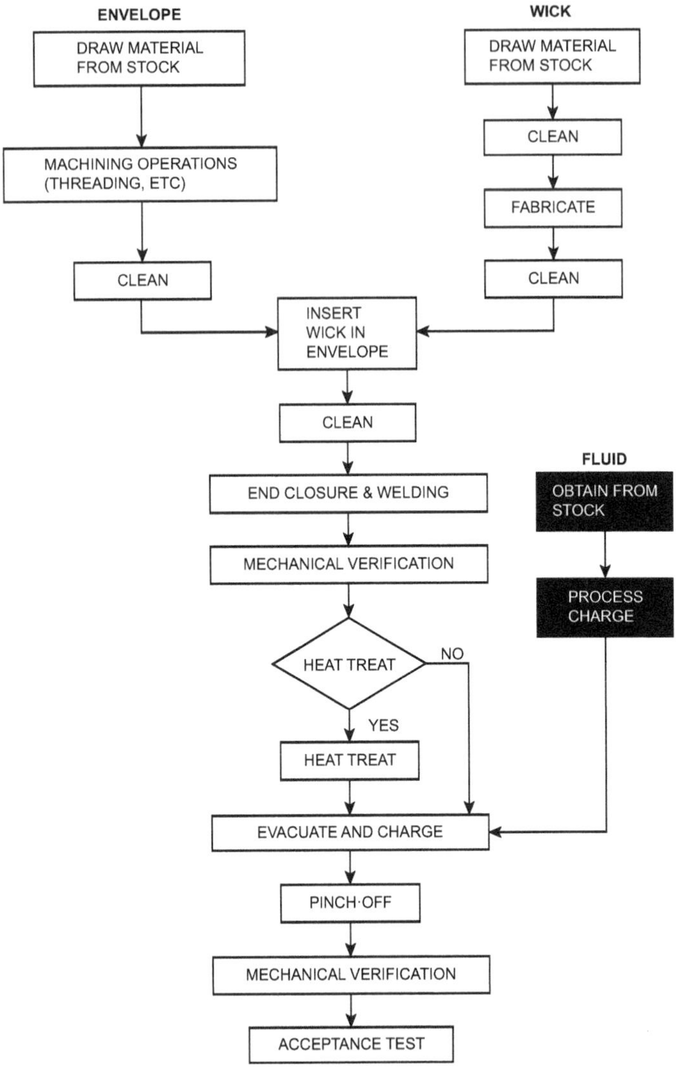

FIG. 4.19 Typical manufacturing steps for charging process [1].

4.8.1 Mechanical sounders

A sound structural design that has been properly verified by non-destructive tests is paramount to reliable, long-term heat-pipe operation. The ASME pressure vessel code is recommended when specifying allowable design stresses, proof pressure, and burst pressure. Simplified methods are also presented by Edelstein and Haslett [1] for including the stress effects due to internal pressure, end caps, thermal expansion, saddle attachments, pipe bends, and dynamic loading. Cost-effective methods of leak detection for ammonia, Freon, and methanol used during pre-charging and post charging operations include: X-ray examination, pressurization under water, helium detection, and copper sulfate/ethylene glycol (for ammonia) [1].

The pipe must be able to withstand the maximum vapor pressure. For this reason, the pipe envelope and end caps have to be properly designed and joints must be made with high quality welding before a design and production procedure can be finalized for large quantity production. Experimental heat pipes may be made and subjected to high vapor pressure by heating the pipe so as to assure that the pipe envelope, the end caps, and the joints will be able to withstand the designed vapor pressure with appropriate margins of safety [2].

4.8.2 Wick wetting

Unfortunately, many of these problems cannot be uncovered until the pipe is charged, sealed and tested. In some cases, a long time can occur until some of these effects are noticed; by then, it is usually too late to provide corrective action. Hence, the objective must be to develop cleaning procedures which will prevent these problems from occurring, and produce, therefore, a more reliable product. Moreover, in keeping with the overall aims of this study, the cleaning procedure should also be simple, inexpensive, and as free as possible from human error [1].

The simplest way to determine if the working fluid has wet the wick is to oscillate the heat pipe along its axis by hand. If the fluid slugs the end caps, it is fairly obvious that the liquid is riding on the wick in the vapor space. This preliminary test works fairly well for pipes having working fluid in the liquid state at room temperature. However, one can shake the pipe with vigor and break the force of surface tension. Liquid within the vapor space will result. This method is limited to qualitative testing of wick wetting. X-ray radiography may sometimes be used to observe uniformity of the liquid inside the wick or the presence of excessive liquid at the bottom of the pipe [2].

4.8.3 Performance versification

If a pipe passes the mechanical and wetting tests, it can be assured that the pipe will operate as a heat pipe under favorable conditions. In order to establish pipe performance such as the maximum heat transfer capability and effective thermal conductance of the pipe, tests using the set-up illustrated schematically in Fig. 4.20 may be used for the moderate-temperature heat pipe. The pipe may be placed at any desired orientation. The heat is supplied by an electrical tape wound around the evaporator section of the pipe [2].

Either the evaporator or the condenser temperature may be maintained constant regardless of the variation of the rate of heat supply by controlling the water inlet temperature and flow rate. Thermocouples installed along the heat

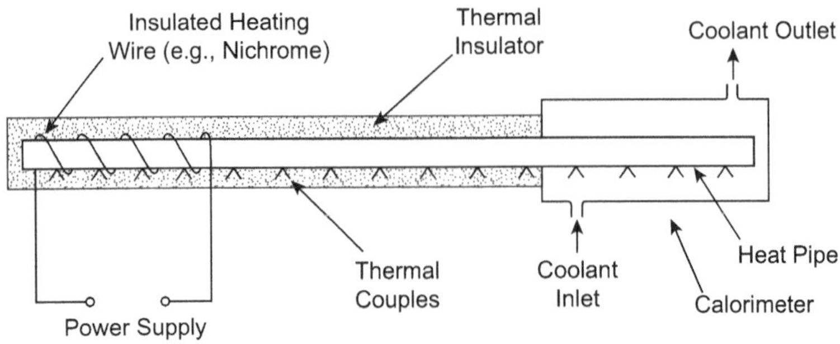

FIG. 4.20 Testing set-up for moderate temperature heat pipe [2].

pipe wall measure the axial wall temperature along the pipe at different heat transfer rates, which are indicated by the electrical power input. As the electrical power input is gradually raised to a certain limit, it may be observed that the temperature indicated by the thermocouple at the end of the evaporator rises suddenly above that indicated by other thermocouples at the evaporator. This sudden rise in temperature at the end of an evaporator indicates partial dry-out of the evaporator. Hence, both heat transport limit and temperature characteristics of heat pipes may be measured using apparatus shown in Fig. 4.20.

The operating temperature and the heat transfer capability for the liquid-metal heat pipes are usually much higher than those for the moderate temperature heat pipes. For increasing the power supply capability, the electrical heating tape shown in Fig. 4.20 can be replaced by an electrical induction coil. When this set-up is used for testing liquid-metal heat pipes, radiation and natural convection of heat to the atmosphere provides a convenient heat sink. In order to control the pipe operating temperature, the thermal resistance of the condenser section may be controlled by providing a concentric annular gap as shown in Fig. 4.21.

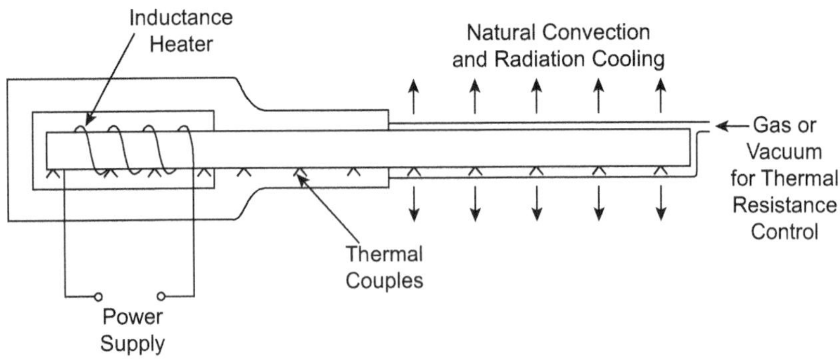

FIG. 4.21 Testing set-up for liquid metal heat pipe [2].

References

[1] F. Edelstein, R. Haslett, Heat Pipe Manufacturing Study, Final Report prepared by Grumman Aerospace Corp. for NASA, Contract No. NASS5-23156, August 1974.

[2] S.W. Chi, Heat Pipe Theory and Practice, Hemisphere Publishing Corporation a McGraw Hill Book Company, 1976.

[3] J.E. Kemme, Heat Pipe Capability Experiment, Report No. LA-3585-MS, Los Alamos Scientific Laboratory, August 1966.

[4] R.A. Rhodes Jr., Procedures for the Construction of a Screen Wick Heat Pipe, A Lecture Note on Heat Pipe, U. S. Army Mobility Equipment R and D Center, October 1973.

[5] D. Reay, P. Kew, Heat Pipes Theory, Design and Application, fifth ed., Elsevier, 2006. Butterworth-Heinemann is an imprint of Elsevier.

Chapter 5

Heat pipe heat exchanger opportunities and industrial applications

Chapter outline

5.1 Introduction 275
5.2 General theory of heat pipe design 277
5.2.1 Capillary limitation 277
5.2.2 Sonic limitation 277
5.2.3 Entrainment limitation 278
5.2.4 Boiling limitation 278
5.3 Holistic approach to heat pipe application 281
5.3.1 Merit number derivation 288
5.3.2 Lowest heat pipe limit driven temperature 289
5.3.3 Heat pipe working fluids 292
5.4 Heat pipe heat exchangers, an innovation for heat transfer management 293
5.4.1 Direct contact heat exchangers highly efficient HVAC systems 294
5.4.2 Variable conductance heat pipe (VCHP) heat exchanger 298
5.4.3 Innovative heat exchanger designs 300
5.5 An overview of the heat pipe technology summary 303
References 304

5.1 Introduction

Heat pipes (HPs) and Thermosyphons (TSs) are passive devices which operate by utilizing the latent heat of an internal working fluid to transfer large amounts of heat, nearly isothermally, with a minimal driving temperature difference through a small cross sectional area and is divided into three segments:

1. Evaporator,
2. Adiabatic and
3. Condenser.

These sections denoted in Fig. 5.1A and B along with its limitations in respect to heat transportation, based on their external thermal boundary conditions. A HP/TS functions when heat is applied to the evaporator section,

Functionality, Advancements and Industrial Applications of Heat Pipes.
https://doi.org/10.1016/B978-0-12-819819-3.00005-5

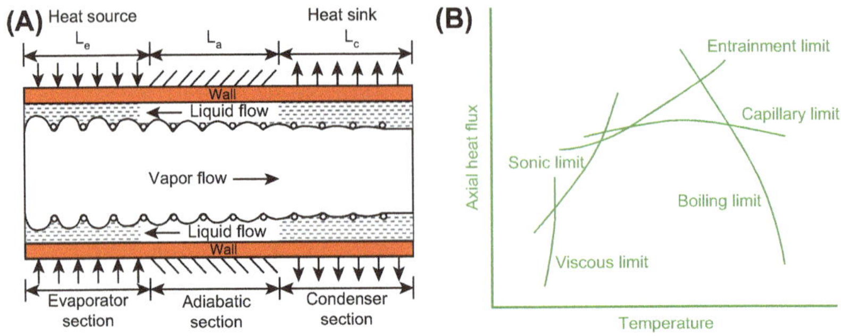

FIG. 5.1 Conventional heat pipe along with heat transport limitation. (A) Conventional heat pipe. (B) Limitations to heat transport in a heat pipe.

which causes vaporization of the working fluid. The vapor flows through the adiabatic section to the lower temperature condenser section, within which condensation of the HP/TS working fluid occurs.

The heat pipe can be of any size or shape, although the constraint of. The main components of heat pipe (Fig. 5.1A) are sealed container with (pipe wall and end caps) with three sections evaporator, condenser and adiabatic sections, with or without a wick structure, and small amount of working fluids. A heat pipe can have multiple evaporators, condensers and adiabatic sections depending upon the type of application and the heat load.

Selection of working fluid also plays an important role for its thermal performance. Within the temperature band, several possible working fluids may exist, and a variety of characteristics must be examined in order to determine the most acceptable of these fluids for the application considered. The prime requirements are compatibility with wall materials, good thermal stability, wettability of the wall materials, optimum vapor pressure over the operating temperature range, high latent heat, high thermal conductivity, low liquid and vapor viscosity, high Surface tension and acceptable freezing or pour point (Fig. 5.1B). Heat pipe fabrication, processing, and testing involve several detailed procedures which are recommended to be strictly followed in order to achieve the highest quality possible.

Longevity of a heat pipe can be assured by selecting a container, a wick and welding materials that are compatible with one another and with the working fluid of interest. Performance can be degraded and failures can occur in the container wall if any of the parts (including the working fluid) are not compatible. For instance, the parts can react chemically or set up a galvanic cell within the heat pipe. Additionally, the container material may be soluble in the working fluid or may catalyze the decomposition of the working fluid at the expected operating temperature. Some details can be found in Chapter One of this book, while More information presented in book published by Zohuri [1].

5.2 General theory of heat pipe design

A straight circular tube type heat pipe containing metallic sodium which is usually used in a high temperature heat pipe has been produced. The compatibility of the tube material (316L, stainless steel) with sodium is well-documented in heat pipe literatures.

Considering a traditional heat pipe such as the one demonstrated in Fig. 5.1A, the maximum heat transport capacity depends strongly on the working fluid and its temperature. The maximum axial heat transfer rate of a particular heat pipe under certain working conditions can be determined by a number of possible heat transfer limitations. The limitations will depend on the geometry of the heat pipe, the wick structure, the vapor channel shape, the working fluid and the operating temperature. Maximum heat transport limitations may be calculated by Eqs. (5.1) through (5.4), which are summarized by Chi [3]. Fig. 5.1B shows that entrainment limitation is one of the most important factors in the present investigation of sodium heat pipe.

5.2.1 Capillary limitation

$$Q_{c,\max} = \frac{2\left(\frac{\sigma\rho_l h_{lv}}{\mu_l}\right)\left(\frac{K}{r_c}\right)(2\pi r_v T_w)}{0.5L_e + L_a + 0.5L_c} \tag{5.1}$$

where,

- σ: Surface Tension, condensing coefficient (N/m)
- ρ_l: Liquid Density Phase (kg/m^3)
- h_{lv}: Heat Transfer Coefficient Liquid, vapor (Latent Heat of Vaporization)
- K: Permeability (m^2)
- r_v: Radius of Vapor Space Section
- t_w: Temperate of Heat Pipe Wall
- μ_l: Liquid Viscosity, (N-s/m^2)
- r_c: Effective Capillary Radius, (m)
- L_e: Length of the Evaporator Section (m)
- L_a: Length of the Adiabatic Section (m)
- L_c: Length of the Condenser Section (m)

5.2.2 Sonic limitation

$$Q_{s,\max} = A_v\rho_v h_{lv}\left[\frac{\gamma_v T_v r_v}{2(\gamma_v + 1)}\right]^{1/2} \tag{5.2}$$

where,

- A_v: Cross-Sectional Area of Vapor Space
- ρ_v: Vapor Density Phase (kg/m^3)

h_{lv}: Heat Transfer Coefficient Liquid, vapor (Latent Heat of Vaporization)
γ_v: Specific Heat Ratio for Vapor Side
r_v: Radius of Vapor Space Section
h_{lv}: Heat Transfer Coefficient Liquid, vapor

5.2.3 Entrainment limitation

$$Q_{e,\max} = A_v h_{lv} \left(\frac{\sigma \rho_v}{2 r_{h,s}} \right)^{1/2} \tag{5.3}$$

where,

A_v: Cross-Sectional Area of Vapor Space
h_{lv}: Heat Transfer Coefficient Liquid, vapor (Latent Heat of Vaporization)
ρ_v: Vapor Density Phase (kg/m^3)
σ: Surface Tension, condensing coefficient (N/m)
$r_{h,s}$: Hydraulic Radius

5.2.4 Boiling limitation

$$Q_{b,\max} = \frac{2\pi L_e k_e T_v}{h_{lv} \rho_v \ln(r_i/r_v)} \left(\frac{2\sigma}{r_n} - P_c \right) \tag{5.4}$$

where,

L_e: Evaporator Length (m)
k_e: Effective Thermal Conductivity of the Liquid-Saturated Wick
T_v: Vapor Temperature
σ: Surface Tension, condensing coefficient (N/m)
ρ_v: Vapor Density Phase (kg/m^3)
h_{lv}: Heat Transfer Coefficient Liquid, vapor (Latent Heat of Vaporization)
r_i: Inner Radius of the Pipe
r_v: Vapor Core Radius
r_n: Initial Radius or Boiling Radius
P_c: Capillary Pressure

Clearly, a heat pipe cannot be operated below the freezing point or above the thermodynamic critical point of its working fluid. Thus, the first criterion for selection of a fluid is that these two thermodynamic conditions fall within the required operating temperature range [1].

These conditions, however, actually represent lower and upper bounds which are seldom approached. Most often, the low end of a given fluid's operating temperature range is established by adverse vapor dynamics (sonic limit, entrainment limit, or simply excessive ΔP_v) due to low vapor densities and corresponding high vapor velocities. The high end of the temperature range is frequently set by the mechanical aspects of containing the fluid vapor pressure [1].

Clearly, a heat pipe cannot be operated below the freezing point or above the thermodynamic critical point of its working fluid. Thus, the first criterion for selection of a fluid is that these two thermodynamic conditions fall within the required operating temperature range [1].

The axial heat transport requirement can have a major impact on the choice of working fluid. Different fluids will yield different capillary pumping limits for the same wick structure. Thus, the case can easily arise where a simple homogeneous wick design can be substituted for a complex arterial wick design by the choice of fluid [1].

To determine the best fluid for a given application, one must theoretically examine the optimal designs for each fluid by integrating the loss equations to determine their respective capillary pumping limit (See Section 2.13.3 book by Zohuri [1]). Sometimes this is actually necessary since, in the general case, there is no simple grouping of fluid properties which serves as a basis for selection. However, there exist such groupings for special cases which at least provide some general guidelines. Thus, for a heat pipe operating in the absence of body forces and for which the vapor pressure drop is negligible, the capillary pumping limit can be shown to be proportional to the grouping $(\sigma \rho_l h_{lv}/v_l)$, sometimes referred to as the "liquid transport factor" or "0 g figure of merit." Fig. 5.2 [1] shows the liquid transport factor for the principal fluids of interest in spacecraft thermal control as a function of operating.

Because heat pipes are effective devices for transporting large amounts of heat with small temperature gradients, they have been considered as elements of heat rejection systems for applications such as space power system, and as heat receivers for solar dynamic space power systems. For high-temperature applications, the working fluids in most liquid metal heat pipes are entirely in the solid state at ambient temperature while pressure in the vapor space is extremely low so that free molecular flow conditions prevail. Unlike for low-temperature heat pipes, these conditions may cause the heat pipe to fail to operate, they also complicate mathematical models of liquid metal heat pipes during the startup period [1].

In summary, High temperature heat pipes are being evaluated for use in energy conversion applications such as fuel cells, gas turbine re-combustors, and Stirling cycle heat sources; with the resurgence of space nuclear power, additional applications include reactor heat removal elements and radiator elements. Long operating life and reliable performance are critical requirements for these applications. Accordingly, long-term materials compatibility is being evaluated through the use of high temperature life test heat pipes.

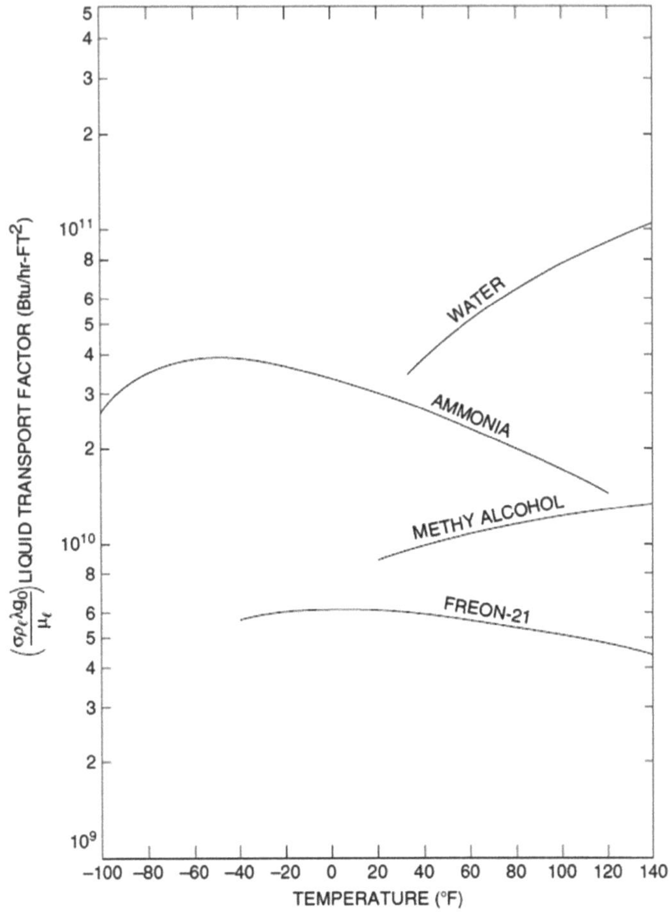

FIG. 5.2 Liquid transport factor for heat pipe working fluids [1].

Thermacore, Inc., has carried out several sodium heat pipe life tests to establish long term operating reliability. Four sodium heat pipes have recently demonstrated favorable materials compatibility and heat transport characteristics at high operating temperatures in air over long time periods. A 316L stainless steel heat pipe with a sintered porous nickel wick structure and an integral brazed cartridge heater has successfully operated at 650–700 °C for over 115,000 h without signs of failure. A second 3 16L stainless steel heat pipe with a specially-designed Inconel 601 rupture disk and a sintered nickel powder wick has demonstrated over 83,000 h at 600–650 °C with similar success. A representative one-10th segment Stirling Space Power Converter heat pipe with an Inconel 718 envelope and a stainless steel screen wick has operated for over 41,000 h at nearly 700 °C. A hybrid (i.e., gas-fired and solar) heat pipe with a Haynes 230 envelope and a sintered porous nickel wick structure was operated for about 20,000 h at nearly 700 °C without signs of degradation. These life test results collectively have demonstrated the potential for high temperature heat pipes to serve as reliable energy conversion system components for power applications that require long operating lifetime with high radiality [4].

However, bear in your mind that, the start of liquid-metal heat pipes from a frozen state involves several highly non-linear and tightly coupled heat and mass transfer processes in the vapor, wick and wall regions. The most challenging tasks have been the modeling of the phase-change in the wick and the free-molecular and transition flow regimes occurring in the vapor region. An analysis has been carried out by Jean Michel Tournier et al. [5] and results presented for the startup from a frozen state were successfully benchmarked using available experimental data for a radiatively-cooled sodium heat pipe by him and his co-author.

Furthermore, Numerical modeling of transient heat pipe behavior is a very complicated problem because it involves the simultaneous solution of a set of time-dependent partial differential equations in different regions of the heat pipe. Coupled nonlinear conditions must be satisfied at the boundaries of these regions. Several investigators have attempted to analyze the transient heat pipe behavior (Bowman and sweeten [6], Jang et al. [7], Chang et al. [8], and Doster and Hall [9]) and different approaches have been used to simplify the problem.

5.3 Holistic approach to heat pipe application

The waste heat recovery driven by heat pipes has been suggested and is accepted as an excellent way of saving energy and preventing possible global warming, while they can be used as part of heat exchanger infrastructure. Heat pipe application as part of sub-system in heat exchangers for the heat recovery is focused on the energy saving and the enhanced effectiveness of the Conventional Heat Pipe (CHP) [1,2], Two-Phase Closed Thermosyphon (TPCT), yet one-dimensional analysis and Oscillating Heat Pipe (OHP) [1] heat exchangers. Thus, we go into the required parameters of effectiveness of the CHP, TPCT and OHP heat exchangers and start to describe them, we need to have some general understanding of heat pies and what they are, however more details can be found in reference by Zohuri [1].

Heat pipes are becoming increasingly popular as passive heat transfer technologies due to their high efficiency. A heat pipe is a passive energy recovery heat exchanger that has the appearance of a common plate-finned water coil except the tubes are not interconnected. Additionally, it is divided into two sections by a sealed partition. Hot air passes through one side (evaporator) and is cooled while cooler air passes through the other side (condenser). While heat pipes are sensible heat transfer exchangers, if the air conditions are such that condensation forms on the fins there can be some latent heat transfer and improved efficiency. See Fig. 5.3.

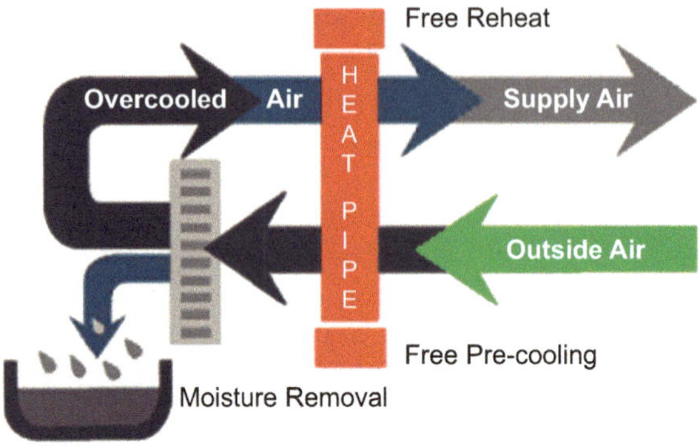

FIG. 5.3 Heat pipe application concept.

Heat pipes are tubes that have a capillary wick inside running the length of the tube, are evacuated and then filled with a refrigerant as the working fluid and are permanently sealed. The working fluid is selected to meet the desired temperature conditions and is usually a Class I refrigerant. Fins are similar to conventional coils - corrugated plate, plain plate, spiral design. Tube and fin spacing are selected for appropriate pressure drop at design face velocity. HVAC systems typically use copper heat pipes with aluminum fins; other materials are available.

Heat pipe heat exchanger enhancement can improve system latent capacity. For example, a 1 °F dry bulb drop in air entering a cooling coil can increase the latent capacity by about 3%. Both cooling and reheating energy is saved by the heat pipe's transfer of heat directly from the entering air to the low-temperature air leaving the cooling coil. It can also be used to precool or preheat incoming outdoor air with exhaust air from the conditioned spaces.

Advantages

- Passive heat exchange with no moving parts,
- Relatively space efficient,
- The cooling or heating equipment size can be reduced in some cases,
- The moisture removal capacity of existing cooling equipment can be improved,
- No cross-contamination between air streams.
- Shock/Vibration tolerant
- Freeze/thaw tolerant

Disadvantages

- Adds to the first cost and to the fan power to overcome its resistance,
- Requires that the two air streams be adjacent to each other,
- Requires that the air streams must be relatively clean and may require filtration.

Limitation

- ➢ Heat pipes must be tuned to particular cooling conditions. The choice of pipe material, size and coolant all have an effect on the optimal temperatures at which heat pipes work.
- ➢ When used outside of its design heat range, the heat pipe's thermal conductivity is effectively reduced to the heat conduction properties of its solid metal casing alone - in the case of a copper casing, around 1/80 of the original flux. This is because below the intended temperature range the working fluid will not undergo phase change; and above it, all of the working fluid in the heat pipe vaporizes and the condensation process ceases.
- ➢ Most manufacturers cannot make a traditional heat pipe smaller than 3 mm in diameter due to material limitations.

As we have indicated Overall, a heat pipe is a heat-transfer device that combines the principles of both thermal conductivity and phase transition to effectively transfer heat between two solid interfaces (Fig. 5.3).

Phase-change processes and the two-phase flow circulation in the Heat Pipe (HP) will continue as long as there is a large enough temperature difference between the evaporator and condenser sections. The fluid stops moving if the overall temperature is uniform but starts back up again as soon as a temperature difference exists. No power source (other than heat) is needed.

In some cases, when the heated section is below the cooled section, gravity is used to return the liquid to the evaporator. However, a wick is required when the evaporator is above the condenser on earth. A wick is also used for liquid return if there is no gravity, such as in NASA's micro-gravity applications. Fig. 5.4.

When a need for application of heat pipe in industry such Heat Pipe Heat Exchanger (HPHX) at high temperature arises, then (i.e., Application of HPHX for Fluoride-Salt Cooled and High-Temperature Advanced Fission Reactor operating in the range of 450−1100°C or some future Fusion Reactor), we need to utilize typical High Temperature Heat Pipe (HTHP) as passive heat transfer media, which is using superalloy envelope and wick/alkali metal working fluids such as cesium, potassium and sodium filled type heat pipes as illustrated in Fig. 5.5.

The working fluid is cesium, potassium, or sodium, depending on the operating temperature range. At higher temperatures, heat pipes use lithium as the working fluids, and refractory metals such as tungsten or molybdenum as the wick and envelope; See Fig. 5.6. The merit number for the alkali metals are much higher than the merit numbers for working fluids that are suitable at lower temperatures.

Ultra-High Temperature Heat Pipes illustrated in Fig. 5.6 have a refractory metal wick and envelope that are using Lithium as working fluid.

Note that, The envelope for this heat pipe is Titanium-Zirconium-Molybdenum (TZM) alloy, which has 0.5% titanium and 0.08%

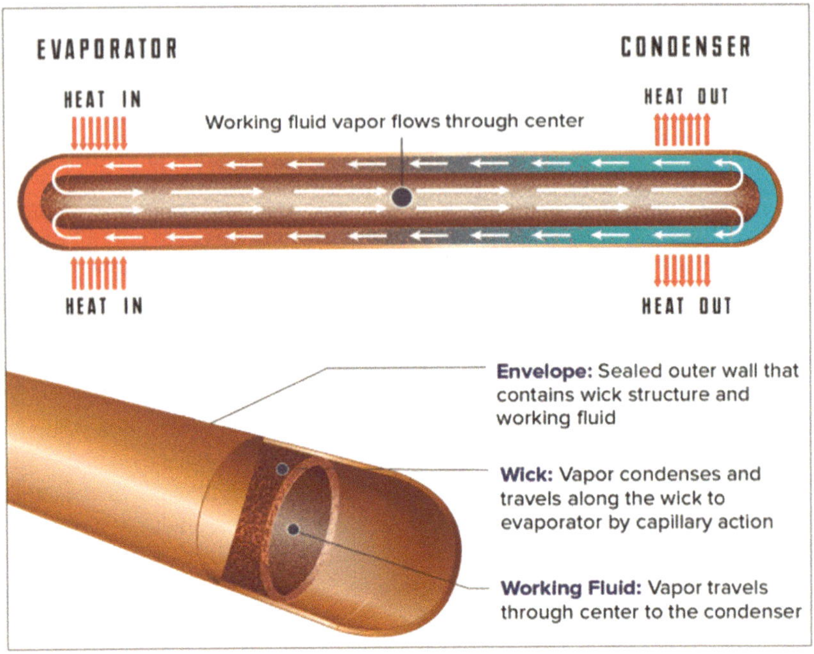

FIG. 5.4 Tope view depiction of heat pipe.

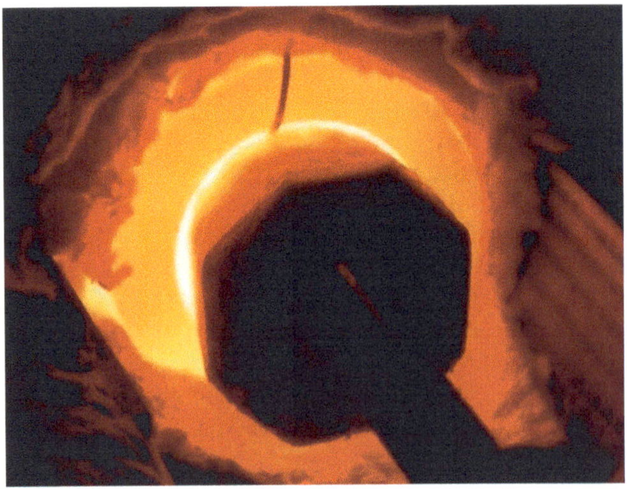

FIG. 5.5 High temperature isothermal furnace liners having a superalloy envelope. *Courtesy of Advanced Cooling Technologies.*

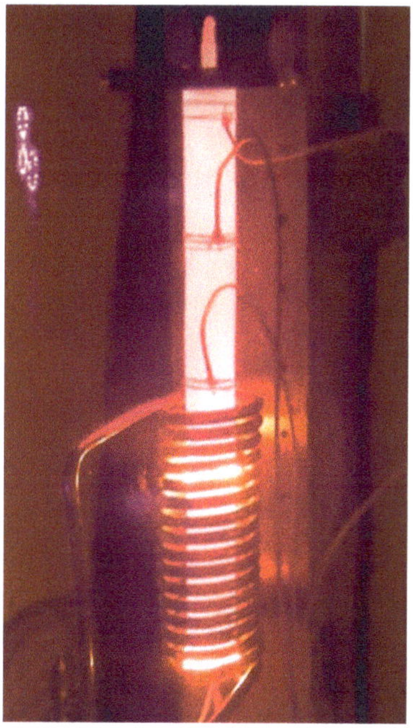

FIG. 5.6 Ultra-high temperature heat pipes. *Courtesy of Advanced Cooling Technologies.*

zirconium. Again, note that, Evaporation and condensation of the working fluid is what gives heat pipes their high effective thermal conductivity, which can be as high as 100,000 W/m K. During heat pipe operation, working fluid is vaporized in the evaporator, and then condensed in the condenser, transferring heat.

The first step in selecting a heat pipe working fluid and envelope/wick material is to determine the operating temperature range. For a heat pipe to operate, it must be at saturated conditions, where the heat pipe contains both liquid and vapor. Because of this, fluids can only operate (theoretically) between the triple (freezing) point and the critical point, where vapor and liquid phases have the same properties. In reality, the operating temperature range for any given fluid is smaller, since the power that the heat pipe can carry drops off sharply near the freezing and critical temperatures.

For example, a water heat pipe will carry some power between the water triple point (0.01 °C) and the critical point (373.9 °C). The practical operating temperature range for a copper/water heat pipe is roughly 25–150 °C. At lower temperature, fluid properties limit the heat transfer. At higher temperatures, a copper envelope to withstand the vapor pressure becomes too thick to be practical. Titanium and Monel envelopes extend the upper temperature limit to 300 °C.

The most commonly used envelope/fluid pairs include:

- Copper/Water for Electronics Cooling
- Copper or Steel/Refrigerant R134a for Energy Recovery
- Aluminum/Ammonia Spacecraft Thermal Control
- Superalloys/Alkali Metals (Cesium, Potassium, Sodium) for High Temperature Heat Pipes

There are a large number of other compatible envelope/fluid pairs that are used at other temperature ranges, or when additional factors must be considered. For example, copper/methanol is often used for electronics cooling when the heat pipe needs to operate near or below 0 °C, when water freezes. Fluid choices in a given temperature range are ranked by the Merit Number as it is presented in Eq. (5.1) and plot in Fig. 5.7 below:

$$N_l = \frac{\rho_l \sigma \lambda}{\mu_l} \tag{5.5}$$

where,

ρ_l: Liquid Density
σ: Surface tension as illustrated in Fig. 5.8 (Zohuri) [1].
λ: Latent Heat
μ_l: Liquid viscosity

High liquid density and high latent heat reduce the fluid flow required to transport a given power, while high surface tension increase the pumping capability. A low liquid viscosity reduces the liquid pressure drop for a given power.

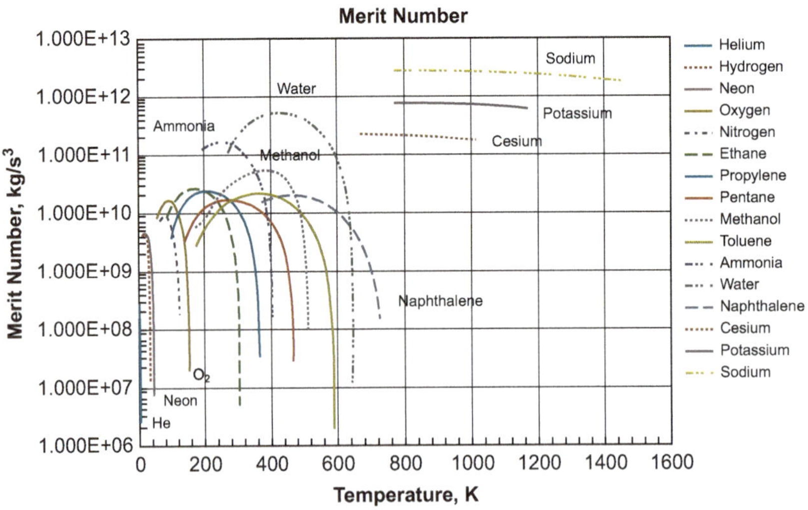

FIG. 5.7 Merit number for commonly used heat pipe working fluids. *Courtesy of Advanced Cooling Technologies.*

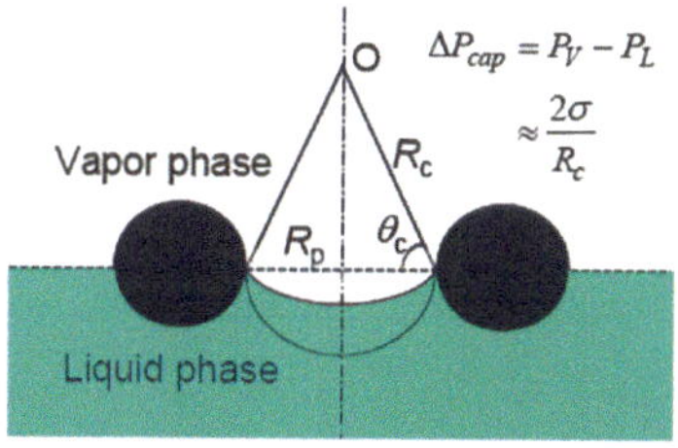

FIG. 5.8 Capillary pumping of the working fluid in heat pipe [1].

The Merit number as a function of temperature is shown in Fig. 5.6 for a number of typical heat pipe working fluids. From the figure, it is very clear why water is chosen as the heat pipe working fluid whenever possible. Its Merit number is ~10 times higher than everything else except the liquid metals, meaning that it will carry ten times more power (in the proper temperature range) than other working fluids. The Merit number is derived below (See Section 5.3.1).

Bear in mind that, High Temperature Heat Pipe (HTHP) materials of construction are typically Alloy 600 for Cesium, Potassium, and Sodium. Companies like Advanced Cooling Technology (ACT) also manufactures a Haynes 230/Sodium heat pipe for extended operation near 1100 °C because of the significantly higher creep strength. Austenitic stainless steels can be used for applications that are at the lower end of the temperature range.

Typical applications for high temperature heat pipes include the following:

- Heat Engine Receivers (Steam, Stirling, Brayton, Rankine)
- Solar Thermal
- Heat exchangers
- Hypersonic wing leading edges
- Waste heat recovery
- Nuclear power
- Thermoelectric Generators
- Isothermalizing furnace elements

High temperature heat pipes can be used to build custom heat transfer devices for both high power throughput and precise temperature uniformity. Power throughput in the 1–100 kW range is typical. Precision heat treating and materials processing furnaces are capable of extraordinarily precise temperature set points and isothermality. Set point accuracy, stability, and isothermality of ±0.1 °C is common with a single heated zone, using an off-the-shelf temperature controller.

Because of the inherent temperature uniformity and stability of high temperature heat pipes, they are an integral component in nearly all of the most precise temperature calibration instruments in the primary calibration laboratories around the world. This technology can also be applied to research, commercial, and industrial applications for processes such as annealing, sintering, crystal growing, brazing, and controlled diffusion.

5.3.1 Merit number derivation

The amount of power that a heat pipe can carry is governed by the lowest heat pipe limit at a given temperature (See Section 5.3.2 below). For a given heat pipe, the Merit number ranks the maximum heat pipe power when the heat pipe is capillary limited. (The capillary limit generally controls the power in the mid-range, while other limits control at higher and lower temperatures).

The capillary limit is reached when the sum of the liquid, vapor, and gravitational pressure drops is equal to the capillary pumping capability:

$$\Delta P_{\text{Capillary}} = \Delta P_{\text{Liquid}} + \Delta P_{\text{Vapor}} + \Delta P_{\text{Gravity}} \tag{5.6}$$

The Merit number neglects the vapor and gravitational pressure drops and assumes that the capillary pumping capability is equal to liquid pressure drop. The equation for the liquid pressure drop in a heat pipe is:

$$\Delta P_l = \frac{\dot{m}\mu_l L_{\text{Effective}}}{\rho_l k_{\text{Wick}} A_{\text{Wick}}} \tag{5.7}$$

where,

ΔP_l: Liquid Pressure Drop, assumed equal to the wick pumping capability
$L_{\text{Effective}}$: Effective Length
k_{Wick}: Wick Permeability
A_{Wick}: Wick Area

and the rest of parameters as defined before.
The mass flow rate is the heat transfer rate divided by the latent heat λ as:

$$\dot{m} = \frac{Q}{\lambda} \tag{5.8}$$

We also have the wick pumping capability using Fig. 5.8, written as:

$$\Delta P_l = \frac{2\cdot\sigma}{r_c} \tag{5.9}$$

Combining the three Eqs. (5.7) through (5.9) and solving for Q, the maximum heat transfer when only the liquid pressure drop is considered becomes:

$$Q = \frac{2A_{\text{Wick}}k_{\text{Wick}}}{r_c L_{\text{Effective}}}\ \frac{\rho_l\sigma\lambda}{\mu_l} \tag{5.10}$$

where the first term consists of heat pipe and wick properties, and the second term is the Merit Number as illustrated in Eq. (5.5) accordingly.

Note that the Merit number only ranks fluids based on the capillary limit. Close to the triple point, the sonic limit or viscous limit control the heat pipe power.

5.3.2 Lowest heat pipe limit driven temperature

The most important heat pipe design consideration is the amount of power a heat pipe is capable of transferring. Heat pipes can transfer much higher powers for a given temperature gradient than the best metallic conductors. The maximum power that the heat pipe can carry can be set either by the heat source and heat sink conditions, or by internal heat pipe limits.

Fig. 5.9 here shows; the temperature drops across a heat pipe.

The vapor space temperature drop is usually small compared to the temperature drops required to conduct heat into and out of the heat pipe.

In a properly designed heat pipe, the maximum power is set by the source and sink conditions. As shown in Fig. 5.9, the temperatures drops across a heat pipe are:

- Conduction through the envelope wall and wick
- Evaporation
- Vapor Space Temperature Drop
- Condensation
- Conduction through the envelope wick and wall

In addition, there are further temperature drops to bring the heat to the heat pipe evaporator, and reject the heat from the condenser, for example, using a finned heat sink and forced convection at the condenser. See Fig. 5.10, where heat pipe assembly used to move heat away from the hot side of Thermoelectric Cooler (ECT) to external heat sink.

Note that: ACT-TEC Thermoelectric Cooler Series Solid State Enclosure Air Conditioning – Solid state Peltier cooling, no compressors or refrigerants – Powder coated aluminum construction. – 316 stainless steel housing for NEMA 4X option – Precise temperature control with built-in adjustable thermostat – Low vibration, long life dual ball bearing fans. See Fig. 5.11.

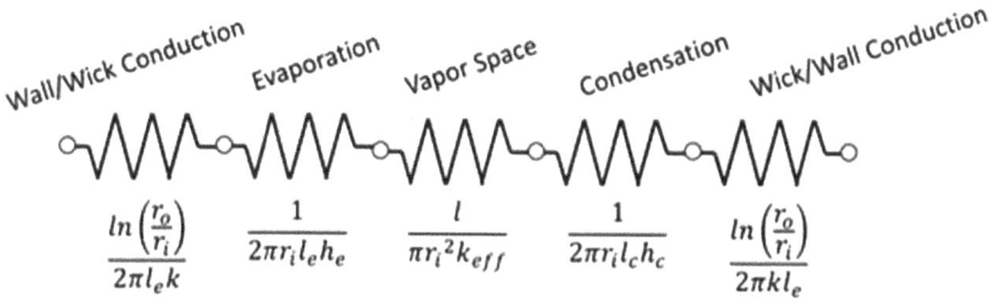

FIG. 5.9 Temperature drops across a heat pipe. *Courtesy of Advanced Cooling Technologies.*

FIG. 5.10 Finned heat pipe assembly used by ECT. *Courtesy of Advanced Cooling Technologies.*

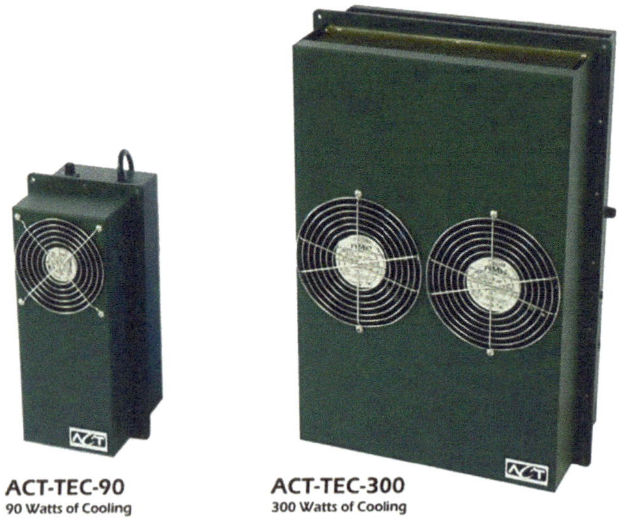

FIG. 5.11 ACT-TEC thermoelectric cooler series solid state enclosure air conditioning. *Courtesy of Advanced Cooling Technologies.*

However, if operating conditions cause the heat pipe to exceed its power capacity, the effective conductivity of the heat pipe will significantly reduce. Therefore, assuring heat pipes meet your maximum system requirements is a critical aspect of design.

As shown in Table 5.1, there are five primary heat pipe transport limitations that must be considered during design: viscous, sonic, capillary, entrainment/flooding and boiling; See Table 5.1. These limits are a function of many variables including operating temperature, wick selection and fluid properties. The most common limit for terrestrial applications is the capillary limit. ACT developed a heat pipe calculator to help customers design accordingly.

TABLE 5.1 Heat pipe and thermosyphon performance limits.

Heat pipe limit	Description	Cause
Viscous (Vapor Pressure)	Viscous forces prevent vapor flow within the heat pipe.	Heat pipe operating near triple point with a very low vapor pressure – need to use a different working fluid.
Sonic	Vapor flow reaches sonic velocity when leaving the evaporator, choking the flow.	Too much power at lower operating temperature. Typically, this is seen at start-up and will self-correct.
Heat Pipe Entrainment	High velocity vapor flow strips liquid from the wick.	Not enough vapor space for the given power requirement. Occurs at low temperatures.
Thermosyphon Flooding	High velocity vapor flow prevents liquid return in a gravity aided thermosyphon.	Not enough vapor space for the given power requirement. Occurs at low temperatures.
Capillary	The capillary action of the wick structure cannot overcome gravitational, liquid, and vapor flow pressure drops.	Power input too high. Wick structure not designed appropriately for power and orientation.
Boiling	Boiling occurs in the wick which prevents liquid return	High radial heat flux into the heat pipe evaporator.

Courtesy of Advanced Cooling Technologies.

To calculate the heat pipe performance limit, the different heat pipe limits are plotted as a function of temperature; See Fig. 5.12 (Top). Note that the viscous limit is not shown, since it is not relevant in the normal operating temperature range. The lowest limit at each temperature is then the heat pipe performance limit curve; See Fig. 5.12 (Bottom).

In Fig. 5.12, the top plot is the individual limits, showing that the entrainment and capillary limits are controlling over certain temperature ranges, while the bottom is the heat pipe performance limit curve, calculated by taking the lowest limit at each temperature.

The viscous, sonic, and entrainment/flooding limits are all related to the vapor velocity and are more significant at lower temperatures. The reason is that the vapor pressure and vapor density decrease as the temperature is lowered, so the vapor velocity must increase to carry the same power.

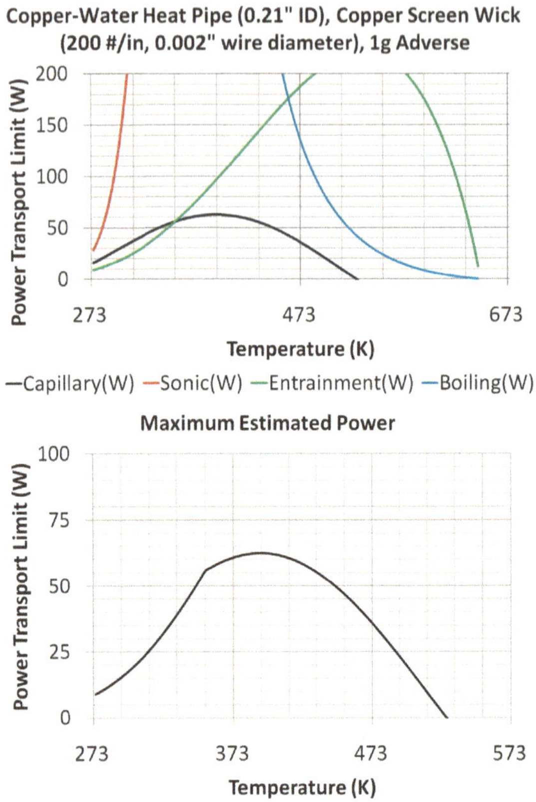

FIG. 5.12 Heat pipe performance limits. *Courtesy of Advanced Cooling Technologies.*

5.3.3 Heat pipe working fluids

Evaporation and condensation of the working fluid is what gives heat pipes their high effective thermal conductivity, which can be as high as 100,000 W/m K. During heat pipe operation, working fluid is vaporized in the evaporator, and then condensed in the condenser, transferring heat.

The first step in selecting a heat pipe working fluid and envelope/wick material is to determine the operating temperature range. For a heat pipe to operate, it must be at saturated conditions, where the heat pipe contains both liquid and vapor. Because of this, fluids can only operate (theoretically) between the triple (freezing) point and the critical point (See Fig. 5.13), where vapor and liquid phases have the same properties. In reality, the operating temperature range for any given fluid is smaller, since the power that the heat pipe can carry drops off sharply near the freezing and critical temperatures.

For example, a water heat pipe will carry some power between the water triple point (0.01 °C) and the critical point (373.9 °C). The practical operating

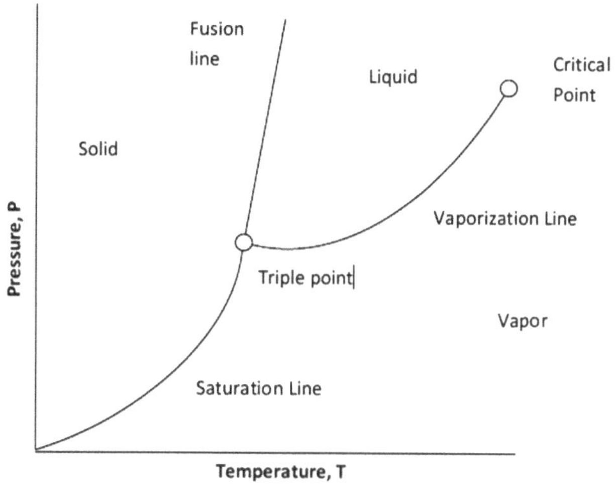

FIG. 5.13 Pressure-temperature diagram.

temperature range for a copper/water heat pipe is roughly 25–150 °C. At lower temperature, fluid properties limit the heat transfer. At higher temperatures, a copper envelope to withstand the vapor pressure becomes too thick to be practical. Titanium and Monel envelopes extend the upper temperature limit to 300 °C.

The most commonly used envelope/fluid pairs include:

- Copper/Water for Electronics Cooling
- Copper or Steel/Refrigerant R134a for Energy Recovery
- Aluminum/Ammonia Spacecraft Thermal Control
- Superalloys/Alkali Metals (Cesium, Potassium, Sodium) for High Temperature Heat Pipes

As discussed in Compatible Fluids and Materials there are a large number of other compatible envelope/fluid pairs that are used at other temperature ranges, or when additional factors must be considered. For example, copper/methanol is often used for electronics cooling when the heat pipe needs to operate near or below 0 °C, when water freezes. Fluid choices in a given temperature range are ranked by the *Merit Number*, which was defined earlier in this chapter.

5.4 Heat pipe heat exchangers, an innovation for heat transfer management

There are recent studies that are taking around Heat Pipe Heat Exchanger in recent past few years and companies like Advanced Cooling Technologies (ACTs) in United States is on the leading edge of heat exchanger technology development engaging in several innovative heat exchanger R&D programs, including:

- Direct-contact heat exchangers for highly efficient Heating, Ventilation, and Air Conditioning (HVAC) systems
- Variable-Conductance-Heat-Pipe (VCHP) heat exchangers that passively maintain outlet temperature in a small range for widely varying inlet temperatures and flow rates
- Innovative heat exchanger designs for vapor compression systems with thermal storage to accommodate varying thermal loads.

These programs utilize different thermal management technologies and demonstrate the breadth of heat exchanger technical expertize at companies like ACT. A brief synopsis of each of these advanced heat exchanger projects follows.

5.4.1 Direct contact heat exchangers highly efficient HVAC systems

ACT is developing a Vertical Direct-contact Heat Exchanger (VDHX) as illustrated in Fig. 5.14 here, for higher efficiency, lower mass HVAC systems. The VDHX is a modification of the momentum-driven vortex phase separator currently under development at ACT for microgravity applications.

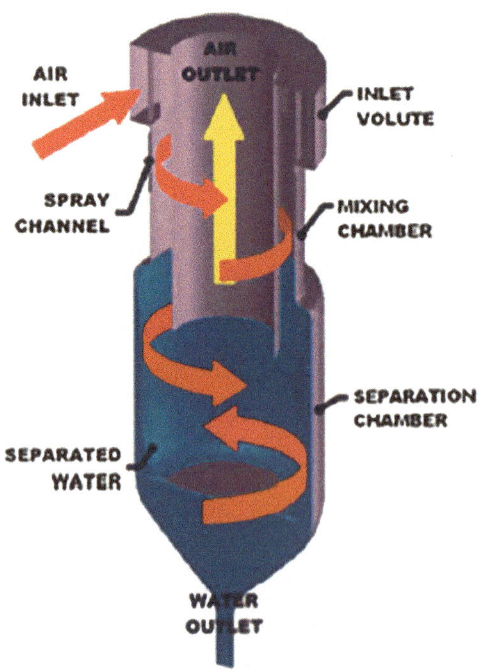

FIG. 5.14 Vertical direct-contact heat exchanger schematic. *Courtesy of ACT Corporation.*

The VDHX is shown in Fig. 5.14. During operation, air enters through the inlet volute, which centripetally accelerates the air flow and forms a high speed, forced air vortex. In addition to the induced vorticial motion, the air flow is directed in the axial direction and moves from the inlet volute into the mixing chamber. Chilled water is introduced in the mixing chamber as droplets generated by the spray channels. The spray channels are oriented such that the droplets enter the spray chamber in cross-flow with the air vortex. During their transit through the spray chamber, the droplets exchange thermal energy with the air stream by direct contact. The length of the spray chamber is designed such that the air reaches thermal equilibrium with the water before exiting this section. This results in significant cooling of the air and slight warming of the water.

Vertical Direct-contact Heat Exchanger (VDHX) provides a compact, lightweight heat exchanger for Heating, Ventilation, and Air Conditioning (HVAC).

Much like a conventional cooling coil system, condensation occurs if the air is cooled below the dew point. However, in a VDHX, condensation occurs at the droplet surfaces rather than on copper fins. In either case, once condensation occurs, the outlet air will achieve almost 100% Relative Humidity (RH). This air-water mixture, which maintains a strong vorticial motion, then flows from the spray chamber into the separation chamber. As the water-air mixture travels through the separation chamber, the centrifugal acceleration field developed within the vortex separates water, including condensate, from the air stream. The centrifugal acceleration experienced by water droplets within the VDHX is over 100 times greater than gravity. As a result, droplet transit occurs within tenths of a second, rather than the tens of seconds typical of a conventional direct contact heat exchanger. This allows the VDHX to minimize volume while maximizing throughput.

Together, these advantages provide an energy-efficient, low-maintenance HVAC heat exchanger with the following benefits compared to conventional finned-tube evaporators.

- Minimum possible temperature potential for heat transfer. This reduces the temperature lift required of and power consumed by the heat pump.
- Air conditioning by evaporative cooling when inlet conditions are appropriate. This allows the heat exchanger to provide cooling by latent heat exchange with the air. Operating in this mode will significantly reduce the heat load and power consumption of the heat pump.
- Freedom of material selection. High thermal conductivity materials are no longer necessary and noncorrosive, lightweight, recyclable, or low cost materials can drive the design instead.
- Continual recycling of the heat transfer surface. Particulate deposition, condensate build up, and biological growth, as well as the performance degradation associated with these, are eliminated.

- Filtration of submicron and larger particles. These particles are filtered by the water recirculation system, which greatly reduces the power consumption of the air recirculation system. Similar water filtration systems have been shown to remove 99% of particles greater than 0.5 μm in diameter, 96% of those 0.3–0.5 μm in diameter, and 86% of those smaller than 0.3 μm.
- Biological filtration. Combined with the intrinsic filtration of the water spray, low-power ultraviolet filtration of the water system allows removal of biologically active material from the recirculating air flow. Typical Ultra Violet (UV) filtration systems can destroy 99% of bacterial growth with less than a minute of exposure.

A Vertical Direct-contact Heat Exchanger (VDHX) sized to provide 2 tons of air conditioning (7 kW) was fabricated and tested. A schematic of the test setup is shown in Fig. 5.15, while Fig. 5.16 is a view through the volute top during operation. As shown in Fig. 5.17, the experimental system provided more than 2 tons of air conditioning.

Fig. 5.16 is the view of the Vertical Direct-contact Heat Exchanger (VDHX) test unit looking down through the volute top.

The following data were used by ACT to evaluate potential VDHX performance at over 70 locations across the United States. The results of this evaluation demonstrate the potential for significant benefits in performance and electrical usage when compared with a conventional system; see Table 5.2.

Table 5.2 was used in a VDHX in an air conditioner booster Coefficient of Performance (COP) from 3.0 to 3.9, reducing during peak times in the summer.

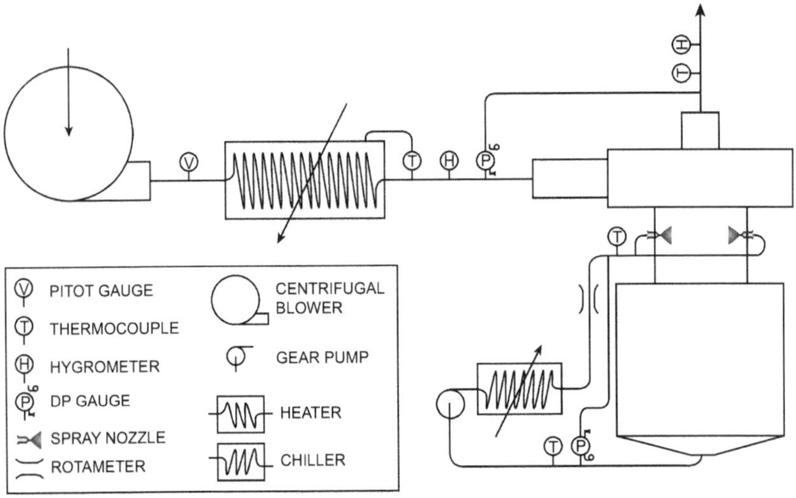

FIG. 5.15 VDHX demonstration test bed schematic. *Courtesy of ACT Corporation.*

FIG. 5.16 VDHX top-view. *Courtesy of ACT Corporation.*

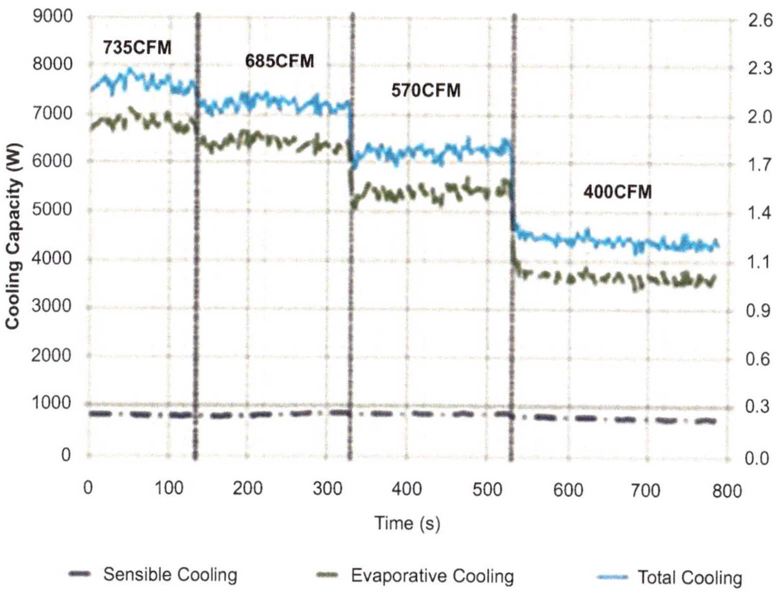

FIG. 5.17 Range of air flow rate cooling capacity. *Courtesy of ACT Corporation.*

Note that the coefficient of performance or **COP** (sometimes CP or **CoP**) of a heat pump, refrigerator or air conditioning system is a ratio of useful heating or cooling provided to work required. Higher **COPs** equate to lower operating costs. The star rating for air conditioners is determined differently to other appliances. For air conditioners, the measure of energy efficiency is the Energy Efficiency Ratio (EER) for cooling and the Coefficient of Performance

TABLE 5.2 VDHX evaluation data.

Result	AC system	AC with VDHX
Coefficient of performance (COP)	3.0	3.9
Annual energy consumption (kWh)	1627	1289
Annual cost ($)	$195.20	$154.71

Courtesy of ACT Corporation.

(COP) for heating. The EER and COP are defined as the capacity output divided by the power input.

5.4.2 Variable conductance heat pipe (VCHP) heat exchanger

VCHP Heat Exchanger for Passively Maintaining Outlet Temperatures in Chemical Reactors are used by industries such as United States Navy (USN), where it is investigating hydrogen fuel cells powered by reformed naval logistic diesel fuel as a means of providing distributed ship service electrical power. Hydrogen fuel cell operation using diesel fuel requires a reforming process to remove sulfur and steam reform the diesel fuel into a hydrogen rich stream. The operating temperature of the reactors must be closely controlled to maintain their chemical equilibrium. Temperature control is made more difficult than typical reforming systems because changes in the fuel cell electrical load and the resulting changes in reactant flow rates occur more frequently and drastically. The fuel reforming system must maintain inlet and outlet temperatures within ±30 °C despite a turndown ratio of 5:1 in reactant flow rate. A passive control scheme is needed to control the reactor temperatures within operational limits over all anticipated reactant flow rates.

Companies in United Sates of America (USA), has developed a Variable Conductance Heat Pipe (VCHP) Heat Exchanger to provide a roughly constant temperature feed to a reactor, despite variations in flow rate and outlet temperature from the previous reactor. A schematic of the VCHP heat exchanger is shown in Fig. 5.18. Heat from the gas stream is transferred by the VCHP to the coolant stream. The non-condensable gas in the VCHP is used to passively control the outlet temperature of the hydrogen. If the hydrogen temperature is too low, the non-condensable gas in the VCHP expands, which blocks more of the condenser, and reduces the heat transfer. Similarly, if the hydrogen temperature is too high, the non-condensable gas in the VCHP expands, which exposes more of the condenser, and increases the heat transfer.

As illustrated in Fig. 5.18 VCHP heat exchanger is used to passively maintain the outlet hydrogen temperature roughly constant over varying inlet flow rate and temperature.

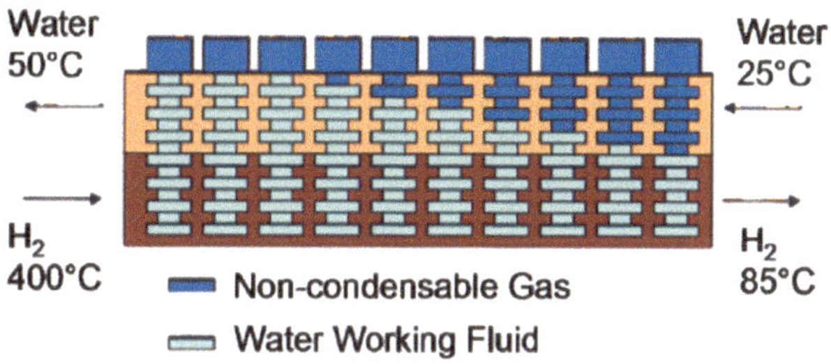

FIG. 5.18 A variable conductance heat pipe heat exchanger cross section schematic. *Courtesy of ACT Corporation.*

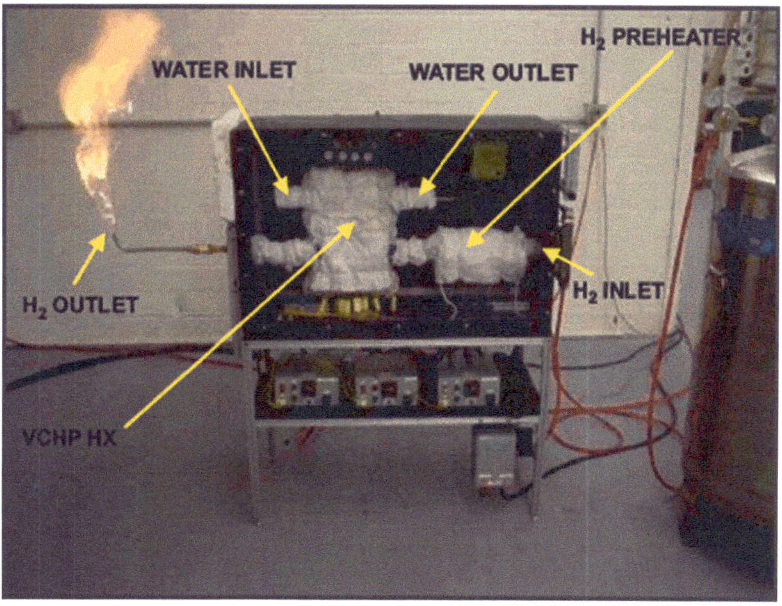

FIG. 5.19 A variable conductance heat pipe heat exchanger during testing condition. *Courtesy of ACT Corporation.*

The VCHP heat exchanger is shown in Fig. 5.19 with some details, while is in testing mode. Hydrogen entering the system first passes through a pre-heater, where it is heated to the desired inlet temperature. The hydrogen then passes through the VCHP heat exchanger, where it is cooled by a counter-current water flow past the top of the VCHPs.

Furthermore, the measured outlet hydrogen temperature is plotted against inlet hydrogen temperature in Fig. 8.20 for the files heat exchanger at a mass flow rate of 2.5 kg/h.

The predicted outlet temperature from the VCHP heat exchanger model is also shown.

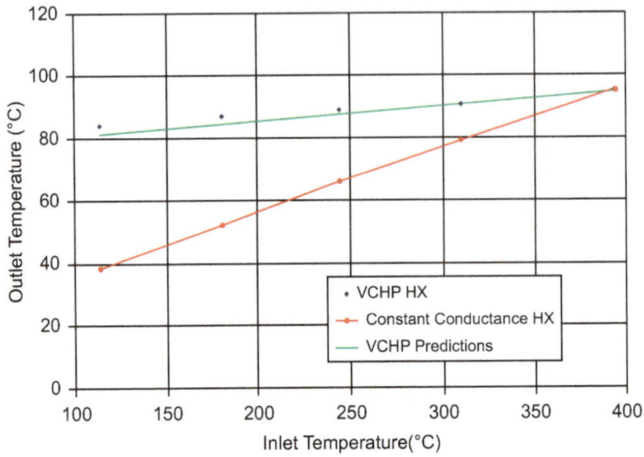

FIG. 5.20 The measure outlet hydrogen temperature plot. *Courtesy of ACT Corporation.*

5.4.3 Innovative heat exchanger designs

Innovative Heat Exchanger Designs for Vapor Compression Systems with Thermal Storage to Accommodate Varying Thermal Loads is discussed here.

There are several cases in which vapor compression systems must accommodate highly variable thermal loads, for example:

- A low, steady-state load must constantly be removed by the vapor compression system
- A transient load that is much higher than the steady state load must be removed for a short period of time. This transient load can be 20 times higher than the steady state load.

The brute force method to provide cooling for these cases is to size the vapor compression system for the highest thermal load experienced during the transient. However, this method has significant penalties in both size, mass, and electrical power. For example, both the required compressor size and the heat exchanger size will increase by a factor of almost 20 over the steady-state system. The approach ACT has taken is to add thermal storage to the system, which allows the compressor and primary heat exchanger to be sized for roughly ten-percent more capacity than the steady state case.

The system that ACT Corporation is developing has two key components (see Fig. 5.21):

- An integral heat exchanger that combines the evaporator, condenser and recuperator into a single heat exchanger. This approach significantly reduces the refrigeration system volume and mass. The use of a recuperator between the cooler vapor at the evaporator outlet and the hotter liquid at the condenser outlet ensures that a superheated vapor enters the

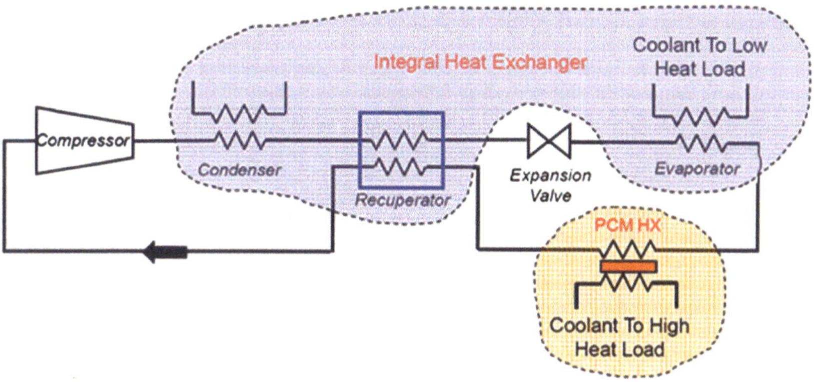

FIG. 5.21 Heat exchanger design with two key components. *Courtesy of ACT Corporation.*

compressor and increases the subcooling of the liquid entering the expansion valve. Both of these improvements contribute to increased system Coefficient of Performance (COP).

- A Phase Change Material (PCM) (See Section 2.10.4 of this book for more details on PCM) heat exchanger that stores the large amount of waste heat generated during the short operating peaks and dissipates the heat during steady state operation. This approach eliminates the need to oversize the compressor and other components in the refrigeration system, which results in significantly reduced system mass, volume and power consumption.

Fig. 5.21 is illustration of ACT corporation heat exchanger design that has two key components as:

1. Reduced-mass, integral heat exchanger that incorporate the condenser, recuperator, and evaporator, and
2. A heat exchanger with Phase Change Material (PCM) for thermal storage.

A more detailed schematic of the integral heat exchanger is shown in Fig. 5.22. Using an integrated heat exchanger:

- Reduces the number of flow connections and lines
- Reduces the system mass and size
- Improves System COP And Reliability

Fig. 5.22 is ACT's Corporation integral heat exchanger reduces mass and size, and improves Coefficient of Performance (COP). And reliability.

A system based on the schematic shown in Fig. 5.8 was modeled, fabricated, and successfully tested. The inclusion of a PCM heat exchanger and a recuperating heat exchanger reduced the overall mass by 36% while providing increased reliability and system efficiency.

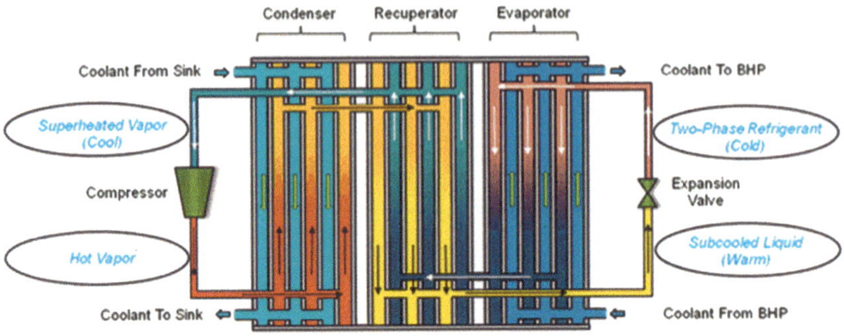

FIG. 5.22 ACT's integral heat exchanger artist schematic. *Courtesy of ACT Corporation.*

For purpose of Heat Pipe Heat Exchanger (HPHX), Some other modeling techniques include using a complex model where you model the liquid vapor interface, envelope wall, wick structure and vapor space separately. This is usually done for smaller models with high heat flux and custom wick structure. The two main resistances associated with heat pipes are the radial resistance and the axial resistance.

The radial resistance is the conduction through the wall, the wick and the two phase heat transfer at the vapor-condensate interface. This value is typically around 0.2 °C centimeters square per watt. The axial resistance is the vapor temperature difference across the length due to the temperature across the length due to the internal pressure difference. This is typically a very low number, about 0.02. This modeling technique is fairly challenging, especially in applications that must include several interfaces (gap pads, greases, etc.) and can lead to long computational run times.

One technique to cut down on computation time is a Mixed Model, which is a method which lumps the interface, wall and wick material into one material and uses a very highly conductive vapor space. This method again simplifies the two phase heat transfer into effective conductivity, but accounts for the radial resistance more accurately than the simplified model. The user can use hand calculations to determine the "lumped" thermal conductivity/ radial resistance. To calculate the $k_{effective}$ of the vapor space, use Fourier's Law as shown in the example below:

Using Fourier's Law to determine $k_{effective} = k_{eff}$ requires having the following information as:

- Known geometry
- Assumed power and ΔT
- Power (Q)
  - ✓ Known to be 25 W
- Effective Length (l_{eff})

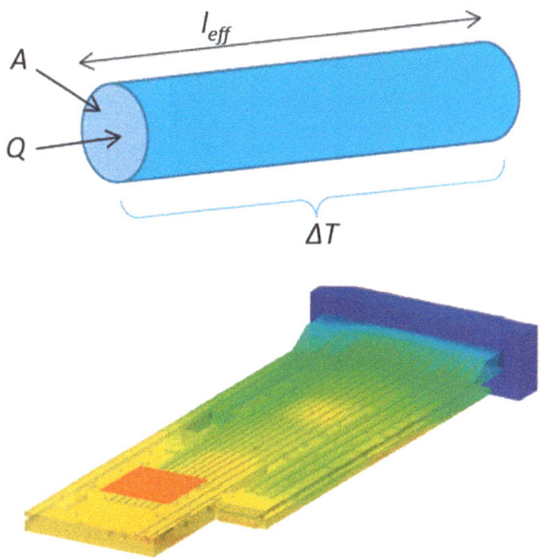

FIG. 5.23 Fourier's law depiction application.

And the relationship is written as Eq. (5.11), utilizing Fig. 5.23 as well.

$$k_{\text{eff}} = \frac{Ql_{\text{eff}}}{A\Delta T} \tag{5.11}$$

where the parameters use in above equation are presented in Fig. 5.23.

Here effective length l_{eff} for purpose of depicting Fig. 5.23, is given as Eqs. (5.12) in below as:

$$l_{\text{eff}} = \frac{l_{\text{evap}} + l_{\text{cond}}}{2} + l_{as} = 2.21 \text{ in} = 0.056 \text{ m} \tag{5.12}$$

and vapor area A is considered to be 4 mm Diameter less the 0.040″ modeled wall, thus $A = 3.004 \times 10^{-6}$ m [2]. For purpose of this model, Vapor ΔT is temperature drop of vapor is only due to pressure drop of vapor from evaporator to condenser (i.e., very low and conservatively assume 2K).

Given above assumptions for the model, the k_{eff} becomes as:

$$k_{\text{eff}} = \frac{25 \text{ (W)} \times 0.056 \text{ (m)}}{(3.004 \times 10^{-6}) \times 2 \text{ (K)}} = 233,000 \frac{\text{W}}{\text{m K}}$$

5.5 An overview of the heat pipe technology summary

A heat pipe is a two phase heat transfer device with a very high effective thermal conductivity. It is a vacuum tight device consisting of an envelope, a working fluid, and a wick structure. As shown in Fig. 5.24, the heat input

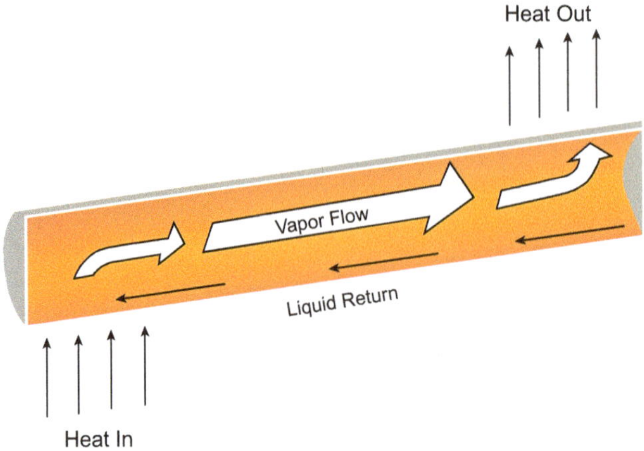

FIG. 5.24 Holistic illustration of heat pipe operation.

vaporizes the liquid working fluid inside the wick in the evaporator section. The saturated vapor, carrying the latent heat of vaporization, flows toward the colder condenser section. In the condenser, the vapor condenses and gives up its latent heat. The condensed liquid returns to the evaporator through the wick structure by capillary action. The phase change processes and two-phase flow circulation continue as long as the temperature gradient between the evaporator and condenser are maintained.

Benefits of these devices include:

- High Thermal Conductivity (10,000−100,000 W/m K)
- Isothermal
- Passive
- Low Cost
- Shock/Vibration tolerant
- Freeze/thaw tolerant

Read more about heat pipes in our Heat Pipe FAQ or download our Heat Pipe Reliability Guide. See a full video and transcription about the basics of heat pipes and their advantages by Advanced Cooling Technologies Corporation.

References

[1] B. Zohuri, Heat Pipe Design and Technology, Modern Applications for Practical Thermal Management, Springer Publishing Company, New York, NY, 2016.

[2] B. Zohuri, Combined Cycle Driven Efficiency for Next Generation Nuclear Power Plants: An Innovative Design Approach, Springer Publishing Company, 2015.

[3] S.W. Chi, Heat Pipe Theory and Practice: A Source Book, second ed., Hemisphere Pub. Corp., 1976.

[4] J.H. Rosenfeld, D.M. Ernst, J.E. Lindemuth, J.L. Sanzi, An Overview of Long Duration Sodium Heat Pipe Tests, Thermacore International, Inc., Lancaster, Pennsylvania 17601, 2004. Available electronically at: http://gltrs.grc.nasa.gov.

[5] J.-M. Tournier, M.S. El-Genk, Transient Analysis of the Startup of a Sodium Heat Pipe from a Frozen State, 1996. https://www.researchgate.net/publication/234935245.

[6] J. Bowman, R. Sweeten, Numerical heat pipe modeling, in: AIAA Paper 89-1705, AIAA 24th Thermophysics Conference, Buffalo, NY, 12-14 June 1989, 1989.

[7] J.H. Jang, A. Faghri, W.S. Chang, E.T. Mahefkey, Mathematical modeling and analysis of heat pipe startup from the frozen state, J. Heat Transf. 112 (1990) 586–594.

[8] M.J. Chang, L.C. Chow, W.S. Chang, M. Morgan, Transient behavior of axially grooved heat pipes with thermal energy storage, in: AIAA Paper 90-1754, AIAA/ASME 5th Joint Thermophysics and Heat Transfer Conference, Seattle, WA, June 1990, pp. 18–20.

[9] J.M. Doster, M.L. Hall, Numerical modeling of high temperature liquid metal heat pipe, in: ASME Paper 89-HT-13, Joint ASME/AIChE National Heat Transfer Conference, Philadelphia, PA, August 1989, pp. 5–8.

Chapter 6

Thermosyphon and heat pipe applications

Chapter outline

6.1 Introduction 308
6.2 Historical development and background of thermosyphon and heat pipe 309
6.3 Heat pipes and thermosyphon 314
6.4 Application of heat pipes and thermosyphon 316
6.4.1 Application in space systems 317
6.4.2 Application in cold regions 317
6.4.2.1 Ground temperature control 318
6.4.2.2 Snow melting and deicing system 319
6.4.3 Application in automobile industry 319
6.4.4 Application in railroad industry 319
6.4.5 Application in electrical, electronics, and nuclear industries 319
6.4.5.1 Electrical industry 319
6.4.5.2 Electronics 320
6.4.5.3 Heating and cooling 320
6.4.5.4 Passive decay heat removal system of the modular HTR 320
6.4.6 Summary statement 320
6.5 Thermosyphon design 321
6.5.1 Geometry 322
6.5.2 Working fluids 323
6.5.2.1 Lithium 324
6.5.2.2 Sodium 325
6.5.2.3 Potassium 325
6.5.2.4 Cesium 325
6.6 Mass flow rate and sonic velocity analysis 326
6.7 Heat transport limitations 327
6.7.1 Sonic limit (choking) of vapor flow 327
6.7.2 Viscous limit 328
6.8 Comparison of alkaline metals thermosyphon with convective loop 329
6.9 Thermosyphon startup 332
6.10 Two-phase instabilities in thermosyphon 333
6.10.1 Surging (chugging) and geysering instability 333
6.10.2 Thermosyphon evaporator instability 334
6.10.3 Fluid superheating (alkaline metals) 334
6.11 Nucleation sites 334
6.12 Inclination effects on a thermosyphon performance 337
6.13 Summary 341
References 342

Functionality, Advancements and Industrial Applications of Heat Pipes.
https://doi.org/10.1016/B978-0-12-819819-3.00006-7

6.1 Introduction

Recent technological developments in next generation nuclear reactors have created renewed interest in nuclear process heat for industrial applications. The Next Generation Nuclear Plant (NGNP) will most likely produce electricity and process heat for use in hydrogen (H_2) production or other technologies such as iron ore extraction, coal gasification, and enhanced oil recovery. A thermal device is needed to use process heat to transfer the thermal energy from NGNP to a hydrogen plant in the most efficient way possible. A conceptual schematic of an advanced nuclear reactor coupled to a hydrogen production plant is shown in Fig. 6.1.

The High Temperature Gas-cooled Reactor (HTGR) supplies thermal energy to drive a Brayton power cycle and process heat for High Temperature Electrolysis (HTE) or a thermochemical Sulfur Iodine (SI) hydrogen production process. Prismatic and pebble-bed advanced gas-cooled reactors are the primary reactor designs being considered to provide the high process heat required for hydrogen production.

As Fig. 6.1 shows, a portion of the hot helium outlet stream serves as the working fluid in a gas-turbine power cycle, and a separate helium stream flows through a high temperature heat exchanger, providing process heat to the hydrogen production plant. The heat transfer system is particularly challenging

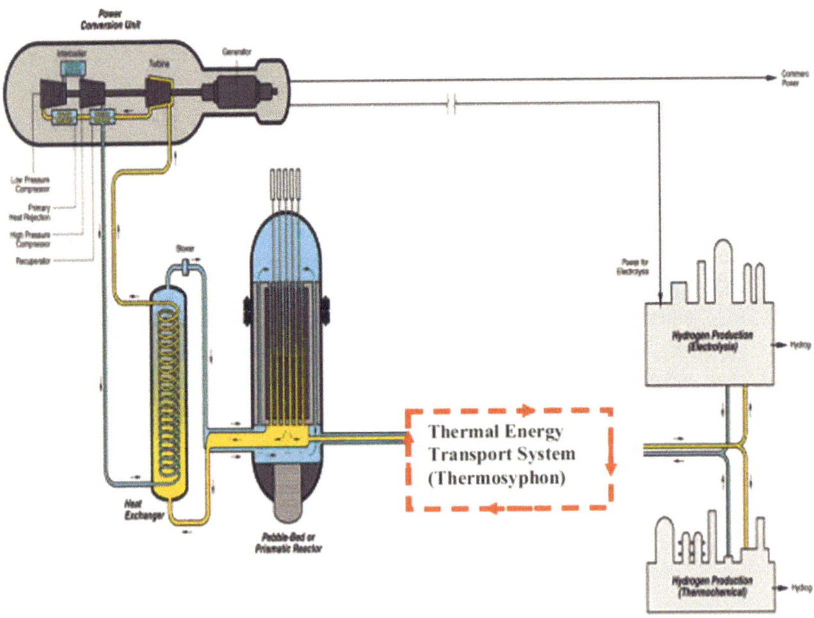

FIG. 6.1 NGNP with process heat transfer [1]. *Modified from NGNP Preliminary Project Plan 2007.*

because of elevated temperatures (up to 1300K), industrial-scale power transport (~50 MW), and the potentially large separation distance between the nuclear and industrial plants (100 + m) dictated by safety and licensing mandates.

Several options could transfer multi-megawatt thermal power over such a distance. One option is simply to produce only electricity, transfer it by wire to the hydrogen plant, and then reconvert the electric energy to heat via Joule heating. However, this option suffers energy losses of 60%–70% because of the thermal-to-electric conversion inherent in the Brayton cycle. A second option is thermal energy transport via a single-phase forced convection loop where a fluid is mechanically pumped between heat exchangers at the nuclear and hydrogen plants. However, the high temperatures required (up to 1300K) in this option present unique materials and pumping challenges. As previously determined by Davis et al. [2], low pressure helium (He) is an attractive option for NGNP but is not suitable for a single purpose facility dedicated to hydrogen production because low pressure helium requires higher pumping power and makes the process very inefficient. A third option is two-phase heat transfer utilizing a high temperature thermosyphon. One of the most significant advantages of heat transfer by thermosyphon is the characteristic of nearly isothermal, phase-change, heat transfer, which makes the thermosyphon an ideal candidate for applications where the temperature gradient is limited, and high delivery temperatures are required; such is the case in thermochemical and high temperature electrolysis hydrogen production.

The nature of isothermal heat transport results in an extremely high thermal conductance (defined as the heat transfer rate per unit temperature difference). The report presents design considerations for a thermosyphon system to transfer thermal energy from NGNP to the hydrogen production facility.

6.2 Historical development and background of thermosyphon and heat pipe

Natural convection refers to the process wherein heat, transferred to a fluid, raises its temperature and reduces its density, giving rise to buoyant forces that lift the fluid (due to density difference) and transport the absorbed heat to some other location where it can be removed. Natural convection occurs in a similar manner in two-phase systems. Here, the application of the liquid phase produces a low-density vapor that is free to rise through the liquid and condense at some other location. In either case, continuous circulation of the heat transfer fluid is maintained [3].

The Perkins tube, a two-phase flow device, is attributed to Ludlow Patton Perkins in the mid nineteenth century. As shown in Fig. 6.2, the Perkins tube, which was actually a single-phase, closed-loop thermosyphon, was used to transfer heat from the furnace to the evaporator of a steam boiler [4]. A demonstration of the excellence of this design is the air expansion tube, which

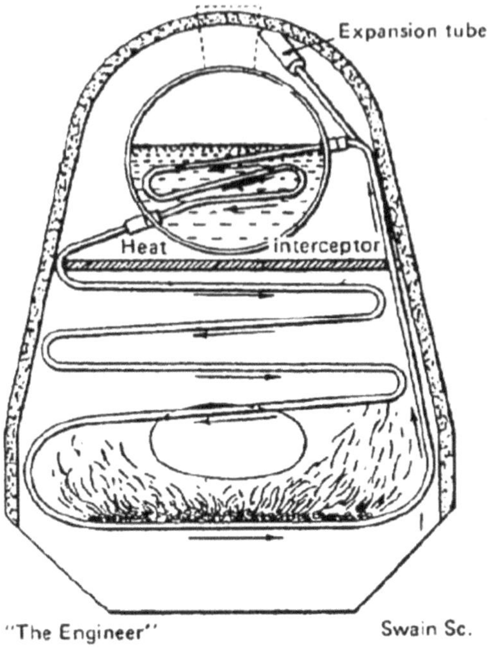

FIG. 6.2 Perkins boiler [4].

provides a space for the air inside the tube when the liquid (water) expands and also functions as a valve for regulating the operating pressure.

Early applications of the Perkins tube include steam generation, domestic heating, warming greenhouses, preventing window fogging, removing heat from dairy products, cooling car engines, and in heat exchangers. The development of modern thermosyphon technology and applications did not start until the 1940s. In 1942 and 1944, Gaugler [5] proposed a two-phase closed thermosyphon tube incorporating a wick or porous matrix for capillary liquid return. In 1963, Grover [6] studied this phase heat transfer device and named it "heat pipe" [7]. Tremendous effort has since been invested in thermosyphon and heat pipe research, resulting in broad applications.

The heat pipe differs from the thermosyphon by virtue of its ability to transport heat against gravity by an evaporation-condensation cycle.

Thermosyphon heat transfer has certain operating and limiting mechanisms that need to be considered before further discussing thermosyphon technologies and applications. Fig. 6.3 illustrates a typical two-phase closed thermosyphon, which consists of a metal pipe with a fixed amount of working fluid sealed inside. During operation, heat is added through the bottom section (evaporator) and the working fluid becomes vapor. The vapor travels through the middle section (adiabatic section) to the top section (condenser) of the tube. In the condenser, the vapor releases the latent heat to the condenser wall

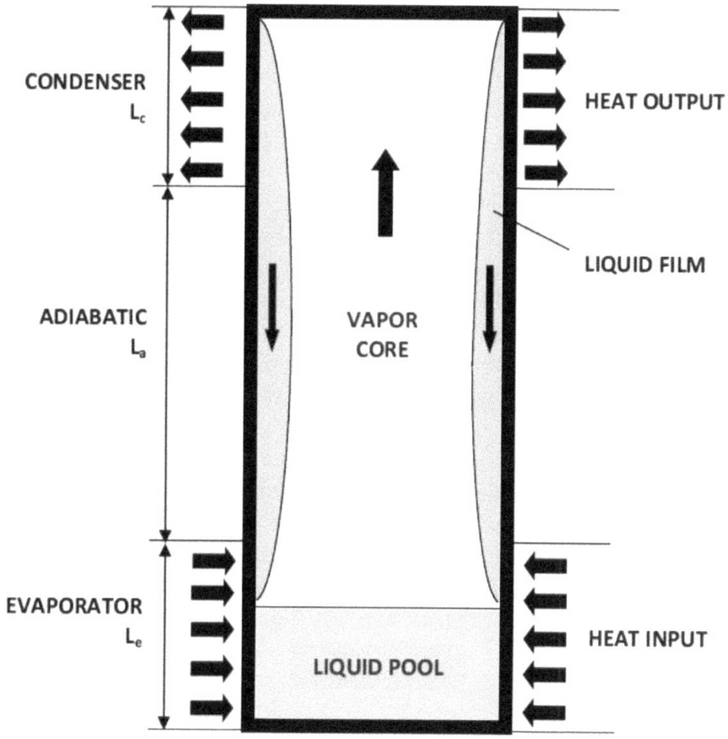

FIG. 6.3 A typical two-phase closed thermosyphon [1].

and becomes liquid. In contrast to a heat pipe, which utilizes capillary forces for liquid return, the thermosyphon relies on gravitational or centrifugal force to return the condensed liquid to the evaporator.

Heat transfer performance of the thermosyphon is a function of many factors, including properties of the working fluid, geometry and orientation of the thermosyphon, gravity field, and operating temperature or pressure. Fundamental heat transfer theory dictates that any mode of heat transfer is driven by a temperature difference and the larger the temperature difference ($\mathbf{T_{hot}} - \mathbf{T_{cold}}$), the higher the heat transfer rate. However, in many applications, it is desirable to transport large amounts of heat over a long distance, but at a relatively small temperature difference ($\mathbf{T_{in}} - \mathbf{T_{out}}$). Inside an operating thermosyphon, the vapor (at constant saturation temperature) carries a large amount of latent heat from the evaporator to the condenser. One of the most significant advantages of the thermosyphon heat transfer is the characteristic of nearly isothermal phase change heat transfer, which makes the two-phase thermosyphon an ideal candidate for applications where the temperature gradient is limited. The characteristics of isothermal heat transfer results in an extremely high transfer coefficient (defined as the heat transfer rate per unit temperature difference).

The adiabatic section, which serves as the passage for vapor and liquid flows, can have various geometric structures. The flexibility in geometry makes the thermosyphons more welcome in applications where the geometric accommodation is restrained by the system. Another important advantage of the thermosyphon is the adjustable heat transfer performance. Unlike conventional heat transfer devices, which usually cannot be regulated and consequently always remain "on" in the presence of a temperature gradient, a thermosyphons' performance can be adjusted by tilting, altering the thermal reservoirs (heat source and sink), and regulating of the condensate return. This characteristic of controllable heat transfer is favored in many applications including thermal radiators in spacecraft where the heat load may drastically vary. The phase change cycle inside the thermosyphon itself is self-actuated and requires no external power (other than the heat supply) or auxiliary equipment, which makes the thermosyphon a reliable and low-cost heat transfer device. Because of these advantages, the thermosyphon is well suited for applications with challenging operating conditions and high reliability requirements, such as transferring process heat from the NGNP to the production facility. There have been many different types of thermosyphons developed for various applications. In addition to the most common circular two-phase thermosyphons (as shown in Fig. 6.4), thermosyphons with

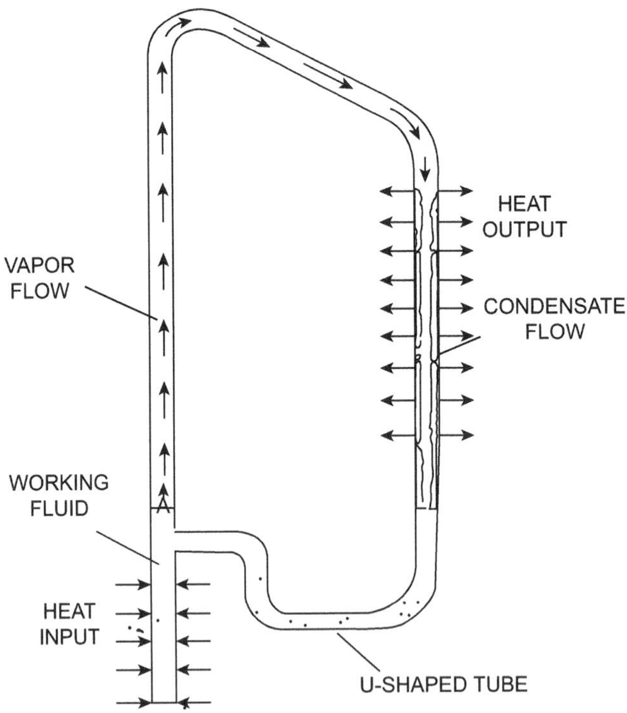

FIG. 6.4 Two-phase closed U tube thermosyphon loop [8].

triangular, rectangular, elliptic and concentric annular cross-sections have been widely used.

The only restriction on the geometric design of the condenser, adiabatic, or evaporator sections is that the condenser must be installed at a higher elevation than the evaporator to utilize gravitational force in returning the condensed liquid to evaporator. Thermosyphons have been designed with a vertical, horizontal, inclined, cranked, or flexible evaporator or condenser for various applications. Each section of the thermosyphon can have a different shape and dimension. Inside a counter-current type thermosyphon, the vapor core and the liquid film flow in opposite directions, resulting in liquid-vapor interfacial shear stress. The liquid-vapor interfacial shear stress can be very large for thermosyphons at high heat transfer levels because of the correspondingly large mass flow rate of the vapor. Once the interfacial shear force overcomes the gravitational force on the liquid film, the liquid flow may be reversed, and the flooding limit is reached. Many novel designs have been put forward to improve the flooding limited thermosyphon heat transfer capacity, which include an internal physical barrier along the adiabatic section by-pass line for liquid return, a cross-over flow separator, and others. The main advantage of these designs is that the liquid and vapor flows have partially separate passages, which can result in a higher flooding-limited heat transfer capacity.

To reduce the reliance of thermosyphon operation on gravitational force and thus promote broader applications, several designs have been patented or published. Feldman [9] invented a passive down-pumping thermosyphon and demonstrated his design later in the Heat Pipe Laboratory at the University of New Mexico. Feldman's system, as shown in Fig. 6.5, uses a three-way Passive Pumping Module (PPM) float control valve that allows switching the hot vapor flow to either the condenser during the heat transfer mode or to the accumulator during the pumping mode. The improvement in this design is that the evaporator can be located higher than the condenser. However, two main disadvantages include the discontinuity of heat transfer and reliability problems of floating control valves.

The work done here addresses industrial-scale recovery options for NGNP process heat. NGNP is a graphite-moderated, gas-cooled reactor intended to deliver high temperature fluids that can be used in generating electricity, producing hydrogen, and other process applications. In contrast to single-phase, forced convective heat transferred by pumping a fluid through a thermosyphon (also called a wickless heat pipe) transfers latent heat through the vaporization/condensation process. It uses a highly efficient, controllable, and nearly isothermal vapor heat transfer process with gravity liquid return and requires no pumps or compressors. It can deliver heat to the industrial plant with essentially no temperature loss. The work done here also analyzes the development of new heat transfer methodology—fluids (**Li** [Lithium], **Na** [Sodium], **K** [Potassium], and **Cs** [Cesium]), material needs for the construction of the thermosyphon, understanding the physics behind

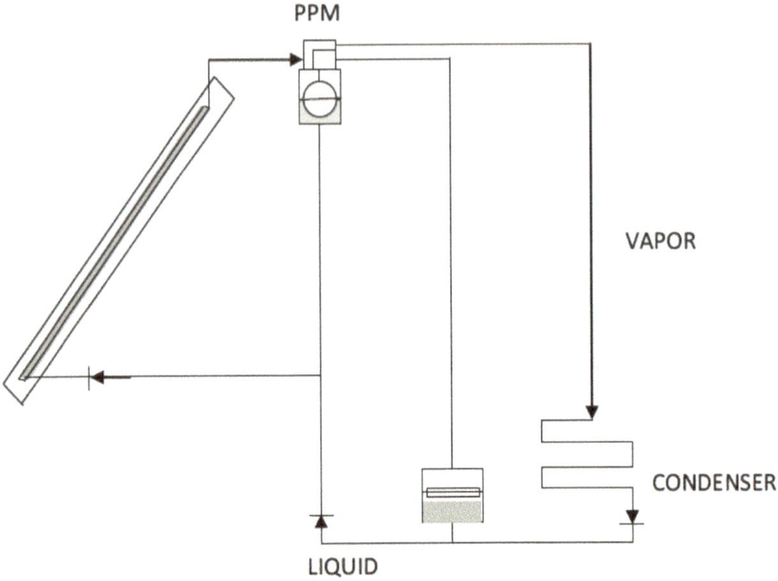

FIG. 6.5 Evaporator–controlled two-accumulator PPM systems [9].

thermosyphon performance based on thermodynamics, analyzing phase change liquid metal Spiral Heat Exchanger (SHE), which enables the thermal system to have enhanced efficiency and economic benefits. This research provides valuable insight and options useful in the design and development of Next Generation Nuclear Plant (NGNP) process heat recovery.

6.3 Heat pipes and thermosyphon

The heat pipe shown in Fig. 6.6 is essentially a constant temperature, heat transfer device. It consists of a closed container in which vaporization and condensation of a fluid takes place. The choice of a fluid depends on the

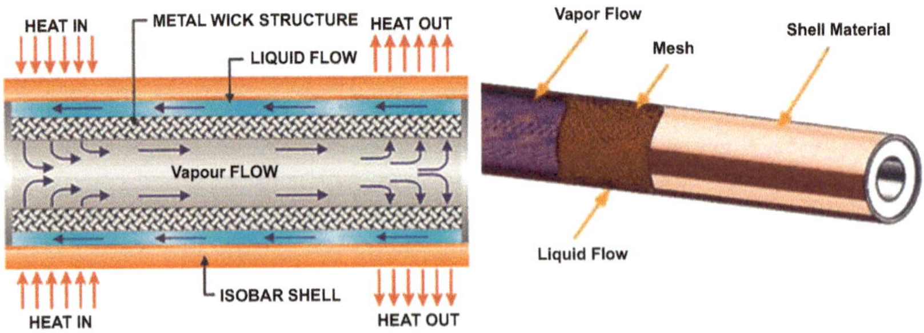

FIG. 6.6 The main regions of the heat pipe. *Courtesy of Acrolab www.acrolab.com.*

temperature range in which the heat pipe will be used. Heat is applied to one end of the heat pipe (evaporator), which raises the local temperature leading to evaporation of the working fluid. Because of the saturation conditions this temperature difference results in a difference in vapor pressure, which in turn causes vapor to flow from the heated section to the cold section of the pipe (condenser). The rate of vaporization is equal with heat absorbed in the form of latent heat of evaporation. The resulting condensate is returned to the heated end (evaporator) of the container by the action of capillary forces in the liquid layer, which is contained in a wick lining inside the cavity. A typical wick might consist of layers of metal screen or some porous metallic structure. A wick is used in the heat pipes to return the working fluid from the condenser to the evaporator.

Wicking material is used in regions to facilitate the path of the vapor to the pipe. Typically, a good wicking material maximizes the movement of the fluid, has uniform porosity, has very small pores such that the wick can generate a large capillary pressure, is resistant to degradation by temperature, and does not react or degrade chemically with the working fluid. Heat pipes can have a number of different geometric configurations. These configurations include cylindrical, spherical, square, or any other geometry such that inner volume of the heat pipe forms a channel from the evaporator section to the condenser section. Metals used to fabricate the heat pipes should be compatible with the working fluid as well as with the external media in contact with the evaporator and the condenser. The outermost shell of the heat pipe is referred to as the container. The container encloses the functioning parts of the heat pipe and provides structural rigidity. The liquid flow takes place in a porous material usually referred to as wick. The interior space of the heat pipe is called the vapor core, which provides passage for the vapor flow. Heat pipes have been used extensively in a variety of energy storage systems such as chemical reactors and space craft temperature equalization. Heat pipes are well suited to thermal storage systems, particularly in the roles of heat delivery and removal, because of their highly effective thermal conductivity and passive operation.

The heat pipe, as can be seen in Fig. 6.7, is similar to a thermosyphon in some respects; the main difference is the mechanism in which the fluid is returned from the condenser to the evaporator section. The operation of a heat pipe relies on the capillary head within the wick, which is sufficient to overcome the pressure drops associated with the liquid and vapor flow and the gravitational head. In a thermosyphon, the condensate is further returned to the hot end (evaporator) by gravity.

Since the latent heat of evaporation is large, a considerable amount of heat can be transported with a very small temperature difference from end to end. The main limitation of the thermosyphon is that the evaporator region has to be located at the lowest point in the system so that the condensate can be returned to the evaporator region by gravitational force. In the case of a heat pipe, a wick is constructed from a few layers of fine gauze fixed to the inside surface, and the

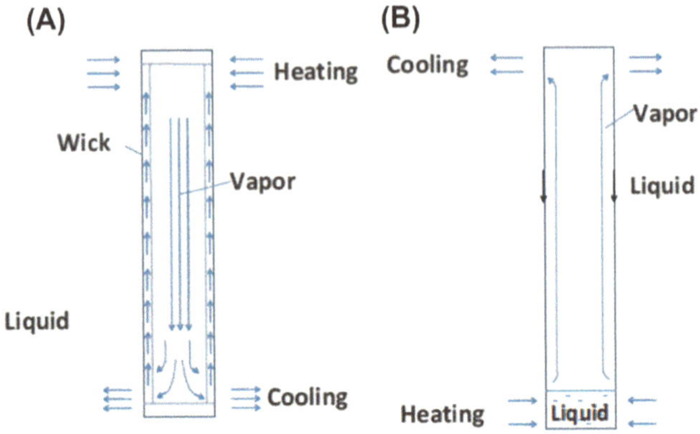

FIG. 6.7 The heat pipe and thermosyphon [4]. (A) Heat pipe. (B) Thermosyphon.

capillary forces return the condensate to the evaporator. The heat pipe has more flexibility in terms of evaporator location but having the evaporator region in the lowest position will allow gravitational forces to assist the capillary forces.

6.4 Application of heat pipes and thermosyphon

Thermal energy devices such as heat pipes and thermosyphons possess many advantages such as high heat recovery effectiveness, high compactness, no moving parts, and high reliability. Since the 1970s, heat pipes and thermosyphons have been extensively applied as waste heat recovery systems in many industries such as energy engineering, chemical engineering, and metallurgical engineering [10]. The heat pipe has been, and currently is being, studied for a wide variety of applications, covering almost the complete spectrum of temperatures encountered in heat transfer processes [4]. The ability of the heat pipe to transport heat over appreciable distances without any need for external power to circulate the heat transfer fluid is one of its most useful properties. Elimination of the fluid pump and power supply leads to greater reliability of the heat transport system and reduced weight, in addition to the saving in power consumption [3].

Heat pipes and thermosyphons have the following unique characteristics:

1. High heat transport capability due to latent heat
2. Small temperature variations
3. The ability to act as a heat flux transformer
4. Heat transfer in one direction only (thermal diodes and switches)
5. Passive heat transfer device
6. Separation of heating (heat source) and cooling (heat sink) parts
7. Heat transport capability through long distance
8. Constant temperature control (variable conductance heat pipe)
9. Heat transport with low temperature drop between heat source and sink.

Heat pipes and thermosyphons have the ability to transport very large quantities of heat with small temperature differences over relatively long distances. The applications of heat pipes and thermosyphons require heat sources for heating and heat sinks for cooling. The original development of the heat pipe and thermosyphon was directed toward space applications. The recent emphasis on energy conservation has promoted the use of the heat pipe and thermosyphon as a component in terrestrial heat recovery units and solar energy utilizations. For a thermosyphon the thermal resistance is smaller, the operating limits are wider (as in a heat pipe the integrity of the wick material might not hold at very high temperatures), and the fabrication cost is lower than that of the capillary heat pipe, which makes a thermosyphon a better heat recovery thermal device. Primarily, the most important aspect of the thermosyphon is that it can easily be turned off when required, whereas a heat pipe cannot be turned off. This added safety feature of a thermosyphon makes licensing of the NGNP process heat transfer system comparatively easier. The results of research and development on heat pipes and thermosyphons over the past 25 years are well documented.

Heat pipes are classified according to operating temperature ranges, as follows [4]:

1. Cryogenic heat pipes, $T < 200K$
2. Low temperature, $200 < T < 550K$
3. Medium temperature, $500 < T < 750K$
4. High temperature, $750 < T < 2800K$.

The physical size and configuration of heat pipes and thermosyphons can vary widely. Thermosyphons have become a unique and versatile heat transfer device with a seemingly unlimited range of applications. Examples of these applications are presented below, but technical details are not included.

6.4.1 Application in space systems

Space systems are one of the most important and original applications of heat pipes and thermosyphons. In 1968, heat pipes were used as components of the thermal control system in the American GEOS-2 satellite. China used heat pipes in a recoverable satellite in 1976 [11]. In the French-German TDF-1 and TV-SAT direct broadcasting satellite, a heat pipe network was incorporated for thermal control [12]. Japan has developed heat pipe payload radiators for its 2-ton geostationary satellite [13]. Heat pipes and thermosyphons were proposed to cool the leading edge of returning hypersonic vehicles to reduce drag forces [14].

6.4.2 Application in cold regions

There are many applications for the heat pipe and thermosyphon that are in existence in industry and few of them are presented in early chapters of this book and we can point out two of them here as a reference and they are presented in the two following sub-sections.

6.4.2.1 Ground temperature control

Heat pipes and thermosyphons were originally designed to transport heat in power plants for space vehicles in the 1960s, but the basic technology was applied to the largest terrestrial application project in the 1970s involving the extensive use of heat pipes for the stabilization of permafrost (perennially frozen ground) in the construction of the Trans-Alaska Pipeline System (TAPS). This well-known application of heat pipes runs.

1285 km long and transports crude oil from Prudhoe Bay in Alaska's North Slope to the ice-free port of Valdez on Alaska's south coast. To ensure foundation stabilization, the permafrost layers must be preserved during any season. Due to their large heat transfer efficiency and high reliability, heat pipes were chosen as the thermal control device for permafrost preservation. As depicted in Fig. 6.8, heat pipes extract heat from the ground during the winter, forming an ever increasing block of frozen soil bulb beneath the support pilings to prevent frost heave and differential thaw settlement. In the summer, the vertical heat pipes act as thermal diodes and prevent heat transfer into the soil.

FIG. 6.8 Typical vertical support member with heat pipes for trans-alaska oil pipeline.

Over 122,000 ammonia-charged closed heat pipes were installed along the entire pipeline [7]. Heat pipes were made of mild carbon steel; ammonia was chosen as the working fluid.

6.4.2.2 Snow melting and deicing system

Thermosyphons have been used to de-ice and melt snow on roofs, roadways, ramps, bridges, airport runways, etc. In 1970, Bienert et al. [15] proposed the use of gravity operated heat pipes to transport low grade energy from the ground beneath runways and highway structures to the surface in order to reduce or possibly prevent their icing. This idea has been tested by placing heat pipes in a small concrete slab at the Fairbank Highway Research Station in McLean, Virginia, and in a 366 m long interchange ramp in Oak Hill, West Virginia. The technical feasibility of the heat pipe extracting heat stored in the ground for de-icing and removing ice from pavement surfaces has been demonstrated [7].

6.4.3 Application in automobile industry

Mercedes-Benz has used heat pipes and thermosyphons in its passenger cars to recover exhaust heat for cabin heating and prevention of window fogging [16].

6.4.4 Application in railroad industry

The Hudson Bay Railroad in Canada runs 820 km from the Pas, Manitoba, to Churchill, Manitoba, on Hudson Bay. The route passes over seasonal, discontinuous, and permafrost terrain. Thaw settlement of localized zones, referred to as sinkholes, has occurred along the line. A total of 40 ammonia charged thermosyphons were installed at four test sites along the railroad right-of-way in 1978. Temperature monitoring has shown a decrease in ground temperature over a 4-year period indicating the successful application of thermosyphons to stabilizing the thaw settlement problem [7].

6.4.5 Application in electrical, electronics, and nuclear industries

The particular purpose of industrial application of heat pipes/thermosyphons is to save energy and material, raise the quality of products by conditioning thermal parameters, and intensify cooling in electro-techniques, mechanical engineering and chemical industry.

6.4.5.1 Electrical industry

So far, research in using heat pipes in electrical industry has been aimed at cooling electric motors, power semiconductor elements, power supply sources, acoustic techniques, magnets, plasma cathodes, etc. [17]. More and more heat pipes and thermosyphons are being used to cool electronic circuits on the micro scale. The original technique was to attach the thermosyphon or heat pipe directly to the hot spots. However, large contact thermal resistance can significantly affect heat transfer capacity. Some laboratory experiments have shown the feasibility of incorporating thermosyphons or heat pipes as integral parts of the circuit substrate by etching the flow channels directly into the silicon wafer to form the thermosyphons or heat pipes [18,40].

6.4.5.2 Electronics

Closed Loop Two-Phase Thermosyphon (CLTPT) involving co-current natural circulation was modeled and consisted for four major components: evaporator, rising tube, condenser, and falling tube [19].

6.4.5.3 Heating and cooling

A novel method for controlling and regulating wickless heat pipes has been developed and tested at the University of Central Florida Heat Transfer Laboratory. However, the unique design presented allows the system to be controlled during operation from the "off" to "fully on" positions, thereby permitting control of the heat transfer rate. In addition, the method of control can be extended so that a single heat pipe system may be used for both heating and cooling without requiring any physical modifications [20].

6.4.5.4 Passive decay heat removal system of the modular HTR

One of the key features of the Modular High Temperature Reactor (HTR) is the ability of passive decay heat removal under the primary system depressurization accident or the loss of forced circulation accident. The passive decay heat removal is performed by the cooling system installed outside the reactor pressure vessel. Decay heat of the core is passively transferred by radiation and convection to the reactor pressure vessel and then to the decay heat removal system. The evaporator part of the heat pipe is installed on the reactor cavity wall.

In the evaporator part, the heat from the reactor pressure vessel is changed into the latent heat of the working fluid and flows to the condenser part outside of the reactor building. The condenser part comprises heat exchanging pipes cooled by the atmospheric air through natural convection strengthened by the stack. Decay heat of the core is transferred to the atmosphere in this condenser part, and condensed working fluid returns to the evaporator part by gravity. In this system there are no active components and the core decay heat is removed by fully passive measures [21].

6.4.6 Summary statement

Based on the above discussion, one can concluded that heat pipes and thermosyphons have become unique and versatile heat transfer devices with seemingly unlimited applications.

6.5 Thermosyphon design

Heat pipes and thermosyphons have been recognized for decades as effective thermal devices for transporting large amounts of heat with small temperature gradients. The heat pipe differs from the thermosyphon only by virtue of its capillary wick and ability to transport heat against gravity; both utilize an evaporation-condensation cycle. The development of modern thermosyphon technology and applications started in the 1940s when Gaugler [5,22] proposed a two-phase closed thermosyphon tube incorporating a wick or porous matrix for capillary liquid return.

Grover [6] studied this heat transfer device and named it "heat pipe" as described by Cheng and Zarling [7]. Since then, considerable effort has been invested in thermosyphon and heat pipe development and has resulted in broad applications. One significant advantage of heat transfer by thermosyphon is the characteristic of nearly isothermal phase change heat transport, which makes the thermosyphon an ideal candidate for applications where the temperature gradient is limited and high delivery temperatures are required, as in the case of thermochemical hydrogen production [22]. The nature of isothermal heat transport results in an extremely high thermal conductance (defined as heat transfer rate per unit temperature difference). A schematic diagram of a thermosyphon system is shown in Fig. 6.9.

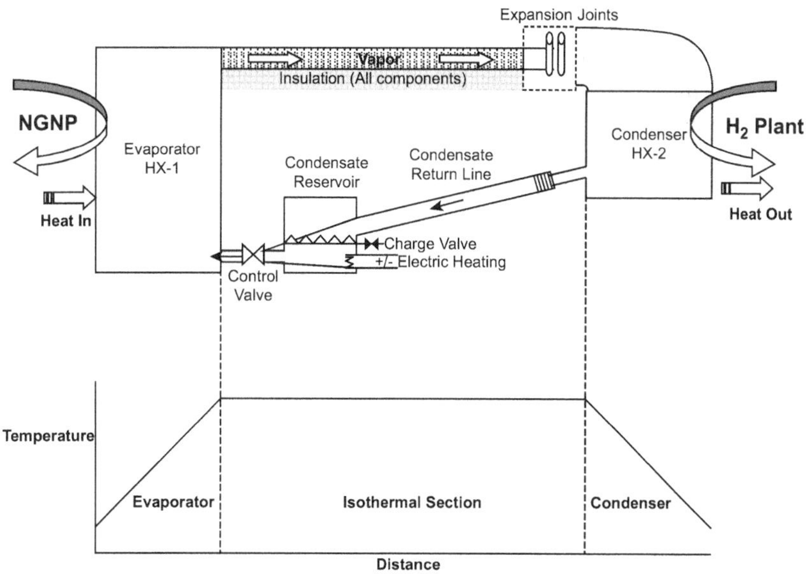

FIG. 6.9 Schematic of a simple controllable thermosyphon [22].

Reported within this section are design considerations for a thermosyphon system that transfers thermal power from NGNP to a hydrogen production facility.

6.5.1 Geometry

A thermosyphon is a two-phase heat transfer device that may be used to transfer process heat from the NGNP plant to the hydrogen plant. A controllable thermosyphon, conceptually illustrated in Fig. 6.9, is a wickless heat pipe with a separate liquid return line. It is an intriguing option to traditional pumped fluid heat transfer. Thermosyphons rely on convection to transport thermal energy inside pipes and high temperature heat exchangers for the evaporation and condensation end processes. Ideally, no pumping power is required in contrast to single-phase gas or liquid loops, which require compressors or pumps, both of which are problematic at very high temperatures.

Heat is transported by saturated or superheated vapor expanded from an evaporative heat exchanger, through a long pipe to a condensation heat exchanger. Liquid condensate returns to the evaporator assisted by gravity through a separate liquid return line with a liquid return control valve. When the thermosyphon is started by applying power (process heat from NGNP) to the evaporator, the working fluid is evaporated, and the latent heat of vaporization is transported ($\sim$isothermally) along the thermosyphon to the condenser region. Expansion joints are added near the condenser section and also at the inlet to the condensate return line in order to accommodate the thermal expansion of the thermosyphon piping at higher temperature. The condensate returns to the evaporator region through a liquid bypass line containing a liquid storage reservoir and a control check valve as shown in Fig. 6.9.

The storage reservoir and part of the liquid lines may incorporate electric resistance heating, if necessary, in order to melt the working fluid and restart the thermosyphon after a long shutdown period. Liquid from the storage reservoir passes into the thermosyphon system evaporator through a control valve, which, as needed, plays a role in controlling the rate of heat transfer and shutting off or isolating the thermosyphon. In order for the thermosyphon system to be shut and completely disabled from heat transfer, the control valve is closed wherein all the working fluid is collected in the liquid storage reservoir and the condensing-evaporating cycle is terminated. When it is desired to resume the thermosyphon action, the control valve is opened to again allow the liquid to flow into the evaporator region of the system. The heat input governs the rate of evaporation and the subsequent rate of heat transfer. The rate of thermal energy exchange can be regulated over a spectrum of conditions from "OFF" to "FULLY ON," hence the term "controllable thermosyphon".

Another salient feature that a thermosyphon construction material must accommodate is thermal expansion. The change in the length with temperature for a solid material can be expressed by the common equation for thermal expansion of a homogeneous material as:

$$\frac{\Delta L}{L_o} = \alpha_l \Delta T \qquad (6.1)$$

where,

ΔT: Superheat Difference
ΔL: Length Change
L_0: Initial Length
α_l: Linear coefficient of thermal expansion (K^{-1})

For a 100 m long thermosyphon made from a high temperature Ni alloy ($\alpha_l = 13.3 \times 10^{-6}$) heated from ambient temperature to 1300K, the expansion is 133 cm. Therefore, in order to accommodate the expansion, either corrugated joints or expansion joints (as shown in Fig. 6.10) must be incorporated.

6.5.2 Working fluids

Depending on the temperature and pressure of operation, favorable working fluids can be identified.

Alkaline metals, for example may be suited for NGNP process heat transfer, because they have the characteristics of:

- High boiling temperature
- Availability and cost effectiveness

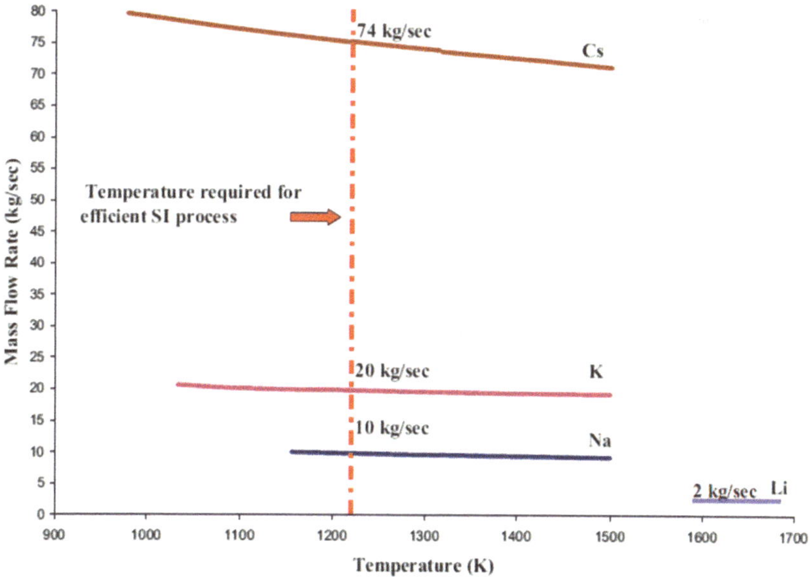

FIG. 6.10 Mass flow rates of alkaline metal vapors, required to transport 50 MW [1].

- Good heat transfer properties (latent and specific heat are both high)
- Typically, good chemical compatibility (except lithium).

Working fluids more suitable than alkali metals may exist, such as molten salts. Corrosive behavior at high temperatures or lack of high temperature thermodynamic properties, especially for superheated vapors, rule out fundamental analysis of many possible thermosyphon working fluids. Thermophysical properties of the alkaline metals at atmospheric pressure are given in Table 6.1.

More salient information on the working fluids is presented below.

6.5.2.1 Lithium

Lithium (Li) is the rarest alkaline metal next to cesium (Cs). It is the least dense of the normally solid elements and is the least typical and most reactive of the alkaline metals. Lithium is harder than other alkaline metals but is softer than lead. Lithium offers interesting characteristics as a heat-transfer fluid in high-temperature systems. It is the lightest metal and has a comparatively high conductivity, high specific heat, high boiling point, and moderately low melting point. Its low density permits high fluid velocities without encountering high pressure drops in the system. At high temperatures, lithium reacts with carbon to form acetylides that hydrolyze to give acetylene, which may be explosively flammable. Near the melting point, lithium may ignite in air and burn with an intense cloud of white smoke, the principal product is the monoxide Li_2O. Lithium cannot be melted in glass or in the usual laboratory ceramics because it severely attacks them. The noble gases, thoroughly freed of oxygen, water, and nitrogen, must be used to prevent contact of lithium with the atmosphere. Stainless steel has been considered unsuitable for handling liquid lithium at high

TABLE 6.1 Properties of alkali metals [23].

Working fluids	Melting temperature (K)	Boiling temperature (K)	h_{fg} (kJ/kg, 1223K)	ρ_v (kg/m^3, 1223K)	Comments
Li	452	1590	21156	2.17E-03	Least favorable
Na	371	1156	3864	4.72E-01	Most favorable
K	337	1033	1794	2.03	Favorable
Cs	301	978	440	12.46	Less favorable

h_{fg}, latent heat of vaporization; ρ_v, vapor density.

temperatures; liquid lithium attacks stainless steel and dissolves nickel [23]. Most likely the material used for the construction of the thermosyphon will be an alloy of nickel, because of its durability and sustainability at high temperatures. In our analysis Li is discarded as a coolant because its boiling point is higher than the operational temperature (1000 °C) of the thermosyphon.

6.5.2.2 Sodium

Solid Sodium (Na) is a relatively soft, silvery-white metal. Molten sodium is a silvery-white metal whose outstanding characteristic is its reactivity with most gases or liquids other than the noble gases and nitrogen. Solid Na tarnishes almost immediately when exposed to air, owing to the formation of a film of oxide. Molten sodium burns readily in the normal atmosphere to form dense flumes of sodium monoxide. With pure oxygen, molten sodium burns with a yellow flame, forming a mixture of sodium monoxide and sodium peroxide. The reaction of sodium with water is energetic and possibly explosively rapid if the contact interfaces are large. Sodium and heavier alkaline metals do not react with nitrogen. Sodium reacts vigorously with halogens, acidic oxides, and mercury, and alloys with lead, tin, zinc, and bismuth. Nitrogen, argon, and helium do not react with sodium, and these gases should be used to prevent the atmosphere from contacting sodium, either in the solid or the liquid state. Sodium does not react with glass at temperatures less than 300 °C; above this temperature, Pyrex glass is rapidly attacked [23].

6.5.2.3 Potassium

Potassium (K) reacts similarly to sodium with the exception that, in general, potassium is more reactive. When exposed to oxygen, potassium oxidizes to superoxide KO_2. The superoxide of potassium will form at low temperatures, but the cause of explosions with combinations of potassium superoxide and potassium metal is not completely understood. When potassium reacts directly with carbon monoxide, an explosive carbonyl is formed, unlike in sodium [23]. Whereas lithium and sodium react only superficially with liquid bromine, potassium detonates when brought into contact with it. Ordinary potassium is the lightest naturally occurring radioactive element; it contains 0.011% beta or gamma emitting 19K40 with a half-life of 2.4×10^8 years.

6.5.2.4 Cesium

Cesium (Cs) is silvery-white in appearance and is very soft and ductile. It is the most electropositive of all the metals and has a high specific gravity. It catches fire in dry air and, in general reacts the same as the other alkaline metals. Cesium absorbs carbon monoxide at room temperature [23].

6.6 Mass flow rate and sonic velocity analysis

A sodium-filled thermosyphon can transport comparable amounts of thermal energy as a single-phase convective loop, within the same diameter pipe. The ideal rate of convective heat transport Q'' through a pipe without losses, modeled in terms of enthalpy, can be written as:

$$Q'' = \Delta h \rho V = \frac{\dot{m}}{A}\Delta h \tag{6.2}$$

where,

V: Average flow velocity (m/s)
Δh: Specific enthalpy change of the transport fluid (kJ/kg)
ρ: Density of the fluid (kg/m^3)
$\dot{m}$: Mass flow rate (kg/s)
A: Cross-sectional flow area (m^2)

Two-phase heat transfer by a thermosyphon has the advantage of high enthalpy transport, which includes the sensible heat of the liquid, latent heat of vaporization, and possible vapor superheat. In contrast, single-phase forced convection transports only the sensible heat of the fluid. Additionally, vapor phase velocities within a thermosyphon can be much greater than single-phase liquid velocities within a forced convective loop. For a working fluid with temperature averaged properties, the rate of transport is:

$$Q'' = \frac{\dot{m}}{A} C_{P,L} \Delta T_L \quad \text{(single phase)} \tag{6.3}$$

$$Q'' = \frac{\dot{m}}{A}\left(\underbrace{C_{P,\text{ Liquid}}\,\Delta T_{\text{Liquid}}}_{\text{Sensible heat of the liquid}} + \underset{\text{Latent heat of vaporization}}{h_{fg}} + \underbrace{C_{P,\text{ Vapor}}\,\Delta T_{\text{Vapor}}}_{\text{Vapor Superheat}} \right) \tag{6.4}$$

where,

$C_{P,\text{Liquid}}$: Heat Capacity for Liquid (kJ/kg·K)
ΔT_{Liquid}: Liquid Temperature Difference (K)
$C_{P,\text{Vapor}}$: Heat Capacity for Vapor (kJ/kg·K)
ΔT_{Vapor}: Vapor Temperature Difference (K)

Q is assumed to be 50 MW (process heat from NGNP for hydrogen production), and the necessary mass flow rate for all the different alkaline metals is compared from:

$$\dot{m} = \frac{Q''A}{C_{P,\text{ Liquid}}\ \Delta T_{\text{Liquid}} + h_{fg} + C_{P,\text{ Vapor}}\ \Delta T_{\text{Vapor}}} \tag{6.5}$$

where

$$\Delta T_{\text{Liquid}} = T_{\text{Boiling}} - T_{\text{Inlet}}$$

$$\Delta T_{\text{Vapor}} = T_{\text{Outlet}} - T_{\text{Boiling}}$$

The inlet temperature of Na is assumed at 393K (based on minimum temperature required to carry out Bunsen reaction in the SI process) and the outlet temperature of Na is assumed at 1223K, which will be needed for better efficiency of the SI process. From Table 2.1, Li can be discarded as the potential working fluid because it does not boil until 1590K at atmospheric pressure. Li is also discarded from consideration by virtue of its high chemical reactivity with Ni alloys; likely material for the thermosyphon system. Fig. 6.10 shows the minimum mass flow rates needed for alkaline metal vapors to supply 50 MW at 1223K as determined by Eq. (6.5), using thermophysical properties from Bystrov et al. [24].

If the alkaline metal vapor reaches sonic velocity, shock waves develop and propagate inside the thermosyphon. In order to avoid sonic shock, a thermosyphon has to operate at vapor velocity lower than sonic velocity, even though lower velocities reduce the heat transport capability. For a thermosyphon to operate near the sonic velocity with Na (Sonic Velocity $V_{\text{Sonic}} \sim 737$ m/s at atmospheric pressure) and transport 50 MW, it would require an internal pipe diameter of approximately 0.19 m. At half the sonic velocity the diameter required is 0.268 m.

For 50 MW, the lowest of mass flow rates is observed in Na, when compared to other alkaline metals. Further, the diameter of the pipe is calculated such that the respective mass flow is attained at the sonic velocity limitation: for Na the diameter obtained is 0.19 m. Similarly, the diameters for K and cesium are 0.15 and 0.16 m, respectively.

6.7 Heat transport limitations

Although the thermosyphon is a very high thermal conductance device, it possesses limitations governed by principles of heat transfer and fluid mechanics. Depending upon operational conditions, heat transport may be limited by sonic limit (choking) of vapor flow or viscous limit [25,38] as discussed below.

6.7.1 Sonic limit (choking) of vapor flow

After continuum flow is established, the evaporator-condenser pressure difference accelerates the vapor until it reaches a maximum velocity at the evaporator exit. The maximum vapor velocity that can exist at the evaporator exit corresponds to sonic velocity, or Mach 1. The limitation of such flow is similar to a converging-diverging nozzle with constant mass flow rate, where the evaporator exit corresponds to the nozzle throat. This choked flow condition is a fundamental limit on the axial vapor flow in a thermosyphon. The maximum axial heat flux at the sonic limit is obtained by calculating the mass flow rate at sonic velocity V_{Sonic}:

$$\frac{Q}{A_{\text{Vapor}}} = \rho_{\text{Vapor}} h_{fg} V_S \tag{6.6}$$

For sodium, the theoretical maximum heat transfer rate is ~1338 MW/m^2 at 1223K. Sodium is one fluid that meets the thermodynamic challenges of NGNP process heat transfer. The speed of sound in the vapor sodium is estimated from the vapor adiabatic compressibility (β_{Soinc}) and the vapor density (ρ_{Vapor}) using the thermodynamic relation given by Fink and Leibowitz [25], which was verified with values given by Bystrov et al. [24] as:

$$V_{\text{Sonic}} = \frac{1}{\sqrt{\rho_{\text{Vapor}} \beta_{\text{Soinc}}}} \tag{6.7}$$

Sonic velocity values for Na vapor in this analysis are from Fink and Leibowitz [25]. At the sonic limit, the mass flow rate per unit area and the corresponding axial heat flux depend only on the properties of the working fluid and, in turn, the operating temperature.

Reay and Kew [4] present an equation that gives the limiting axial heat flux at sonic conditions ($V_{\text{Sonic},0}$−806 m/s) in terms of the stagnation temperature:

$$\frac{Q}{A_{\text{Vapor}}} = \frac{\rho_0 \; h_{\text{fg},0} \; V_{\text{Sonic},0}}{\sqrt{[2(\gamma + 1)]}} \tag{6.8}$$

Where, γ is Ratio of Specific Heat or some books or literature is called Adiabatic Index.

All the fluid properties in Eq. (6.8) are stagnation properties and the maximum heat transfer rate obtained is 1097 MW/m^2. Once the sonic limit is reached, further increases in the mass flow rate or the heat transfer rate can occur only by increasing the stagnation pressure upstream of the choking point.

This limitation is similar to the viscous limitation in that it is typically encountered during startup or over-power transients but does not represent the failure of the system. This limitation was developed for heat pipes with internal annular wicks but should also apply to vapor flows within a thermosyphon.

6.7.2 Viscous limit

At startup for liquid metals, the vapor pressure difference between the evaporator and the condenser is zero or very small. In such cases, the viscous forces may be larger than the vapor pressure gradients and thus prevent vapor flow. This condition is the viscous limitation. It is a function of heat pipe dimensions and vapor conditions in the evaporator [26] and can be expressed as:

$$\frac{Q}{A_{\text{Vapor}}} = \frac{\rho_0\ h_{\text{fg},0}\ P_0}{64\mu L} \tag{6.9}$$

Where, μ is Dynamic Viscosity (N s m^{-2}; or lbft-1 h^{-1}) and P_0 is Vapor initial pressure.

This limitation is mainly for heat pipes with a wick but should also apply to a thermosyphon. The maximum viscous limit heat transfer rate per unit area with Na is much higher than the sonic limit mainly because of the low viscosity (μ ~1.86_10_5 N s/m^2) of Na vapor. Therefore, the viscous limit should not be a concern within a large-scale sodium thermosyphon.

The vapor pressure of a given working fluid plays a major role in determining the maximum allowable length of a thermosyphon. The following analysis determines the length of the thermosyphon such that it can transfer the process heat from the nuclear plant to the process plant for a given thermal load of 50 MW. The distance between the nuclear plant and the process plant is directly proportional to the operating temperature and thermal properties of the working fluid, specifically the vapor pressure of the fluid. The thermosyphon length is therefore determined as:

$$L = \frac{2\ P_{\text{Vapor}}\ D}{4 f \rho_{\text{Vapor}}\ V^2} \tag{6.10}$$

Where higher vapor pressure yields a longer possible thermosyphon, thus more separation distance is possible between the nuclear and process heat plant. Fig. 6.11 shows the plot between half of sonic velocity for various alkaline metals with respect to temperature and the allowable length between the NGNP and hydrogen plant. Half sonic velocity is selected to avoid propagating sonic shock waves in the system and to reduce erosion potential. Also, in this plot, no pressure drop is assumed, i.e., ($P_{\text{Vapor}} \cong \Delta P$). Fig. 6.12 shows the plot of varying length with the change in temperature for an allowable pressure drop.

6.8 Comparison of alkaline metals thermosyphon with convective loop

Alkaline metal thermosyphons and alkaline metal forced convective loops can both deliver comparable rates of heat transfer through a given size pipe. This can be demonstrated by considering the ideal rate of convective heat transport through a pipe without losses, modeled in terms of enthalpy by Eq. (6.2).

Two-phase heat transfer by a thermosyphon has the advantage of high enthalpy transport when compared with single-phase forced convection.

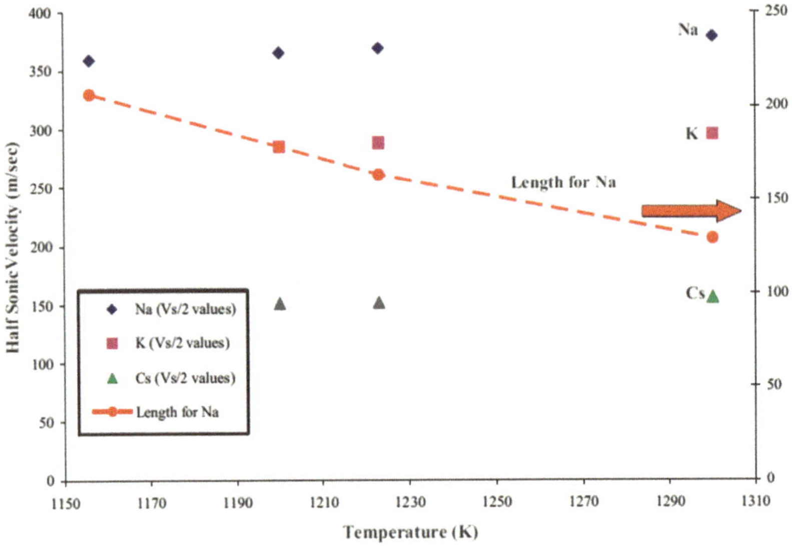

FIG. 6.11 Maximum separation distance possible between NGNP and hydrogen production facility [1].

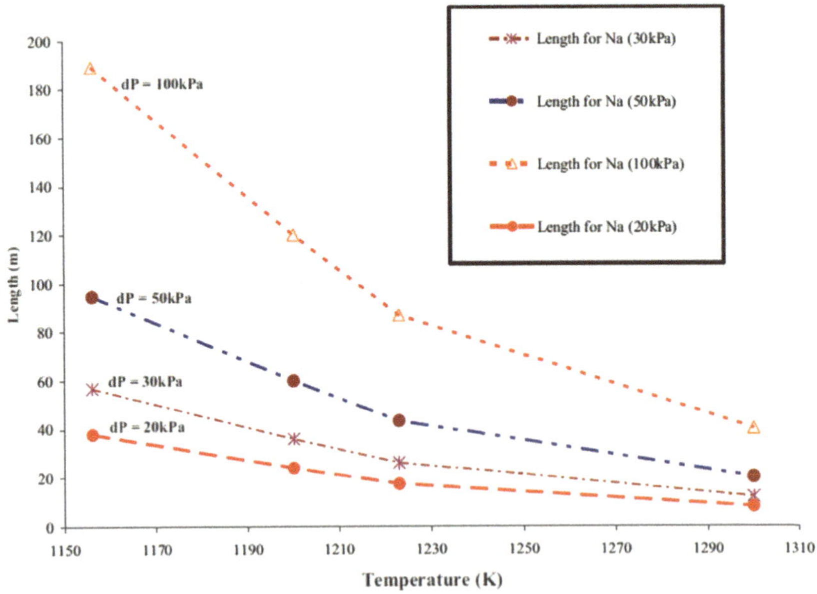

FIG. 6.12 Maximum separation distance possible for an allowable pressure drop [1].

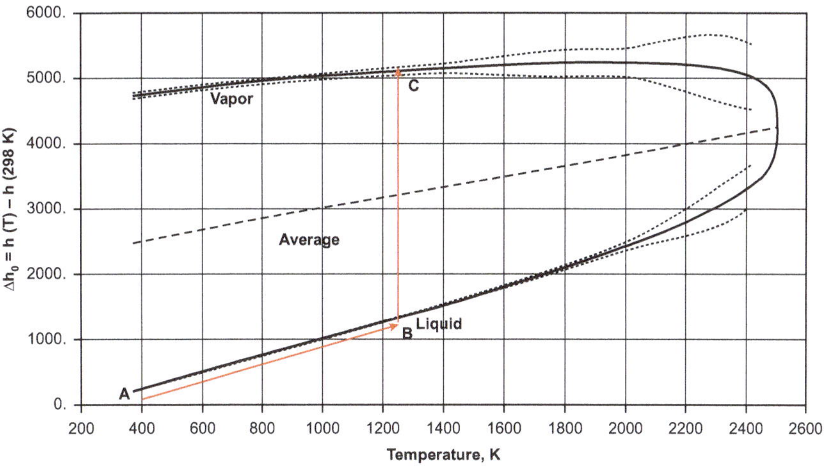

FIG. 6.13 Enthalpy for saturated sodium: liquid and vapor (Thermodynamic data from Fink and Leibowitz [25], Gunnerson et al. [39]).

Additionally, vapor-phase velocities within a thermosyphon can be much greater than single-phase liquid velocities within a forced convective loop.

Fig. 6.13 exemplifies the enthalpy enhancement in heat transfer afforded by a two-phase thermosyphon versus a single-phase convective loop with sodium as the working fluid. The specific enthalpy (Δh) of saturated liquid and vapor, relative to the solid at 298.15K, is illustrated as a function of temperature. Assuming heat transfer from an HTGR to an industrial facility at 1223K, the maximum single-phase heat transfer is given by the enthalpy gain from points A to B in Fig. 6.13, or approximately 1190 kJ for each kilogram of sodium. Compare this with two-phase heat transfer from points A to B to C where the enthalpy gain is approximately 3864 kJ per kilogram with no vapor superheat—over three times more heat per kilogram of sodium than the single-phase. The saturation pressure of sodium at 1223K is only 0.188 MPa, thus pressure and stress forces are minimized. Vapor flow through a pipe is limited by compressible choke flow when the vapor reaches its sonic velocity. The sonic velocity for sodium vapor is approximately 737 m/s at 1223K as given by Bystrov et al. [24].

The limiting heat transfer rate for an ideal sodium thermosyphon operating at 1223K can therefore be estimated by Eq. (6.2):

$$Q'' = h_{fg}\rho_{Vapor}V_{Sonic} = (3864)\text{kJ/kg}(0.47)\text{kg/m}^3(737)\text{m/s} = 1338\text{MW/m}^2$$

Similarly, single-phase liquid sodium could transport the same rate of thermal energy with an average flow velocity of about 2.2 m/s, well within the capabilities of advanced liquid metal pumps. This simple analysis for sodium as the working fluid theoretically illustrates that both a thermosyphon and a forced convective loop can deliver comparable rates of heat transfer through comparable diameter pipes. The thermosyphon, however, has the luxury of controllable heat transfer without the need for high temperature pumping and can deliver the heat at the same approximate temperature as the source. For sodium, the enthalpy gain that can be achieved by two-phase heat transfer, instead of a single phase, is about 3.7 times greater. Ideally, a well-insulated 19 cm diameter sodium thermosyphon could transport ~50 MW of power at 1223K to a hydrogen production facility some distance (100 + m) away. In practice, a larger diameter thermosyphon would reduce the vapor velocity below the limiting sonic velocity and be beneficial in reducing the adverse effects from shock wave prorogation and erosion within the thermosyphon.

6.9 Thermosyphon startup

The charging of a thermosyphon requires a transfer station wherein molten working fluid under an inert environment or vacuum is transferred to the evacuated thermosyphon pipe(s). Pure fluids without condensable gases are required for proper thermosyphon operation. Otherwise, the impurities, which are more volatile than the fluid itself, will be driven to the condenser section of the thermosyphon and less volatile impurities will be collected in the evaporator causing hot spots and reducing heat transfer. Non-condensable gases will accumulate within the condenser. Although conceptually simple, the startup of a large high-temperature thermosyphon is difficult to accurately predict. If the working fluid solidifies at ambient temperature, it must first be melted to flow into the evaporator. The subsequent boiling behavior will be dictated by the thermophysical nature of the working fluid and the nucleation characteristics of the evaporator. Fig. 6.14 describes the procedure commonly practiced for filling up the thermosyphon. The liquid Na valve is opened till 20% by volume limit is reached for filling up the thermosyphon. The 20% by volume is a good approximation for the coolant as described by Gunnerson and Sanderlin [20]. Melting and boiling temperatures define the operating limits of a thermosyphon fluid. Clearly defined upper and lower operating temperature bounds are therefore required for proper selection of the working fluid. Upper temperature limits can challenge the pressure containment design and may influence working fluid degradation and materials compatibility. The minimum nonoperating temperature can affect the thermosyphon startup behavior, especially if operation is to be initiated from a frozen state.

For a thermosyphon to startup efficiently and effectively the working fluid has to initially be in a molten state. The liquid reservoir (as shown in Fig. 6.14) has provisions for external heating. During normal start-up for both heat pipes and thermosyphons, the temperature of the evaporator section increases by a few degrees until the thermal front reaches the end of the condenser as described in Reay and Kew [4]. At this point, the condenser temperature will increase until the pipe structure becomes almost isothermal.

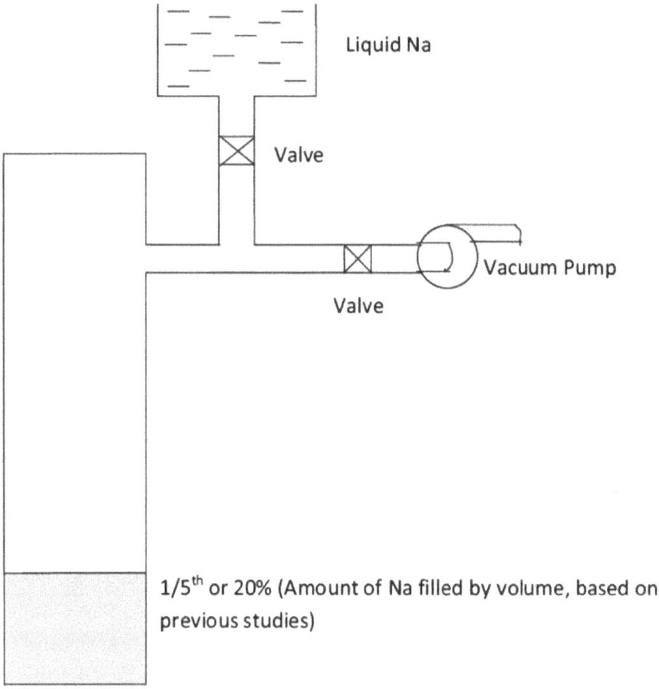

FIG. 6.14 Na Filling procedure for a thermosyphon [20].

6.10 Two-phase instabilities in thermosyphon

Instabilities are common to both forced and natural circulation systems; the latter is more unstable than forced circulation systems. If the system stabilizes to a new steady state or oscillates with increasing amplitude, the system is considered unstable as explained by Sabharwall et al. [41]. Instabilities can cause operational problems in process heat transfer. Thus, it is important to classify mechanisms that can lead to unstable operational behavior of the thermosyphon. The following processes can lead to an unstable behavior for the thermosyphon.

6.10.1 Surging (chugging) and geysering instability

Surging and geysering occur mainly because of liquid superheat. Surging occurs when boiling is initiated in the evaporator, but because of no uniformity in the temperature at the wall and bulk fluid temperature, the vapor being generated becomes trapped, eventually resulting in vapor expulsion as described by Bergles et al. [27]. This mechanism is not destructive but can trigger other types of instability in the system. This phenomenon occurs more readily with liquids having good wettability such as sodium on steel at high temperatures. Geysering is a similar phenomenon that occurs when the heat flux is sufficiently high, and boiling is initiated at the bottom. In low pressure systems this results in a sudden increase in vapor generation due to the reduction in hydrostatic head, and usually causes an expulsion of vapor. The liquid then returns, the subcooled, nonboiling condition is restored, and the cycle starts over again. Geysering is also a naturally occurring phenomenon. A greater number of active nucleation sites in the evaporator section would help avoid surging and geysering in the system.

6.10.2 Thermosyphon evaporator instability

If the evaporator section of the thermosyphon system is not sufficiently long for vapor superheat, then instability could occur such that the fluid at the outlet of the evaporator experiences a static pressure decrease, leading to the onset of fluid condensation within the thermosyphon. Slight vapor superheat from the evaporator should reduce this concern.

6.10.3 Fluid superheating (alkaline metals)

Alkaline metals have relatively high boiling temperatures at atmospheric pressure. If the heater surface does not have enough active nucleation sites, then boiling may not occur near saturation temperature but rather require significant superheat. The maximum superheat temperature for liquid sodium at low pressure has been estimated at 2140K based on a van der Waals equation of state, as described by Gunnerson and Cronenberg [28]. At high superheat temperatures one could expect a vapor burst expulsion upon phase change, which could lead to structural damage and flow excursions. Since the superheat limit is above NGNP temperatures, alkaline metal evaporator and condenser designs should promote active nucleation sites through enhanced surface roughness and reduced interfacial wettability.

6.11 Nucleation sites

A primary condition affecting boiling temperature of the alkali coolants is the density of nucleation sites. A nucleation site is a microscopic reentrant (much smaller opening than the inside of the cavity) that will retain small gas bubbles in the surface to initiate boiling at low wall superheats. It is well known that a large number of nucleation sites promote a lower superheat for a fluid. The alkaline metals like any other metals boil at the boiling point, provided nucleation sites are developed. For instance, water normally boils at 100 °C at normal atmospheric pressure, but if there are not enough nucleation sites, water boiling point may reach a temperature up to 120 °C. Therefore, understanding the nucleation phenomenon is important, especially for the evaporator section of the thermosyphon. Nucleation sites can be thought of (in the case of crystallization) as actual physical locations where energy is drawn off more easily because of the greater surface to volume ratio (high surface area) of the site. In the case of vaporization, the nucleation site can be thought of as an actual physical location where temperatures can be locally higher, and where energy transfer is more efficient—again because of the high surface area.

Nucleation normally occurs at nucleation sites on surfaces containing liquid or vapor. Suspended particles without preferential nucleation sites is homogeneous nucleation. Homogeneous nucleation occurs spontaneously and randomly, but it requires superheating or supercooling of the medium. It is found that the liquid is superheated near the heating surface, and a nucleus of microscopic size is necessary for the birth of a bubble in the liquid. A spherical bubble has a radius such that the energy required for a pressure difference to change the volume is balanced by the energy required to change the area, i.e.,

$$(P_{\text{Vapor}} - P_{\text{Liqure}})dV = \sigma dA \tag{6.11}$$

All the parameters used in Eq. (6.11) are defined as before and for a sphere, $dV = 4\pi r^2$ and $dA = 8\pi r dr$. Therefore,

$$(P_{\text{Vapor}} - P_{\text{Liqure}})4\pi r^2 dr = \sigma 8\pi r dr \tag{6.12}$$

Boiling is nucleated when

$$(P_{\text{Vapor}} - P_{\text{Liqure}}) = \frac{2\sigma}{r} \tag{6.13}$$

At $r = 0$, $(P_{\text{Vapor}} - P_{\text{Liqure}}) = \infty$, which means that the initial pressure difference must be infinitely large for a bubble to develop without a nucleus. So, it is impossible for a bubble to grow without a nucleus. From Eq. (6.13), for a bubble to exist, vapor pressure must be higher than liquid pressure; this pressure differential is required to overcome surface tension, which tends to bring the liquid molecules together.

The presence of dissolved gas, for example air, in the liquid necessitates incorporating the gas partial pressure into account, thus Eq. (6.13) reduces to:

$$P_{\text{Vapor}} - P_{\text{Liqure}} = \frac{2\sigma}{r} = P_{\text{Gas}} \tag{6.14}$$

Where P_{Gas} is gas pressure in cavity at nucleation due to dissolved gas, such as air in Pascal (Pa).

The prediction of 'r,' remains a concern; i.e., is 'r' big enough to meet the coolant boiling requirement.

The mechanism of bubble formation depends on the wetting characteristics of the heating surface. The effect of wetting on bubble formation is shown in Fig. 6.15. Bubbles form most readily if the surface is nonwetting. In addition to wetting, nucleation sites are necessary for bubble formation.

Since steels are wetted by sodium at high temperatures, i.e., a drop of sodium placed on the surface will spread out, with a low or zero contact angle.

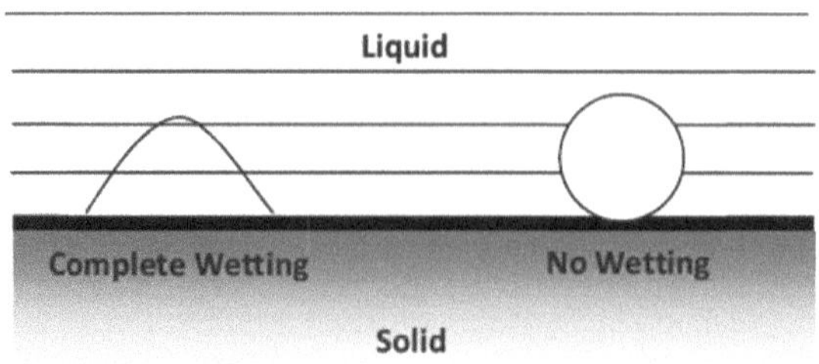

FIG. 6.15 Effect of wetting [1].

A consequence of this is that the sodium will tend to penetrate and fill any cracks or cavities in the surface. However, a cavity filled with liquid cannot act as a nucleation site, so it is necessary to assume that locally the surface is nonwetted.

The contact angle θ as shown in Fig. 6.16, is used as the wettability index. For θ <90 degrees the liquid wet the wall (e.g., water on glass), for >90 degrees the liquid does not wet the wall (e.g., mercury on glass) and if $\theta = 0$ degrees the liquid perfectly wets the wall. The contact angle between the surface and the liquid can be represented by a factor Φ [29] such that

$$\Phi = \frac{2 + 2\cos\theta + \cos\theta\sin^2\theta}{4} \tag{6.15}$$

If the liquid completely wets the surface $\theta = 0$ degrees, $\Phi = 1$ and there is no reduction of the free energy of formation of the embryo. If the surface is completely nonwetting $\theta = 180$ degrees, then $\Phi = 0$ and no superheat is required for nucleation at the surface. Bubbles form more readily if the surface is nonwetting. In practice most solid-liquid systems lie in the range between 0 and 90 degrees and #lies in the range of 1−0.5. In addition to wetting, nucleation sites are necessary for bubble formation. For the nucleus to continue to grow, the liquid temperature must be progressively increased above the saturation temperature to exceed the equilibrium superheat corresponding to the radius of curvature of the interface.

This radius of curvature decreases as the nucleus grows until the contact angle with the flat surface is established, further growth then tends to increase the radius of curvature and the bubble grows spontaneously in the superheated

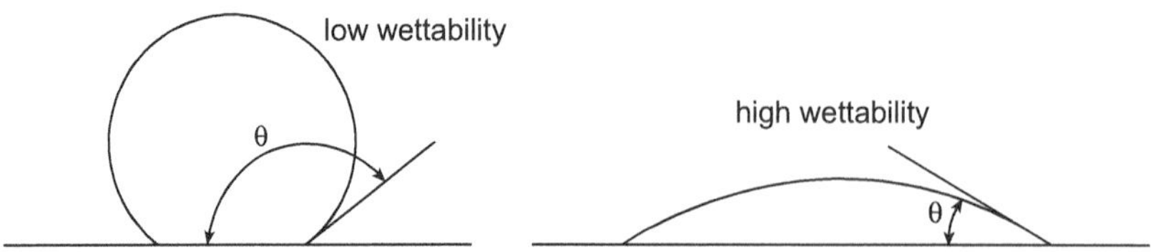

FIG. 6.16 Sketch of the contact angle between surface and liquid [1].

liquid; i.e., the size of the cavity determines the superheat at which a vapor bubble will be nucleated at that site. For large cavity sizes, the growth rate decreases, the dynamic forces become small, and the bubble size at departure is set by a balance between buoyancy and surface tension forces [30].

$$\text{Bubble departure diameter} = 0.0208\,\theta\left[\frac{\sigma}{g\left(\rho_{\text{Liquid}} - \rho_{\text{Vapor}}\right)}\right]^{0.5} \tag{6.16}$$

When surface tension forces are dominant, the departing bubbles tend to be spherical. With inertial forces dominant the bubble tends to be hemispherical and when both forces are significant the bubble has an oblate shape [30].

The main objective of treatment of the evaporative section is to attain consistent nucleation in order to ensure proper heat transfer, which does not occur on polished or smooth surfaces. In order to enhance thermal nucleation, more active nucleation sites are needed (a vapor pocket is needed).

If the nucleation site becomes filled with liquid, it is referred to as an inactive nucleation site ($T_{\text{Wet}} - T_{\text{Sat}}$) and becomes very large (the liquid must be superheated). For an active nucleation site, vapor must be in the pocket as shown in Fig. 6.17.

6.12 Inclination effects on a thermosyphon performance

The performance of a two-phase, closed thermosyphon depends strongly on the gravity field. In the internal phase change circulation, gravitational force plays the role of returning the liquid from the compensation chamber to the evaporator. Any change of the gravitational influence, such as thermosyphon inclination or the gravity field itself, will have a significant impact on the heat transfer characteristics of the thermosyphon. Understanding the inclination dependence is therefore critical to both thermosyphon design and application. Inclining the thermosyphon from the horizontal plane permits buoyancy forces to play a greater role, which is reflected in thermal performance. As the thermosyphon is tilted from horizontal, the indirect effects weaken along with some direct (end) effects, while other direct effects strengthen. However, this strengthening occurs over the long walls and therefore reinforces some parts of the primary flow while opposing others. Around 5 degrees, a stagnation point

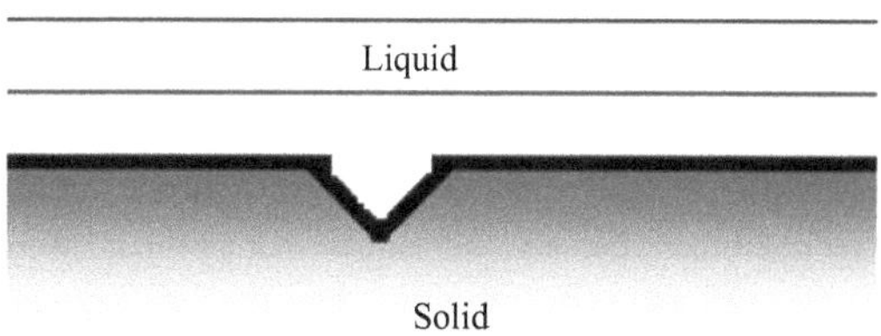

FIG. 6.17 Nucleation site (pocket) [1].

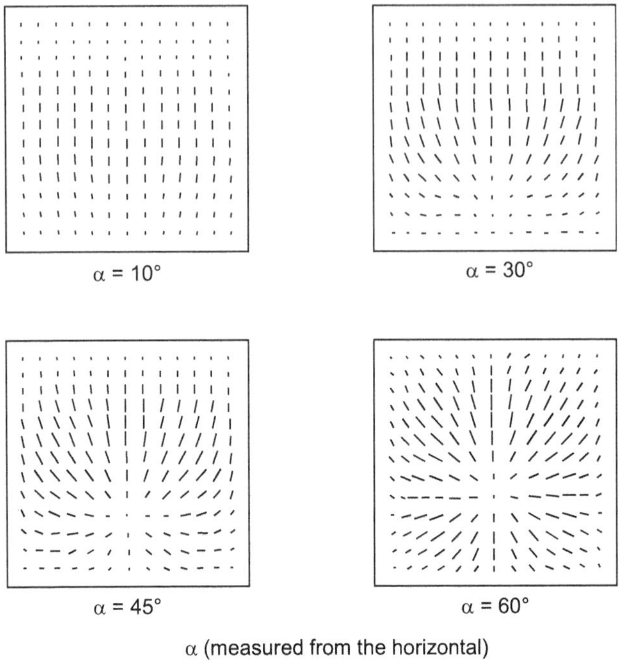

FIG. 6.18 Progressive Displacement of the Stagnation Point at the Hot end of the Tilted Thermosyphon [31].

appears at the bottom corner of the heated end, this is caused by alterations in the direct buoyancy forces operating over the end surface and the bottom surface near the end [31]. As the tilt increases, the stagnation point thus created moves toward the axis of the tube, as indicated in Fig. 6.18. When the thermosyphon is vertical, the stagnation point lies near the axis and the flow exhibits the familiar core annulus pattern.

For many applications, thermosyphons need to be tilted to accommodate the geometric structure of the entire system. Many studies have been conducted to investigate the effects of the inclination angle on thermosyphon heat transfer performance characteristics and limits. A comprehensive review of previous work (mostly experimental) reveals the efforts and progress made over the past 15 years and points out the necessities of this work. Nguyen—Chi and Groll [32] modified flooding correlations for inclined thermosyphons by introducing a function that has little physical meaning and was empirically developed from curve-fitting the experimental data.

Negishi and Sawada [33] conducted experiments to investigate the interactive influence of the inclination angle and working fluid inventory upon heat transfer performance of an inclined two-phase, closed thermosyphon. The experimental setup was designed for both visual observation and measurement of thermosyphon performance. They concluded that in order to obtain a high heat transfer rate, the working fluid inventory needs to be more than 25% of the evaporator volume, and the inclination angle should be almost 85 degrees (from vertical position). The maximum heat flux was found at an inclination angle between 50 and 70 degrees (depending upon working fluid inventory). The heat transfer coefficient, a measure of thermosyphon efficiency, was found to slightly decrease with inclination angle.

Bontemps et al. [34] experimentally studied evaporator and condenser heat transfer coefficients at various inclination angles and working fluid inventories. They found that evaporator heat transfer coefficients slightly decrease with inclination angle, while condenser heat transfer coefficients increase with inclination angle. It was also concluded that the critical heat flux is almost independent of inclination angle.

Terdtoon et al. [35] found that there are four regions of the inclination angle as shown in Fig. 6.19:

a. Vertical position to about 20 degrees (from vertical position)—heat transfer capacity increases with inclination angle and is independent of working fluid inventory;
b. 20–40 degrees—heat transfer capacity increases with inclination angle and then reaches the maximum value; the working fluid inventory has an influence on the heat transfer capacity;

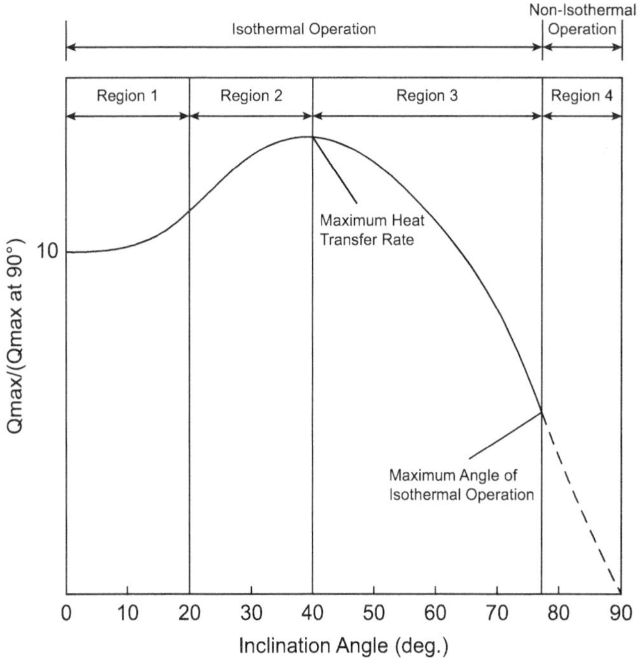

FIG. 6.19 Operation regimes at various inclination angles [35].

c. 40 degrees to minimum angle of isothermal operation—heat transfer capacity decreases with inclination angle and working fluid inventory has a large influence on both heat transfer rate and the minimum angle of isothermal operation; and
d. minimum angle of isothermal operation to limiting angle of operation—part of the evaporator section is not wetted and the thermosyphon is in a non-isothermal operational state.

For increasing tilt angles above the horizontal, the secondary vortices shrink in length. Near a tilt of 45 degrees, the vortices subside and are virtually absent for angles between this and vertical, the heat transfer rate then decreases as the vertical position is approached [31]. Due to differences in operating conditions, previous studies have resulted in diverse values of optimum inclination angle (corresponding to the maximum heat transfer rate). A few theoretical and numerical analyses have been carried out by Ma et al. [36] and Wang and Ma [37] starting from first principles and deriving an analytical expression for condensation heat transfer coefficient at various inclination angles. By comparing their analytical results with experimental data, they proposed a semi-empirical correlation to include the effects of vapor pressure and working fluid inventory. Zuo and Gunnerson [8] developed numerical models based on first principles. The two-dimensional liquid film flow and one-dimensional vapor flow are coupled by appropriate boundary conditions and auxiliary equations. By numerically solving the model, performance parameters (such as temperature along the pipe wall, etc.) and performance limits (such as dry-out and flooding) were studied. For a thermosyphon with a large aspect ratio and not close to horizontal position, flooding is more important once the minimum working fluid inventory is satisfied. The minimum working fluid inventory increases with inclination angle, especially for large inclination angles.

There is a range of inclination angles at which the thermosyphon shows better performance than at a vertical position, i.e., the critical heat transfer rate is higher. When the inclination angle increases from 0 degrees vertical to 90 degrees horizontal, the mean heat transfer coefficient will increase until it reaches the maximum value and then decrease [8]. When the inclination angle increases, the condenser heat transfer coefficient increases because of a thinning up-side liquid film, whereas the evaporator heat transfer coefficient decreases because of the enlargement of the liquid pool heating area and the comparatively poor liquid pool heat transfer, which has also been reported by Bontemps et al. [34]. When the thermosyphon is tilted to a small angle from a vertical position, the increase in condenser heat transfer coefficient dominates and the mean heat transfer coefficient of the thermosyphon increases. At large inclination angles, the liquid pool heating area expands, and thus the decrease in evaporator heat transfer coefficient dominates, resulting in lower mean heat transfer coefficients.

6.13 Summary

Recent technological developments in next generation nuclear reactors have created renewed interest in nuclear process heat for industrial applications. To utilize process heat, a thermal device is needed to transfer the thermal energy from NGNP to the hydrogen plant in the most efficient way possible. One of the most significant advantages of heat transfer by thermosyphon is the characteristic of nearly isothermal phase change heat transfer, which makes the thermosyphon an ideal candidate for applications where the temperature gradient is limited and high delivery temperatures are required, as in thermochemical hydrogen production. Inside an operating thermosyphon, the vapor (at constant saturation temperature) carries a large amount of latent heat from the evaporator to the condenser.

Thermal energy devices such as heat pipes and thermosyphons possess many advantages, including high heat recovery effectiveness, high compactness, no moving parts, and high reliability. Because of these advantages, heat pipes and thermosyphons have been extensively applied in many industries.

Developing very high temperature reactor technologies for the production of hydrogen and other energy products and developing technologies to close the nuclear fuel cycle are high priority research and development issues for a successful nuclear future. Both High Temperature Electronic (THE) and the SI processes are in the early stages of development, but both will require technology to transfer thermal energy (heat) from the Next Generation Power Plant (NGNP)q to the hydrogen production facility. From an efficiency standpoint, thermochemical cycles may have theoretical advantages over the alternative of high-temperature electrolysis, because their efficiency is not burdened with the detraction of the efficiency of electricity.

The success of the next generation of nuclear reactors will depend, in part, on successfully utilizing process heat and on the selection of heat transport devices, such as the thermosyphon as discussed in this chapter. For high temperature applications requiring the transport of large amounts of thermal power at a small temperature difference between the heat source and the heat sink, the thermosyphon is one option.

Thermosyphon working fluids should have a high latent heat of vaporization such as found for liquid metals, which in turn makes them a preferable choice for transporting thermal energy. Although considerable data already exists on the thermophysical properties of possible working fluids, data, especially in the vapor phase, is rarely available at the high temperatures of NGNP. Accurate performance predictions of thermosyphon startup and operational behavior is required to design and build reliable and efficient systems for high temperature process heat transfer from NGNP to a hydrogen plant.

Instabilities are common in both forced and natural circulation systems; the latter is more unstable than the former. The mechanisms that can lead to unstable operational behavior of the thermosyphon have been discussed along with the importance of having active nucleation sites in the evaporator section. Due to differences in operating conditions, previous studies have resulted in diverse values of the optimum inclination angle (corresponding to the maximum heat transfer rate) for a thermosyphon, the optimum inclination angle when the heat transfer capacity reaches its maximum is reported to be around 40 degrees [35]. The information in this section provides insights and options useful in the design and development of an NGNP process heat recovery system using a thermosyphon.

References

[1] P. Sabharwall, Engineering Design Elements of a Two-Phase Thermosyphon to Transfer NGNP Thermal Energy to a Hydrogen Plant, INL Report INL/EXT-09-15383 (Revision 0), July 2009.

[2] C. Davis, C. Oh, S. Barner, D.W. Sherman, Thermal-hydraulic Analyses of Heat Transfer Fluid Requirements and Characteristics for Coupling A Hydrogen Production Plant to a High-Temperature Nuclear Reactor, INL/EXT-05-00453, 2005.

[3] C.C. Silverstein, Design and Technology of Heat Pipes for Cooling and Heat Exchange, Hemisphere Publishing Corporation, 1992.

[4] D. Reay, P. Kew, Heat Pipes: Theory Design and Applications, fifth ed., Butterworth-Heinemann, Elsevier, 2006.

[5] R.S. Gaugler, US patent 2350348. Appl, December 21, 1942. Published 6 June 1944.

[6] G.M. Grover, Evaporation-Condensation Heat Transfer Device, U.S. Patent 3229759, December 2, 1963.

[7] K.C. Cheng, J.P. Zarling, Applications of Heat Pipes and Thermosyphons in Cold Regions, vol. II, Begell House, 1993, pp. 1–32.

[8] Z.J. Zuo, F.S. Gunnerson, Numerical Study of the Thermosyphon Flooding Limit, Fundamentals of Heat Pipes, ASME HTD-Vol. 278 (1994) 47–55 (ii).

[9] K.T. Feldman, Investigation of passive pressure-pumped thermosyphons, in: Proceedings of the 6th International Heat Pipe Conference, Vol. II, Grenoble, France, May 25–29, 1987, pp. 665–670.

[10] S.H. Noie, Heat transfer characteristics of a two-phased closed thermosyphon, Appl. Therm. Eng. 25 (March 2005) 495–506.

[11] Z. Hou, Y. Wen, Heat pipe operating in a spin satellite, in: Proceedings of the 6th International Heat Pipe Conference, Vol. II, Grenoble, France, May 25–29, 1987, pp. 439–442.

[12] M. Amidieu, S. Aucel, R. Giovannini, B. Moschetti, The heat pipe network of TDF1/TV-SAT: review of the development at completion, in: Proceedings of the 6th International Heat Pipe Conference, Vol. II, Grenoble, France, May 25–29, 1987, pp. 386–393.

[13] H. Mitsuma, R. Imai, H. Suzuki, Y. Ido, J. Kawashima, T. Ishibashi, Y. Kuriyama, Development of light weight payload radiator with heat pipes for a 2-ton geostationary satellite, in: Proceedings of the 6th International Heat Pipe Conference, Vol. II, Grenoble, France, May 25–29, 1987, pp. 368–373.

[14] M.M. Chen, A. Faghri, An analysis of the vapor flow and the heat conduction through the liquid-wick and pipe wall in a heat pipe with single or multiple heat sources, Int. J. Heat Mass Transf. 33 (9) (1990) 1945–1955.

[15] W.B. Bienert, Nuclear Wastes in Comparison with Other Heat Sources for Deicing Bridges, Ramps and Pavements, FHWA-DTM-70-6, 1970.

[16] W.D. Munzel, Heat pipes for heat recovery from exhaust gas of a diesel engine in a passenger car, in: Proceedings of the 6th International Heat Pipe Conference, Vol. II, Grenoble, France, May 25−29, 1987, pp. 740−743.

[17] A. Gerak, L. Horvath, F. Jelinek, V. Zboril, Examples of heat pipe application in chemical, electrical and other industries, in: Proceedings of the 6th International Heat Pipe Conference, Vol. II, Grenoble, France, May 25−29, 1987, pp. 676−684.

[18] G.P. Peterson, B.K. Bage, Entrainment limitations in thermosyphons and heat pipes, J. Energy Resour. Technol. 113 (1991) 147−153.

[19] S.I. Haider, Y.K. Joshi, W. Nakayama, A natural circulation model of the closed loop, two-phase thermosyphon for electronics cooling, J. Heat Transf. 124 (2002) 881−890.

[20] F.S. Gunnerson, F.D. Sanderlin, A controllable, wickless heat pipe design for heating and cooling, in: Fundamentals of Heat Pipes, ASME HTD, vol. 278, 1994.

[21] K. Ohashi, H. Hayakawa, M. Yamada, T. Hayashi, T. Ishii, Preliminary study on the application of the heat pipe to the passive decay heat removal system of the modular HTR, Prog. Nucl. Energy 32 (1998) 587−594.

[22] P. Sabharwall, F. Gunnerson, Engineering design elements of a two-phase thermosyphon for the purpose of transferring NGNP thermal energy to a hydrogen plant, J. Nucl. Eng. Des. (2008) (Submitted).

[23] R.N. Lyon, Liquid Metals Handbook, the Committee on the Basic Properties of Liquid Metals, second ed., Office of Naval Research, Department of Navy, June 1952.

[24] P.V. Bystrov, D.N. Kagan, G.A. Krechetova, E.E. Shpilrain, Liquid-Metal Coolants for Heat Pipes and Power Plants, Hemisphere Publishing Corporation, 1990.

[25] J.K. Fink, L. Leibowitz, Thermodynamic and Transport Properties of Sodium Liquid and Vapor, ANL/RE-95/2, 1995.

[26] T. Dickinson, Performance Analysis of a Liquid Metal Heat Pipe Space Shuttle Experiment, Masters' Thesis, Air Force Institute of Technology, Ohio, 1996.

[27] A.E. Bergles, J.G. Collier, J.M. Delhaye, G.F. Hewitt, F. Mayinger, Two-Phase Flow and Heat Transfer in the Power and Process Industries, Hemisphere Publishing Corporation, 1981.

[28] F.S. Gunnerson, A.W. Cronenberg, On the thermodynamic superheat limit for liquid metals on its relation to the leidenfrost temperature, J. Heat Transf. 100 (November 1978) 734−737.

[29] J.G. Collier, J.R. Thome, Convective Boiling and Condensation, third ed., Oxford University Press, 1996.

[30] M.A. Johnson, J. Pena, R.B. Mesler, Bubble shapes in nucleate boiling, in: Chemical Engineering Program Symposium Services, vol. 62, 1966.

[31] G.S.H. Lock, The Tubular Thermosyphon, Oxford University Press, 1992.

[32] H. Nguyen − Chi, M. Groll, Entrainment or flooding limit in a closed- two phase thermosyphon, in: D.A. Reay (Ed.), Advances in Heat Pipe Technology, Pergamon Press, New York, 1981, pp. 147−162.

[33] K. Negishi, T. Sawada, Heat performance of an inclined two-phase closed thermosyphon, Int. J. Heat Mass Transf. 26 (8) (1983) 1207−1213.

[34] A. Bontemps, C. Goubier, C. Marquet, J.C. Solecki, C. Nardi, Performance limit of a toulene loaded, closed two-phase thermosyphon, in: Proceedings of the 6th International Heat Pipe Conference, Vol. II, Grenoble, France, May 25−29, 1987, pp. 634−640.

[35] P. Terdtoon, M. Shiraishi, M. Murakami, Investigation of effect of inclination angle on heat transfer characteristics of closed two-phase thermosyphon, in: Proceedings of the 7th International Heat Pipe Conference, Vol. II, Minsk, May 1990, pp. 517–524.

[36] T. Ma, X. Liu, J. Wu, Flow patterns and operating limits in two-phase closed thermosyphon, in: Proceedings of the 6th International Heat Pipe Conference, Vol. II, Grenoble, France, May 25–29, 1987, pp. 576–581.

[37] J.C.Y. Wang, Y. Ma, Condensation heat transfer inside vertical and inclined thermosyphons, J. Heat Transf. 113 (1991) 777–780.

[38] P. Sabharwall, M.W. Patterson, F. Gunnerson, Theoretical design of a thermosyphon for process heat transfer from NGNP to hydrogen plant, J. Nucl. Eng. Des. (Special Issue) (2009) (i) (Accepted).

[39] F.S. Gunnerson, P. Sabharwall, S. Sherman, Comparison of sodium thermosyphon with convective loop, in: Proceedings of the 2007 AIChE Conference, Salt Lake City, November 2007.

[40] G.P. Peterson, A.B. Duncan, M.H. Weichold, Experimental investigation of micro heat pipes fabricated in silicon wafers, J. Heat Transf. 115 (1993) 751–756.

[41] P. Sabharwall, J.N. Reyes, B. Woods, R. Peterson, Q. Wu, Effects of fluid axial conduction on liquid metal natural circulation and linear stability, Masters Thesis, Oregon State University (2004).

Chapter 7

Thermodynamic analysis of thermosyphon

Chapter outline

7.1 Introduction 345
7.2 General model (vertical thermosyphon) and flooding 347
7.3 Two-phase thermosyphon thermodynamic analysis with spiral heat exchanger 350
7.4 Summary 352
References 352

7.1 Introduction

With the world energy crisis, where petroleum become more expensive and scarcer, where hydroelectric resources are exhausting and where new technologies that enable the use of Solar and Eolic (Wind) energies are not economically viable, the importance of procedures to improve the energetic efficiency of industrial processes is growing, leading to the development of new solutions. High temperature streams (above 600 °C), released to atmosphere from furnaces, represent good examples of recoverable thermal energy, available in many plants. In this paper, the technology of high-temperature thermosyphon is considered to be applied in regenerative heat exchangers in petroleum plants.

What is eolic energy?

Functionality, Advancements and Industrial Applications of Heat Pipes.
https://doi.org/10.1016/B978-0-12-819819-3.00007-9

Wind energy is the energy obtained from the wind. It is one of the oldest energy resources exploited by humans and is today the most mature and efficient energy of all renewable energies. The term "wind" comes from the Latin "aeolicus", pertaining to or related to Eolo, God of the winds in the Greek mythology.

High temperature thermosyphons work at temperatures above 600 °C. The working fluid consists of a liquid metal such as sodium, lithium or potassium. The tube material (metal) must be chemically compatible with the working fluid, to avoid chemical reactions, which could produce undesirable non condensing-gases. The material of the tubes must also resist to the corrosion, while it keeps its mechanical properties at the high working temperatures. The manufacture of this device is also challenging and demands careful, well determined procedures.

In recent years, extensive studies have been conducted to provide a thorough understanding of thermosyphon operation and appropriate design schemes for practical use. Additionally, there are several surveys which illustrate state-of-the-art research on two-phase thermosyphons. Gross [11] surveys condensation heat transfer inside a two-phase thermosyphon including a detailed description of fluid flow and a review of published experimental and theoretical investigations.

Grief [1] reviewed research on natural circulation loops including the two-phase thermosyphon loop. A recent book by Lock [2] provides a thorough discussion on various types of thermosyphons. Dobran [3] summarizes heat pipe research and applications. A comprehensive review on flooding and entrainment inside heat pipes and thermosyphons was provided by Peterson and Bage [4].

The first comprehensive experimental investigation on thermosyphon performance was conducted by Lee and Mital [5]. Since then, numerous experiments concentrating on various aspects of thermosyphon operation have been conducted. These studies include: Casarosa and Dobran [6], Chen et al. [7], Lock and Fu [8], Gunnerson and Sanderlin [9], and many others in the proceedings of the International Heat Pipe Conference.

The experimental thermosyphons used in previous studies were from vertical to inclined, from circular, annular to rectangular cross-section, from extra-long [23] to a very small size. The heating and cooling methods were from constant-heat flux, constant-temperature, forced-convective heating/cooling. The measurement techniques involved in these experiments ranged from conventional thermocouples, laser, thermal imaging techniques to neutron radiography [10].

Previous experimental studies have been able to provide useful insight into thermosyphon operation, set references for validation of theoretical models and provide database for design purposes. Additionally, many theoretical analyses incorporated empirical or semi-empirical correlations to simplify the models and the solution processes.

Since thermosyphon operation involves a series of complex processes related to phase change heat transfer, two-phase flow and interfacial phenomena, an exact theoretical analysis is difficult (if not impossible). Many previous studies have focused on isolated phenomena or regions inside the thermosyphon such as condensation heat transfer in the condenser [11] and flooding in the condenser [4,11].

The first model considering the thermosyphon as an integrated system was provided by Shiraishi et al. in 1981 [12]. Dobrin [13] developed a system model based upon conservation of mass, momentum, and energy and including thermal hydraulics of the vapor core, liquid film, and the liquid pool. Reed and Tien [14] presented a one-dimensional model to predict steady-state and transient performance of the two-phase thermosyphon.

Harley and Faghri [15] developed a transient two-dimensional model which includes conjugated heat transfer through the wall and the falling condensate film. In their study, vapor was assumed to be in a laminar flow regime. The effects of working fluid inventory and liquid pool were not included. Zuo and Gunnerson [16] provided a numerical model to describe thermosyphon performance under various heating and cooling condition as well as the effects of working fluid inventory.

Generally, previous theoretical models consist of a set of highly nonlinear partial differential equations with several empirical or semi-empirical correlations specified for shear stresses and heat transfer coefficients. The two-dimensional model provided by Harley and Faghri [15] avoided the use of empirical correlations for interfacial mechanism because of the assumption of laminar vapor and liquid flows. However, to obtain solutions from these models, numerical techniques must be incorporated, and significant programming efforts and computational time are required. Changes in geometry or boundary conditions may invalidate the models.

Recently, Gunnerson et al. [17] performed convective loop analysis and comparison with Na as the coolant and stated two-phase benefits when compared with single phase for transferring the Next Generation of Nuclear Power (NGNP) process heat, which enabled the author to carry out further analysis with the thermosyphon. This chapter provides a different but more practical view into the physics behind thermosyphon operation. The continuous phase change circulation inside an operating thermosyphon is considered as a thermodynamic cycle. Several important factors such as temperature difference (superheated vapor and saturated liquid) between the evaporator and the condenser, heat exchange between the liquid and the vapor are illustrated by classical thermodynamic diagrams and require only simple engineering calculations.

7.2 General model (vertical thermosyphon) and flooding

The circulation of the working fluid in a thermosyphon is the consequence of a pumping process that results from the conversion of potential thermal energy into kinetic energy in a well-defined thermodynamic cycle.

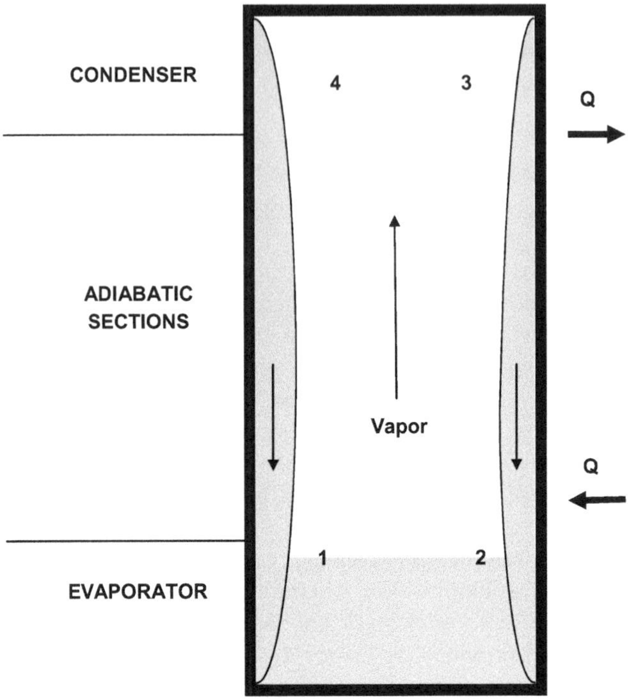

FIG. 7.1 Working fluid circulation inside a vertical two-phase thermosyphon.

As illustrated in Fig. 7.1, thermosyphon operation consists of four processes:

1. In the evaporator section, the liquid receives heat and becomes vapor (process 1–2)
2. The vapor travels through the adiabatic section due to pressure difference between the evaporator and the condenser (process 2–3)
3. In the condenser, the vapor releases heat and becomes liquid (process 3–4)
4. The condensed liquid returns to the evaporator through the adiabatic section due to gravitational force (process 4–1). These four processes form an internal thermodynamic cycle. The working fluid at state 2 can be slightly superheated for some cases (such as in the lateral case). Entropy increases during processes 2–3 and 4–1 due to irreversibilities from liquid-vapor interfacial and liquid-wall shear forces.

In order for liquid at state 4 (in condenser) to go back to the original state 1 (in evaporator), gravitational force must overcome frictional forces and move the liquid from the condenser to the evaporator. Pressure difference between the evaporator and the condenser depends on the heat transfer rate, geometric dimensions of the adiabatic section and thermophysical properties of the working fluid. Generally, the larger the heat transfer rate and the L/D (L being

Length and D is Diameter of the pipe size) of the adiabatic section, the larger the pressure difference.

The flooding phenomenon in thermosyphon operation has been extensively studied [4,14,18,19]. Whether flooding occurs when liquid flow is reversed or when liquid-vapor interface becomes unstable has been discussed [24]. The ultimate cause of flooding is that the downward gravitational force is unable to overcome upward frictional forces.

In order for the condensate to return to the evaporator, the following condition must be satisfied:

$$\rho_L g L_a \geq \Delta P_f \tag{7.1}$$

As mentioned earlier, ΔP_f corresponds to the frictional forces the liquid flow experiences and can be calculated when the thermosyphon geometry and heat transfer rate are known and ρ_L id liquid density as well as L_a being the Length of diabatic section and g is gravity. For thermosyphon loops, Eq. (7.1) may be viewed as a gravity limiting criterion (analogous to capillary limit in a wicked heat pipe).

From thermodynamics, a difference in saturation pressure is directly related to a difference in saturation temperature. A common P-T diagram is shown in Fig. 7.2. For thermosyphon operation, the vaporization line as well as liquid and vapor zones are of interest.

Close inspection of the vaporization line shows the slope of the P-T curve (dP/dT) increasing with temperature, indicating that the same temperature difference can provide a larger ΔP (more pumping power) if the thermosyphon is operating at a higher temperature.

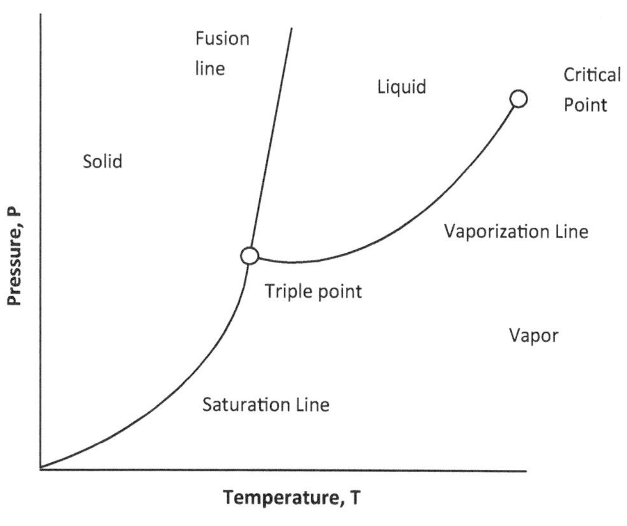

FIG. 7.2 Pressure-temperature diagram [20].

The same conclusion has been drawn by Richter and Gottschild [21] for heat pipe operation. Additionally, with a higher operating temperature (increase in temperature) the liquid viscosity reduces and causes a higher flooding limit.

7.3 Two-phase thermosyphon thermodynamic analysis with spiral heat exchanger

Fig. 7.3 shows a schematic diagram of the thermosyphon design, with Spiral Heat Exchangers (SHEs) and each corresponding thermodynamic state of the working fluid (Na) [22]. The thermodynamics of thermosyphon design are illustrated in Fig. 7.4. The area under T-s curve gives the net amount of heat being added to the working fluid (in the evaporator), which is ideally the same as the net amount of heat removed by the condenser. Thus, the T-s curve and P-v behavior is represented as just a single line, i.e., the net-work in the closed system is zero, as shown in Fig. 7.4.

The heat transferred to the working fluid in the evaporator region is given by

$$Q_{in,}\text{evaporator} = h_3 - h_1 = \Delta h_{1\ 3} \tag{7.2}$$

The heat rejection in the condenser is given by

$$Q_{ou,}\text{condenser} = h_{3\prime} - h_5 = \Delta h_{5\ 3\prime} \tag{7.3}$$

The continuous phase change circulation inside an operating thermosyphon may be viewed as a thermodynamic cycle. The working fluid (subcooled state) enters the evaporator at a temperature T_1 (slightly higher temperature than its melting temperature) and is raised to the upper operating temperature $T_{\prime}^{1}$

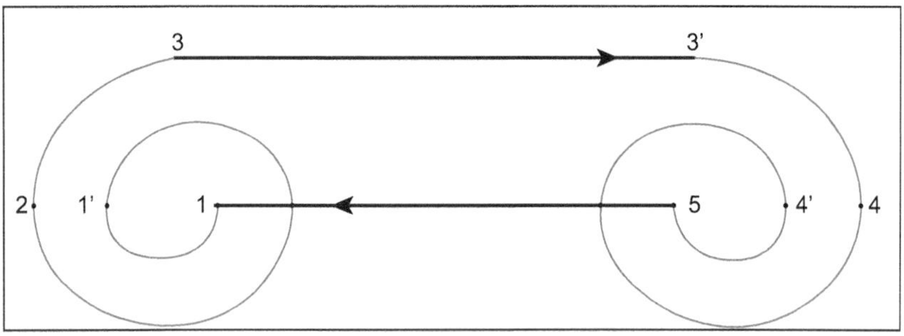

Position Number	Thermodynamic State of the Fluid
1 , 5	SubCooled Liquid
1' , 4'	Saturated Liquid
2 , 4	Saturated Vapor
3 , 3'	Superheated Vapor

FIG. 7.3 Schematic diagram of the thermodynamic process.

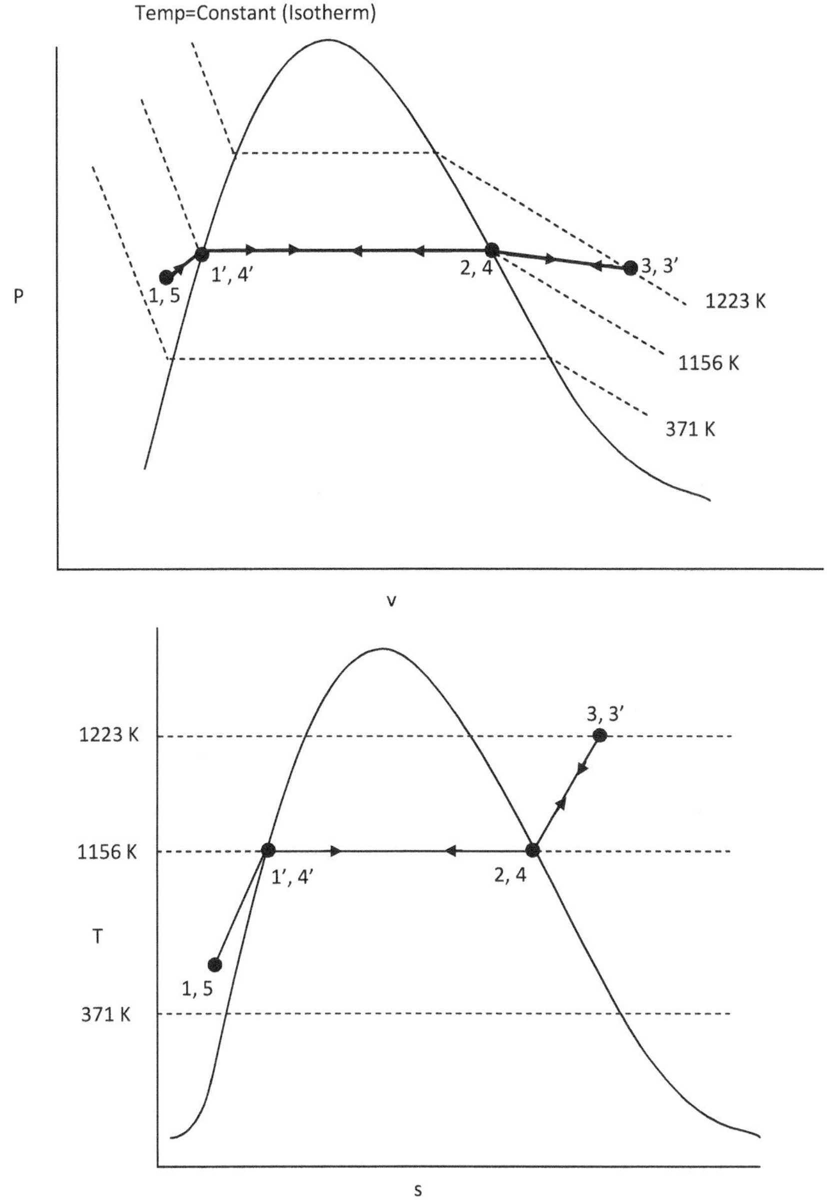

FIG. 7.4 Pressure-specific volume and temperature-specific entropy diagrams.

(saturated liquid state) along the path 1−1′. Evaporation and expansion from the liquid volume to the vapor volume occurs along the path 1′−2 (saturated vapor state). In process 1′−2, heat is being added at constant temperature, which raises enthalpy of the working fluid from $h_{1'}$ to h_2

In process 2–3, the vapor is being superheated, which further leads to the increase in enthalpy from h_2 to h_3 but exhibits a little drop in pressure due to the frictional pressure drop, which is proportional to the square of the velocity. The maximum velocity and losses due to friction occur at the evaporator exit.

$$\frac{\Delta P_e}{L_e} = \frac{4 f_e \rho_V V_e^2}{2 D_e} \tag{7.4}$$

For sodium vapor at half sonic velocity, $\frac{\Delta P_e}{L_e} \sim 10^{-3}$ MPa/m. In process 3′–4, the working fluid releases its thermal energy in the form of heat rejection, which drops the temperature. By stage 4′, the vapor is all condensed and a saturated liquid state remains. In process 4′–5 the fluid is all in liquid state and even after the temperature drops, the temperature at 5 is always higher than the melting point temperature of the fluid, which is now in subcooled state. If the thermosyphon is being operated after a long nonoperational period, external heating will be provided such that the fluid temperature never goes below the melting temperature (as can be seen from Fig. 7.4). The subcooled liquid Na returns to the evaporator through the condensate return line by gravity. Thermosyphons can be operated without having a phase change, but with a phase change, a thermosyphon has the advantage of high-enthalpy transport, which includes the sensible heat of the liquid, the latent heat of vaporization, and possible vapor superheat.

7.4 Summary

Empirical mathematical expressions and numerical schemes are helpful, but sometimes they may mask the real physics from a design engineer's point of view. This section provides a different view into the physics behind thermosyphon operation and performance based on thermodynamics. The continuous phase change circulation inside an operating thermosyphon is considered as a thermodynamic cycle. Several important factors such as temperature difference (superheated vapor and saturated liquid) between the evaporator and the condenser, heat exchange between the liquid and the vapor are illustrated by classical thermodynamic diagrams and require only simple engineering calculations.

References

[1] R. Grief, Natural circulation loops, J. Heat Transf. 110 (1988) 1243–1258.
[2] G.S.H. Lock, The Tubular Thermosyphon, Oxford University Press, 1992.
[3] F. Dobran, Heat pipe research and development in Americas, Heat Recovery Syst. CHP 9 (1989) 67–100.
[4] G.P. Peterson, B.K. Bage, Entrainment limitations in thermosyphons and heat pipes, J. Energy Resour. Technol. 113 (1991) 147–153.

[5] Y. Lee, U. Mital, A two-phase closed thermosyphon, Int. J. Heat Mass Transf. 15 (1972) 1695–1707.
[6] C. Casarosa, F. Dobran, Experimental investigation and analytical modeling of a closed two-phase thermosyphon with imposed convection boundary conditions, Int. J. Heat Mass Transf. 31 (1988) 1815–1833.
[7] K.S. Chen, Y.Y. Chen, S.W. Shiao, P.C. Wang, An experimental study of steady-state behavior of a two-phase natural circulation loop, Energy Convers. Mgmt. 31 (1991) 553–559.
[8] G.S.H. Lock, Fu, Observations on an evaporative elbow thermosyphon, J. Heat Transf. 115 (1993) 501–503.
[9] F.S. Gunnerson, F.D. Sanderlin, A controllable, wickless heat pipe design for heating and cooling, in: Fundamentals of Heat Pipes, vol. 278, ASME HTD, 1994.
[10] M. Tamaki, A. Yoneyama, Y. Ikeda, K. Ohkubo, G. Matsumoto, Observation of working fluid in a heat pipe by neutron radiography, in: Proceedings of the 7th International Heat Pipe Conference, Vol. II, Minsk, May 1990, pp. 67–74.
[11] U. Gross, Reflux condensation heat transfer inside a closed thermosyphon, Int. J. Heat Mass Transf. 35 (2) (1992) 279–294.
[12] M. Shiraishi, K. Kikuchi, T. Yamanishi, Investigation of heat transfer characteristics of a two-phase closed thermosyphon, in: D.A. Reay (Ed.), Advances in Heat Pipe Technology, Pergamon Press, New York, 1981, pp. 95–104.
[13] F. Dobrin, Heat pipe research and development in Americas, Heat Recovery Syst. CHP 9 (1989) 67–100.
[14] J.G. Reed, C.L. Tien, Modeling of the two-phase closed thermosyphon, J. Heat Transf. 109 (1987) 722–730.
[15] C. Harley, A. Faghri, Complete transient two-dimensional analysis of two-phase closed thermosyphons including the falling condensate film, J. Heat Transf. 116 (1994) 418.
[16] Z.J. Zuo, F.S. Gunnerson, Numerical modeling of the steady-state two-phase closed thermosyphon, Int. J. Heat Mass Transf. 37 (17) (1994) 2715–2722 (i).
[17] F.S. Gunnerson, P. Sabharwall, S. Sherman, Comparison of sodium thermosyphon with convective loop, in: Proceedings of the 2007 AIChE Conference, Salt Lake City, November 2007.
[18] Z.J. Zuo, F.S. Gunnerson, Numerical study of the thermosyphon flooding limit, in: Fundamentals of Heat Pipes, vol. 278, ASME HTD, 1994, pp. 47–55 (ii).
[19] B.H. Kim, G.P. Peterson, Theoretical and physical interpretation of entrainment phenomenon in capillary-driven heat pipes using hydrodynamic instability theories, Int. J. Heat Mass Transf. 37 (17) (1994) 2647–2660.
[20] O.P. Single, Engineering Thermodynamics, Macmillan India Limited, 1990.
[21] R. Richter, J.M. Gottschild, Thermodynamic aspects of heat pipe operation, J. Thermophys. Heat Transf. 8 (2) (1994) 334–340.
[22] B. Zohuri, Compact Heat Exchangers: Selection, Application, Design and Evaluation, first ed., Springer Publisher, New York, NY, September 21, 2016.
[23] F.D. Haynes, J.P. Zarling, G.E. Gooch, Performance of a Thermosyphon with a 37-meter-long Horizontal Evaporator, Cold Reg. Sci. Technol. 20 (3) (1992) 261.
[24] S. Roesler, M. Groll, Measurement of the Condensate Structure in Closed Two-Phase Thermosyphons, Proceedings of the 7th International Heat Pipe Conference Vol. I (May 1990) 69–80. Minsk.

Chapter 8

Thermosyphon & heat pipe dimensionless numbers in boiling fluid flow

Chapter outline

8.1 Introduction 355
8.2 Thermosyphon 355
8.3 Heat pipe 358
8.4 Results and discussion 359
8.5 Summary 363
References 363

8.1 Introduction

Bridgman [1] has by far presented the most extensive proof of the mathematical principles underlying dimensional analysis. Because it operates only upon the dimensions of the variables, it does not directly produce numerical results from the variables, but instead yields a modulus by which the observed data can be combined, and the relative influence of the variables established [1]. Dimensional Analysis [7] and ([2] and [3]) is a mathematical technique used in research work for design and for conducting model tests. It deals with the dimensions of the physical quantities involved in the phenomenon. All physical quantities are measured by comparison, which is made with respect to an arbitrarily fixed value. Fundamental dimensions are directly measurable quantities such as mass (M), length (L), time (T), and temperature (θ). Derived or secondary dimensions are those expressed in terms of fundamental dimensions, e.g., velocity is denoted by distance per unit time (L/T) and density by mass per unit volume (M/L^3). Then velocity and density become derived quantities.

8.2 Thermosyphon

It is useful to relate the hydraulic characteristics and thermo-physical properties of flow to the ratio of the various forces encountered in the flow to improve the heat transfer characteristics of a thermosyphon for NGNP process heat transfer (Sabharwall et al. 2009 (iv)). The physical quantities and their

Functionality, Advancements and Industrial Applications of Heat Pipes.
https://doi.org/10.1016/B978-0-12-819819-3.00008-0

respective fundamental dimensions used for the analysis are: l (L), V_S (LT^{-1}), C_p ($L^2\,T^{-2}\,\theta^{-1}$), h_{fg} ($L^2\,T^{-2}$), ρ_l (ML^{-3}), ρ_v ($M\,L^{-3}$), μ_l ($M\,L^{-1}T^{-1}$), μ_v ($M\,L^{-1}T^{-1}$), D (L), K ($M\,L\,T^{-3}\theta^{-1}$), m· ($M\,T^{-1}$), Q ($M\,L^2\,T^{-3}$).

Repeating Variables: ρ_l, D, μ_l, C_p

The repeating variables were chosen such that in the analysis we have the effect of the fundamental property of the fluid (ρ_l, μ_l, C_p) and also the characteristic property of the heat transfer device (D).

Number of physical quantity (n): 12.

Fundamental Dimensions (k): M, L, T, and θ i.e., 4.

Therefore, the number of dimensionless pi terms: n − k: 12−4 = 8

$$\Pi_1 = (\rho_L)^a(D)^b(\mu_L)^c(C_P)^d(l)^1 \tag{8.1}$$

$$\Pi_1 = \frac{1}{D}\text{(First pi term)} \tag{8.2}$$

The first pi term is just the geometric ratio or an aspect ratio of the given thermosyphon.

$$\Pi_2 = (\rho_L)^a(D)^b(\mu_L)^c(C_P)^d(V_s)^1 \tag{8.3}$$

$$\Pi_2 = \underbrace{\frac{\rho_L\,V_S\,D}{\mu_L}}_{Re}\text{(Second pi term)} \tag{8.4}$$

The second pi term is the Reynolds number, which is the ratio of inertial forces ($\rho_L V_S$) to viscous forces $\left(\frac{D}{\mu_L}\right)$, and it quantifies the relative importance of these two types of forces for sonic flow conditions and is also used to identify different flow regimes, such as laminar or turbulent flow. In this case, the Reynolds number can be taken as the limiting Reynolds number as maximum velocity is limited by the sonic velocity of vapor.

$$\Pi_3 = (\rho_L)^a(D)^b(\mu_L)^c(C_P)^d(h_{fg})^1 \tag{8.5}$$

$$\Pi_3 = \frac{\rho_L^2 D^2 h_{fg}}{\mu_L^2} \tag{8.6}$$

$$\Pi_3 = (Re)^2\underbrace{\left(\frac{h_{fg}}{V_S^2}\right)}_{E_r}\text{(Third pi term)} \tag{8.7}$$

E_r is the ratio of thermal to kinetic energy. Thermal energy is the energy given out by the evaporator to convert the liquid into its vapor state, which is traveling with sonic velocity V_S.

$$\Pi_4 = (\rho_L)^a(D)^b(\mu_L)^c(C_P)^d(\mu_V)^1 \tag{8.8}$$

$$\Pi_4 = \frac{\mu_V}{\mu_L} \text{(Fourth pi term)} \tag{8.9}$$

The fourth pi term is the ratio of the viscosities of vapor to liquid. Both the viscosities vary inversely with each other with respect to increase in temperature. Vapor viscosity tends to increase making the vapor more viscous, whereas liquid viscosity tends to decrease with the increase in temperature.

$$\Pi_5 = (\rho_L)^a (D)^b (\mu_L)^c (C_P)^d (K)^1 \tag{8.10}$$

$$\Pi_5 = \frac{K}{C_P \mu_L} = \frac{1}{Pr} \text{(Fifth pi Term)} \tag{8.11}$$

The fifth pi term is the inverse of the Prandtl number, which approximates the ratio of thermal diffusivity to momentum diffusivity (kinematic viscosity).

$$\Pi_6 = (\rho_L)^a (D)^b (\mu_L)^c (C_P)^d (\dot{m})^1 \tag{8.12}$$

$$\Pi_6 = \frac{\dot{m}}{\mu_L D} \tag{8.13}$$

$$\Pi_6 = (Re) \underbrace{\left(\frac{A}{D^2}\right)}_{\text{Dimensionless Geometric Parameter}} \text{(Sixth pi term)} \tag{8.14}$$

The sixth pi term has both dependence of fluid flow properties (Re) and geometric parameters and can simply be taken as a modified Reynolds number. For a noncircular duct, 'D' should be replaced by 'D_h' which is the hydraulic diameter. The geometric parameter for a circular duct is equal to $\frac{\pi}{4}$, for a noncircular duct such as a square duct it is equal to 1, and for any other geometrical shape of the thermosyphon duct this value can be easily obtained by using the following expression:

$$\frac{A}{D^2} = \frac{(\text{Perimeter})^2}{16(\text{Area})} \tag{8.15}$$

$$\Pi_7 = (\rho_L)^a (D)^b (\mu_L)^c (C_P)^d (\rho_V)^1 \tag{8.16}$$

$$\Pi_7 = \frac{\rho_V}{\rho_L} \text{(Seventh pi term)} \tag{8.17}$$

The seventh Pi term is the density ratio of vapor to liquid.

$$\Pi_8 = (\rho_L)^a (D)^b (\mu_L)^c (C_P)^d (Q)^1 \tag{8.18}$$

$$\Pi_8 = \frac{\rho_L^2 D Q}{\mu_L^3} \tag{8.19}$$

$$\Pi_8 = \left(\frac{\rho_L V_S D}{\mu_L}\right)^3 \left(\frac{Q}{\rho_L D^2 V_S^3}\right) \tag{8.20}$$

$$\Pi_8 = (\mathrm{Re})^3 \underbrace{\left(\frac{h_{fg}}{V_S^2}\right)}_{E_r} \text{(Eighth pi term)} \tag{8.21}$$

E_r is the ratio of thermal to kinetic energy. Thermal energy is the energy given out by the evaporator to convert the liquid into its vapor state, which is traveling with sonic velocity V_S.

8.3 Heat pipe

Carrying out similar dimensional (Buckingham's-π) analysis for a heat pipe, the physical quantities and their respective fundamental dimensions that are used for this analysis are the same as those used previously for the thermosyphon, the only additional physical quantity is surface tension, i.e., σ (M T $^{-2}$).

Repeating Variables: ρ_L, D, μ_L, C_p

Number of physical quantity (n_1): 13.

Fundamental Dimensions (k_1): M, L, T and θ, i.e., 4.

Therefore, number of pi terms: $n_1 - k_1$: $13-4=9$.

In this case, the first 8 pi terms are the same as those obtained for the thermosyphon, the only additional pi term in this case is the Π_9 (ninth pi term) due to the effect of surface tension, which cannot be ignored in this case because of the wick. Although the Π_8 term for both the thermosyphon and the heat pipe are the same, by rearranging some terms we can also obtain the Π_8 term for a heat pipe as:

$$\Pi_8 = \frac{\rho_L^2 DQ}{\mu_L^3} \tag{8.22}$$

$$\Pi_8 = \left(\frac{\rho_L V_s D}{\mu_L}\right)^2 \left(\frac{Q}{V_S^2 D \mu_L}\right) \tag{8.23}$$

$$\Pi_8 = (\mathrm{Re})^2 \underbrace{\left(\frac{\sigma}{V_S \mu_L}\right)}_{\frac{1}{Ca}} \left(\frac{Q}{DV_S \sigma}\right) \tag{8.24}$$

$$\Pi_8 = (\mathrm{Re})\left(\frac{1}{\mathrm{Ca}}\right)\left(\frac{\frac{h_{fg}}{\sigma}}{\rho_L D}\right) \tag{8.25}$$

$$\Pi_8 = (\mathrm{Re})\left(\frac{1}{\mathrm{Ca}}\right)\underbrace{\left(\frac{\frac{h_{fg}}{\sigma D^2}}{\rho_L D^3}\right)}_{E_M} \text{(Eighth pi term)} \quad (8.26)$$

The Reynolds, which is the ratio of inertial forces to viscous forces, also quantifies the relative importance of these two types of forces for given flow conditions. $\left(\frac{1}{\mathrm{Ca}}\right)$ is the inverse of the capillary number, which represents the relative effect of surface tension forces versus viscous forces for a fluid moving with velocity V_S. E_M is the ratio of two energies. The numerator represents latent heat of vaporization, which can further be referred to as latent energy per unit mass, and the denominator represents surface tension energy per unit mass.

$$\Pi_9 = (\rho_L)^a (D)^b (\mu_L)^c (C_P)^d (\sigma)^1 \quad (8.27)$$

$$\Pi_9 = \frac{\rho_L D \sigma}{\mu_L^2} \quad (8.28)$$

$$\Pi_9 = \underbrace{\left(\frac{\rho_L \sigma}{\mu_L \rho_V V_S}\right)}_{F_N}\left(\frac{D \rho_V V_S}{\mu_V}\right)\left(\frac{\mu_V}{\mu_L}\right) \quad (8.29)$$

$$\Pi_9 = \underbrace{\left(\frac{\rho_L \sigma}{\mu_L \rho_V V_S}\right)}_{F_N}(\mathrm{Re})\left(\frac{\mu_V}{\mu_L}\right) \text{(Ninth pi term)} \quad (8.30)$$

where F_N refers to the heat pipe fluid number [4] and is independent of any geometric parameters, the magnitude of F_N normally decreases with an increase in temperature, for all fluids.

8.4 Results and discussion

Most of the pi-groups that were obtained during the analysis are commonly used in many heat transfer applications. In this section, mainly E_r and E_M, which are derived dimensionless numbers, are explained in detail. E_r is the ratio of thermal to kinetic energies, thermal energy is the energy used up by the fluid for phase transformation from liquid to vapor state. The square of the velocity represents the kinetic energy head.

$$E_r = \frac{h_{fg}}{V_S^2} \quad (8.31)$$

On rearranging Eq. (8.31) by multiplying the numerator and the denominator by ρ_V,

$$E_r = \frac{h_{fg}\rho_V}{V_S^2\rho_V} \tag{8.32}$$

Eq. (8.32) is the ratio of thermal energy per unit volume to kinetic energy per unit volume. Because the pressure drop ΔP is one of the key parameters in any heat exchanger optimization, the denominator of Eq. (8.32) has been related to ΔP (Sabharwall et al. 2009 (iii)) [5].

$$\frac{\Delta P}{l} = \frac{4f\rho_V V_S^2}{2D} \tag{8.33}$$

From Eqs. 8.32 and 8.33,

$$E_r = \frac{h_{fg}\rho_V}{\frac{\Delta P}{2f}\frac{1}{\left(\frac{l}{D}\right)}} \tag{8.34}$$

For a given aspect ratio $\left(\frac{l}{D}\right)$ and assuming negligible change in the friction factor for a given turbulent flow range, the following appears:

$$\frac{E_{r1}}{E_{r2}} = \frac{\left[\frac{(h_{fg}\rho_V)}{\Delta P}\right]_1}{\left[\frac{(h_{fg}\rho_V)}{\Delta P}\right]_2} \tag{8.35}$$

Eq. (8.35) is the ratio of volumetric latent energy per unit pressure drop for two different conditions. These may be different alkaline metals at the same temperature or be for the same fluid, but at different temperatures. The above equation could be used to determine an optimum working fluid for a given set of operating conditions. For example, ratios of E_r of Li with other alkaline metals at 2,000 K shows 1.5, 1.8, and 2.6 times volumetric energy gain when compared with Na, K, and cesium respectively.

Eq. (8.35) could also be used for the same coolant but at different temperatures; the volumetric latent energy decreases because of the decrease in latent heat, which could optimize the working fluid selection based on operating temperature. Thus, at lower temperatures the volumetric latent energy per unit pressure drop is higher because of higher latent heat and lower pressure drop (because V_S increases with increase in temperature, though the density decreases but the net effect due to square dependence of pressure drop on velocity leads to a lower pressure drop), as can be seen in Fig. 8.1. Li, on the other hand, shows higher thermal energy than the rest of the alkaline metals, as shown in Fig. 8.1, but can be discarded as it does not boil until 1,590 K at atmospheric pressure, which is higher than the available temperature from NGNP.

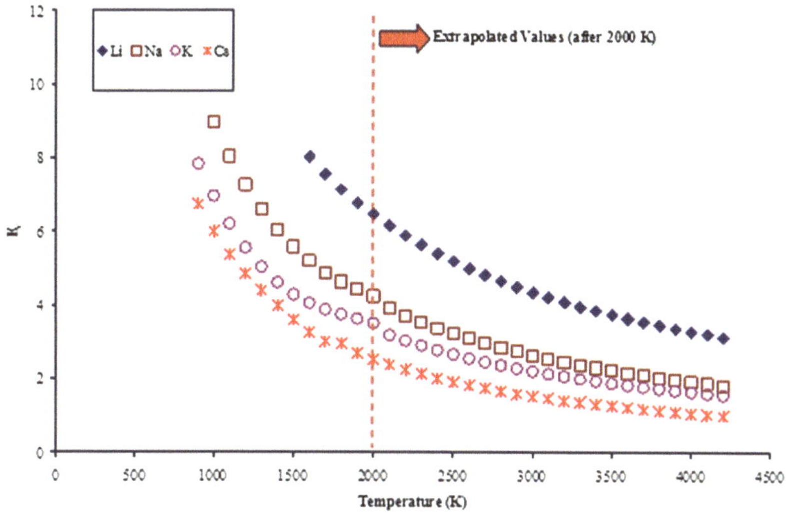

FIG. 8.1 Variation in E_r with respect to temperature for alkaline metals.

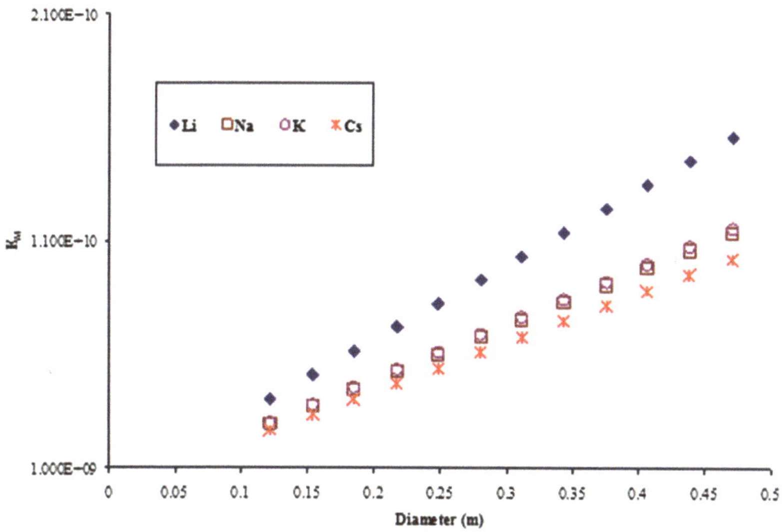

FIG. 8.2 Variation in E_M with respect to diameter for alkaline metals.

In Fig. 8.2, the dependence of is shown with varying diameters at 1,300 K. E_M is the ratio of the latent heat of vaporization per unit mass to the surface tension energy per unit mass.

For a calculated value for Na of E_M equal to 2.422×10^{10}(at 1,300 K) and at 1 m diameter, in order to obtain the same $E_M$value for Li, K, and Cs, the diameter obtained was equal to 0.7 3 m, 0.98 m, and 1.11 m, respectively.

The narrower the pipe diameter, the greater the pressure drop; therefore Li has the highest pressure drop among all alkaline metals for a given value of E_M. Capillary pressure [4] (P_c) that draws the alkaline metal into the wick is determined by porosity of the wick and surface tension of the alkaline metal, as can be seen from the following equation:

$$P_c = \frac{2\sigma}{D_w} \tag{8.36}$$

The wick for Li does not have to be very fine: it could be coarse, because Li has the highest surface tension when compared with other alkaline metals, thus reducing the manufacturing cost.

From Eqs. 8.26 and 8.36, E_M could be rewritten in terms of P_c such that:

$$E_M = \frac{h_{fg}}{\frac{\sigma}{D_w}\frac{D_w}{D\rho_L}} \tag{8.37}$$

$$E_M = \frac{\text{latent heat energy/mass}}{(\text{height of liquid column})(g)\left(\frac{D_w}{D}\right)} \tag{8.38}$$

From Eq. (8.38), we can conclude that for an optimum working fluid for a given D and D_w the value of E_M should be as low as possible, such that greater height of the liquid can be achieved through the wick in a heat pipe. As discussed previously, for E_r the magnitude should be as high as possible such that the volumetric energy gain is maximized for a given heat pipe or thermosyphon.

It is important to recognize the fundamental limitations [6] and strengths of the dimensional analysis techniques from which the above written equations have been derived. Also, the dimensionless numbers would vary were different repeated variables chosen. The weakness of dimension analysis is that it sheds no light on the physical nature of a phenomenon, nor does it provide any explicit formulation for the unknown dimensionless function. Conversely, the strength of this technique is its generality, since it does not depend on any particular physical model of the phenomenon, a factor that is most important in complex situations such as the chaotic analytically intractable turbulent flow regimes. Once the conclusions derived from dimensional analysis are empirically validated, which should be done in the near future for both heat pipe and thermosyphon devices, an optimum working fluid can be chosen. The correlations are perfectly general and are therefore not restricted to only alkaline metals or any kind of fluid or the range of velocities measured, or even the particular kind of heat pipe or thermosyphon employed.

8.5 Summary

Dimensional Analysis is a valuable mathematical technique useful in research work for design and conducting model tests. This analysis yielded two terms—E_r and E_M—particular to the operation of these devices in addition to those commonly used in many heat transfer applications. $E_r$relates the latent heat of vaporization to the pressure drop across the device, while E_M relates the latent heat to the capillary pressure. The significance of these two terms is discussed. The universal nature of these numbers should be useful in increasing the fundamental understanding of both thermosyphons and heat pipes.

References

[1] D.Q. Kern, Process Heat Transfer, McGraw-Hill International Book Company, 1982.

[2] B. Zohuri, Dimensional Analysis and Self-Similarity Methods for Engineers and Scientists, Springer Publishing Company, New York, April 16, 2015.

[3] B. Zohuri, Dimensional Analysis beyond the Pi Theorem, Springer Publishing Company, New York, November 3, 2016.

[4] K.C. Sockalingam, Performance Characteristics and Optimization of Water Heat Pipes, PhD Thesis, University of California, Berkeley, 1972.

[5] P. Sabharwall, V. Utgikar, F. Gunnerson, Effect of mass flow rate on the convective heat transfer coefficient: analysis for constant velocity and constant area case, J. Nuc. Technol. 166 (May 2009) (iii).

[6] H. Soumerai, Practical Thermodynamic Tools for Heat Exchanger Design Engineers, Wiley & Sons, 1987.

[7] G.I. Barenblatt, Scaling, Self-Similarity, and Intermediate Asymptotic, Cambridge University Press, 1996.

Index

Note: 'Page numbers followed by "f" indicate figures, "t" indicate tables and "b" indicate boxes'.

A

Activated Brazing Alloy (ABA), 163
Adiabatic, 9–11, 275
Advanced High-temperature Reactor (AHTR), 56
Aircraft Nuclear Propulsion (ANP), 95–96
Alkali Metal Thermal-to-Electric Conversion (AMTEC), 78–80, 87
Alkaline metals, 361f
  thermosyphon, 323–324, 324t, 329–332, 331f
AlSiC HiK™ Plates, 194, 195f
Aluminum Silicon Carbide (AlSiC), 194
American Society for Testing and Materials (ASTMs), 211
Annular heat pipes, 225f
  cross-section, 225, 225f
  Isothermal Furnace Liners, 225, 226f
  temperature calibration industry, 226–227
Apollo Moon missions, 81
Apollo spacecraft, 78
Army Air Forces (AAF), 95
Atomic Energy Commission (AEC), 95
Austenitic stainless steels, 287
Axially grooved wick, 34, 35f
Axial Power Rating (APC), 24

B

Boiling limitation, 28
  adverse vapor dynamics, 278–279
  axial heat transport requirement, 279
  liquid transport factor, 279, 280f
  sodium heat pipes, 280–281
  thermodynamic conditions, 278
Brayton cycle, 98, 309

C

Capillary limitation, 28, 277
Capillary Pumped Loop (CPL), 12
Carbon fiber filaments, 20
Cassini–Huygens mission, 64
Cassini spacecraft, 62, 67
Central Processing Units (CPUs), 49, 109
Cesium, 192, 283, 325
Class I refrigerant, 2
Closed Loop Pulsating Heat Pipes (CLPHPs), 197
Closed Loop Two-Phase Thermosyphon (CLTPT), 320
Coefficient of Performance (COP), 297–298, 300–301
Coefficient of Thermal Expansion (CTE), 194
Cold Reservoir Variable Conductance Heat Pipe, 15, 15f
Compatible fluids/materials
  cesium and potassium heat pipes, 192, 192f
  freeze– thaw and thermal cycling, 193f
    ACT Corporation, 192–194
    AlSiC HiK™ Plates, 194, 195f
    Aluminum HiK™ Plates, 194
    Aluminum Silicon Carbide (AlSiC), 194
    Heat Pipes, 194
  working fluid and envelope compatibility, 189, 190t–191t
Compressed Air Energy Storage (CAES), 124–125, 125f
Computational Fluid Dynamics (CFDs), 229–231, 230f
Concentrated Solar Power (CSP), 123, 123f, 128, 161
Conduction, 3–4, 189, 302
Conductively coupled multicell (CC/MC) thermionic fuel element, 71–72
Constant Conductance Heat Pipes (CCHPs), 14, 16
  axial groove extrusion, 202, 202f
  excess-liquid heat pipe, 204, 205f
  gas-loaded heat pipe, 203–204, 203f–204f
  liquid flow control, 205–206, 206f
  manufacture, 202
  vapor flow control, 206, 207f
  variable conductance, gas-loaded heat pipes, 206–207, 208f

Controllable thermosyphon, 322
Convection
natural, 309
single-phase forced, 329–331
Conventional heat pipe (CHP), 9–11, 9f, 14, 281
heat-radiation characteristics, 17–18, 18f
structural comparison, 17, 17f
Copper pipe system, 186
Corrosivity, 131–133
Counter-flow heat pipe heat exchanger, 118, 120f
Cryogenic heat pipes, 213–214
Cylindrical Inverted Multicell (CIM), 72–74, 73f

D

Defense Advanced Research Projects Agency (DARPA), 91
Defense/avionics
air application, 116, 117f
F-22 Fighter, 113–114, 114f
ground application, 114–115
sea application, 115, 116f
space application, 116–117, 117f
Defense Threat Reduction Agency (DTRA) program, 76
Deicing system, 173–174, 319
Demonstration Using Flattop Fissions (DUFF), 104
Dimensional analysis, 355
Diode Heat Pipe (DHP), 224
collecting heat, 222
rejecting heat, 222–223, 223f
role of, 223–224
Direct Methanol Fuel Cell (DMFC), 163–164
DuoFin, 50

E

Electric blankets, 13
Electromagnetic Interference (EMI), 107
Electron beam welding (EBW), 253
Electronics, 107
and electrical equipment cooling, 109–113, 110f–112f
Electromagnetic Interference (EMI), 107
heat pipe heat sinks, 107, 108f
high end CPU heat pipe heat sink, 107–108, 108f
high power IGBT heat pipe heat sink, 108–109, 109f
End cap installation, 253–254, 254f
Energy conservation, 54
Energy-dependent boundary equations
Heat Pipe-Operated Mars Exploration Reactor (HOMER), 58
Heat Pipe Power System (HPS), 59
heat pipe space reactor, 58, 59f
Los Alamos National Laboratory (LANL), 59
power conversion technique, in space, 59, 60f
space nuclear power development, 59, 60f
Energy Efficiency Ratio (EER), 297–298
Energy storage methods
electrical storage, 124–125, 125f
Thermal Energy Storage (TES), 125–126
Entrainment limitation, 28, 277–278
Eolic energy, 345b–346b
Evacuated Tube Heat Pipe Solar Collectors (ETHPSC), 152, 152f

F

Fibrous materials, 20
Fin efficiency, 36
Finned heat pipe, 51f
general use, 50, 51f
green house application, 50, 50f
Fission systems
heat, 68–69
propulsion, 69
Fixed Conductance Heat Pipes (FCHP), 14, 16–17. *See also* Constant Conductance Heat Pipes (CCHPs)
Flooding phenomenon, 347–350, 348f–349f
Fluid charging, 262
Fourier's Law, 302, 303f

G

Galileo, 63–64
Gas-loaded heat pipe, 14, 14f
GASPIPE, 15
Gas turbine engines, 166–169, 168f
General Purpose Heat Source–Radioisotope Thermoelectric Generator (GPHS-RTG), 62, 65

Geostationary communications satellites, 61
Geysering instability, 333–334
Goddard Space Flight Center (GSFC), 98–99
Graphics Processing Units (GPUs), 49, 109
Grooved tube, 21
Ground temperature control, 318–319, 318f

H

Heat exchanger, 165, 166f
counter-flow heat pipe heat exchanger, 118, 120f
deigns, 118
Heat Pipe Heat Exchanger (HPHE), 120, 121f
quoted effectiveness, 119–120
staging heat pipe tube, 118–119, 120f
types, 118, 119f
Heat flux transformation, 52–53
Heating, Ventilation, and Air Conditioning (HVAC) systems, 2–4, 41b–44b, 186
Heat pipe-cooled nuclear reactor, 58
Heat Pipe Driven Heat Exchanger (HPHX), 54–55
Heat pipe fabrication, 276
Heat pipe heat exchanger
benefits, 304
conventional heat pipe, 275–276, 276f
heat pipe application, holistic approach, 304f
advantages, 282
capillary pumping, 287f
compatible envelope/fluid pairs, 286
concept, 281, 282f
disadvantages, 282
envelope/fluid pairs, 285–286
heat pipe working fluids, 292–293, 293f
high liquid density and high latent heat, 286
high temperature heat pipes, 287
high temperature isothermal furnace liners, 283, 284f
inherent temperature uniformity and stability, 287–288
limitation, 283
lowest heat pipe limit driven temperature, 289–291
Merit number, 286–289, 286f
phase-change processes, 283
Titanium-Zirconium-Molybdenum (TZM) alloy, 283–285
tope view depiction, 283, 284f
two-phase flow circulation, 283
ultra-high temperature heat pipes, 283, 285f
heat pipe design
boiling limitation, 278–281, 280f
capillary limitation, 277
entrainment limitation, 277–278
metallic sodium, 277
sonic limitation, 277–278
Innovative Heat Exchanger Designs, 300–303
limitations to heat transport, 275–276, 276f
longevity, 276
Variable Conductance Heat Pipe (VCHP), 298–299
Vertical Direct-contact Heat Exchanger (VDHX), 294f
air flow rate cooling capacity, 297f
Coefficient of Performance (COP), 297–298
demonstration test bed schematic, 296, 296f
Energy Efficiency Ratio (EER), 297–298
evaluation data, 296, 298t
top-view, 296, 297f
working fluid, 276
Heat Pipe Operated Mars Exploration Reactor (HOMER), 58, 78–80
Heat Pipe Operated Mars Exploration Reactor-15 (HOMER-15), 80–82, 80f, 82t
baseline design, 85–86
Heat Pipe Operated Mars Exploration Reactor-25 (HOMER-25)
characteristics, 83, 84t
designs, 80–82
fuel pins and heat pipes configuration, 85, 85f
mass summary, 83, 84t
Heat Pipe Power System (HPS), 59, 77–78
Heat pipes (HPs), 275
advantages, 41b–44b
applications, 41b–44b, 316–320
to avoid, 41b–44b
electronic devices, 38
fin efficiency, 36
high-capacity power electronics cooler, 36
industrial, 51–58
Loop Heat Pipe (LHP), 37–38

Heat pipes (HPs) (*Continued*)
possible, 41b–44b
sounding rocket, 39
thermal energy storage systems, 123–152
benefits, 184, 185f, 186
concept, 3–4, 4f
constraints
axially grooved wick, 34, 35f
basic heat pipe, 29, 29f
benefits, 30
fluid names and contents, 33–34
latent heat to specific heat, 30–31, 32t
liquid transport factor, 30–31, 33f
material composite, 33t, 34
physical elements, 29
sections, 29
surface tension, 30–31, 30f
symbols, 32
viscosity, 30–31, 31f
description and types
closed loop, 5–6
Cold Reservoir Variable Conductance Heat Pipe, 15, 15f
components and principle, of operation, 9–11, 9f
Constant Conductance Heat Pipe (CCHP), 16
conventional, 9–11, 9f
electric blankets, 13
Fixed Conductance Heat Pipes (FCHPs), 16–17
gas-loaded heat pipe, 14, 14f
heat-radiation characteristics, 17–18, 18f
heat transfer, 6–7
isothermaliser, 7
liquid, 6
Loop Heat Pipe (LHP), 7, 10f, 11–12
Mercury Heat Pipe, 9–11
NASA's New Millennium Space Program, 11, 11f
parts and functions, 7, 8f
sodium/molybdenum, 7, 10f
structural comparison, 17, 17f
super-conductors, 5
temperature regulation, 13, 13f
thermal conduction, 6
traditional, 7, 7f
Variable Conductance Heat Pipe (VCHP), 15–16, 16f
wick, 5–6
disadvantages, 41b–44b
efficiency, 41b–44b
fundamental dimensions, 358
heat configuration, 1, 2f
history, 2–5
impacts, 35
manufacturers, 41b–44b
operating ranges, 27–29, 28f
operation principles
assemblies design guidelines, 24
container, 18–19
forming/shaping, 26
grooved tube, 21
heat removal, 25
length and pipe diameter, 26
reliability, 25–26
respect to gravity, 24–25
screen mesh, 21–23, 22f–23f
sintered powder, 21
temperature limits, 25
wick/capillary structure, 20–21, 26–27, 27f
working, 23–24, 24f–25f
working fluid, 19, 20f
phase-change processes, 185
physical quantity, 358
principles, 1–4
questions and answers, 187–188
repeating variables, 358
surface tension energy, 359
technology types, 41b–44b
and thermosyphon, 314–316, 314f, 316f
two-phase flow circulation, 185
types, 195
Heat pipe technology (HPT), 1
Heat pipe testing techniques
mechanical sounders, 270–271
performance versification, 271–272, 272f
wick wetting, 271
Heat pumps, 165
Heat Transfer Fluid (HTF), 134, 143
High-capacity power electronics cooler, 36
High Power Electric Propulsion (HiPEP), 98
High Temperature Electrolysis (HTE), 308
High Temperature Gas-cooled Reactor (HTGR), 308
High-temperature heat pipe furnace, 56–57
High temperature heat pipes, 211–213, 213f
High temperature thermosyphons, 346

HiK™ heat pipe plates
Computational Fluid Dynamics (CFDs), 229–231, 230f
conductivity structures, 227
configuration, 227–228, 228f
cooling embedded VME and VPX systems, 231–233, 231f, 233f
3-dimensional, 229, 230f
position, 227–228, 228f
thermal analysis, 228–229, 229f
Home energy system, 159–164, 160f–164f

I

Inclination effects, 338f
experimental setup, 338–339
gravitational influence, 337–338
inclination angle, 339–340, 339f
interactive influence, 338–339
Industrial application
die-casting and injection molding, 53
electronic components, cooling, 53
energy conservation, 54
groups, 51–52
heat flux transformation, 52–53
Heat Pipe Driven Heat Exchanger (HPHX), 54–55
heat pipe inserts, 56
heat source and sink, 52
high-temperature heat pipe furnace, 56–57
miscellaneous, 57–58
permafrost preservation, 55
snow melting and deicing, 55–56
spacecraft, 53–54
temperature control, 53
temperature flattening, 52
Innovative Heat Exchanger Designs, 301f
ACT Corporation, 300–301, 302f
brute force method, 300
components, 301
Fourier's Law, 302, 303f
Heat Pipe Heat Exchanger, 302
radial resistance, 302
Insulated-Gate Bipolar Transistors (IGBTs), 108–109
Interface Heat Spreaders (IHS), 112
Intermediate Temperature Thermal Management System (ITTMS), 221
International Space Station (ISS), 11–12, 100
Isothermal Furnace Liner (IFL), 225
Isothermaliser heat pipes, 7, 21–23

K

Kilopower Reactor Using Stirling Technology (KRUSTY), 101–104, 101f, 103f–104f
Kuiper Belt Objects (KBOs), 64–65

L

Latent Heat Thermal Energy Storage (LHTES)
arrangements of heat pipe, 136, 139f
charging process, 141f
effect of heat input, 149, 149f
Evacuated Tube Heat Pipe Solar Collectors (ETHPSC), 152, 152f
experimental setup, 136, 137f
flow configuration, 138, 141f
gravity heat pipes, 143, 146f
heat pipe heat exchanger, 134, 135f
Heat Transfer Fluid (HTF), 143
heat transfer mechanisms, 152, 153f
heat transfer pathways, 143–147, 146f
one heat pipe and with three heat pipes, 136, 138f
Phase Change Material (PCM), 134–136, 143, 148, 151f
contours of molten, 143, 145f
embedded heat pipes, 138, 141f
liquid fraction, 136, 140f
physical model and computational domain, 136, 137f
solar thermal system, 151, 152f
surface energy balance, 142
thermal resistance network, 140, 142f, 147
unit with heat pipe, 134, 134f
Levelized Cost of Energy (LCOE), 124
Liquid Controlled Heat Pipe (LCHP), 15
Liquid flow control, 205–206, 206f
Liquid Metal Fast Breeder Reactor (LMFBR), 9–11, 56
Liquid metal working fluids, 3–4
Liquid transport factor, 30–31, 33f, 279, 280f
Liquid Trap Diode, 217–218, 217f
Liquid-vapor interfacial shear stress, 313
Lithium, 188, 324–325
Longevity, 276
Loop Heat Pipe (LHP), 12, 37–38, 111–112
application, 7, 10f
operation, 11
two parallel condensers, 7, 10f
two parallel evaporators, 7, 10f

Loop heat pipes/capillary pumped loop, 196, 196f
Los Alamos National Laboratory (LANL), 59, 77, 183
Lowest heat pipe limit driven temperature
ACT-TEC Thermoelectric Cooler Series Solid State Enclosure Air Conditioning, 289, 290f
finned heat pipe assembly, 289, 290f
heat pipe and thermosyphon performance limits, 290, 291t
heat pipe performance limits, 291, 292f
temperature drops, 289, 289f
vapor space temperature drop, 289
Luna lander, 153–159
Lunar base mission, 81

M

Manufacturing techniques, 239–240
assembly
end cap installation, 253–254, 254f
end closure and welding, 255, 255f–257f
forming/shaping, 258b–259b
heat removal, 258b–259b
length and pipe diameter, 258b–259b
reliability, 258b–259b
respect to gravity, 258b–259b
summary, 255–259
temperature limits, 258b–259b
wick forming and insertion, 253
wick structures, 258b–259b
components, 240
end cap, 244f
aluminum alloy 6061, 242
fully mechanized fusion, 242
fusion welding, 242
geometry, 243
joint designs, 242, 244f
joint efficiency, 242–243
stainless steel alloy 304L, 242
thickness, 242
envelope, 241, 241f
evacuation and charging
equipment, 259–262, 261f
fluid charging, 262
fluid purity and inventory, 262–264, 263f
gas blockage, 264–267, 264f, 266f, 268f–270f
over-all evacuation and procedure, 259, 260f
setup, 259, 261f
fill tube, 243
flow chart, 245, 246f
full tube closure, 267
heat pipe testing techniques, 267–272
parts cleaning, 247f
chromated deoxidizer solutions, 251t
contaminants, 249
machining, 248
non-etch alkaline cleaners, 250t
operation, 248
passivating solutions, 251t
problems, 249
procedures, 249–252, 250t
wick cleaning and pretreatment, 248
restrictions/assumptions, 245
wick, 244
working fluid, 240–241, 245
Mars One mission, 99–101, 99f–100f
Martian surface, 80b–81b
Massachusetts Institute of Technology (MIT), 95
Maximum Expected Operating Pressures (MEOPs), 34–35
Mean Time Between Failure (MTBF), 186
Mechanical sounders, 270–271
Medicine/human body temperature control, 169–171, 170f–171f
Mercury Heat Pipe, 9–11
Merit number derivation, 286–287, 286f
capillary limit, 288
capillary pumping capability, 288
mass flow rate, 288
Micro heat pipe (MHP), 199f
chip speed, 198, 199f
components, 198
defined, 197
example, 200, 201f
laptops, 200
micro machining technology (MEMS), 198
power density, 198, 200f
power dissipation, 198, 200f
super-thin Micro-Heat Pipe, 200, 201f
Military radar systems, 114–115, 115f
Modular High Temperature Reactor, 320
Monel, 189
Moon Age and Regolith Explorer (MARE), 153–154
configuration, 154, 154f

thermal concept, 155, 155f–156f
Thermal Control System (TCS), 155
thermal mathematical model, 156, 157f
Multi-layered Insulation (MLI), 156–157
Multi-Mission RTG (MMRTG), 62

N

NASA's New Millennium Program, 7, 11f
National Nuclear Security Administration (NNSA), 101–102
Natural convection, 309
Navy antenna cooling, 115
Neutron shielding, 93–94
New Horizons, 64–68, 66f–67f
Next Generation Nuclear Plant (NGNP), 347, 360
alkaline metals, 323–324
industrial-scale recovery options, 313–314
process heat transfer, 308, 308f
thermosyphons, 355–356
Non-adherence, 35–36
Non-condensable gas (NCG), 25–26, 29–30, 157–158, 218, 332
Non-etch alkaline cleaners, 250t
Nuclear Energy Propulsion of Aircraft (NEPA), 95–97, 97f
Nuclear Engine for Rocket Vehicle Application (NERVA), 61
Nuclear heat conversion, 74
Nuclear reaction control, 102
Nuclear reactor industry, 9–11
Nuclear reactor power system
Defense Advanced Research Projects Agency (DARPA), 91
masses of current on-board nuclear reactors, 88, 89t–90t
Space Power 100 (SP-100), 87, 87f–88f
Nuclear Thermal Rocket (NTR), 61, 69
Nuclear thermionic technology development
conductively coupled multicell (CC/MC) thermionic fuel element, 71–72
Cylindrical Inverted Multicell (CIM), 72–74, 73f
Nucleation sites
contact angle, 336, 336f
dissolved gas, 335
inactive nucleation site, 337
pocket, 337, 337f
spherical bubble, 335
wetting, 335, 336f

O

Oak Ridge National Laboratory (ORNL), 96
Oscillating Heat Pipe (OHP), 215f–216f, 281
adiabatic section, 216
condensing section, 216
evaporating section, 216
feature, 215
types, 216–217
working fluid, 215
Ovens/furnaces, 171–172

P

Passive heat transfer device, 5
Passive thermal technical discipline lead (TDL)
Moon Age and Regolith Explorer (MARE), 153–154
configuration, 154, 154f
thermal concept, 155, 155f–156f
thermal control system (TCS), 155
thermal mathematical model, 156, 157f
Small Business Innovative Research (SBIR), 153–154
Variable Conductance Heat Pipes (VCHPs), 158–159, 158f
Warm Electronic Box (WEB), 158–159
Pentium processor, 110
Perkins boiler, 309–310, 310f
Permafrost stabilization, 172, 173f
Phase Change Materials (PCMs), 159–160, 301
high-melting-point, 128
Latent Heat Thermal Energy Storage (LHTES), 134–136, 143, 148, 151f
contours of molten, 143, 145f
embedded heat pipes, 138, 141f
liquid fraction, 136, 140f
thermal storage applications
classification, 129, 131t
corrosivity, 131–133
High-Temperature Reactors (HTR), 133
inorganic, 129, 131t–132t
shuttle heat pipe, 129, 130t
Very-High Temperature Reactors (VHTR), 133
Phase change medium, 123–124
Photovoltaics (PV), 123, 159
Planar Array Antenna Assemblies (PAAAs), 115
Polyethylene (PEX), 195

Potassium, 325
Potential space nuclear thermionic missions
Defense Threat Reduction Agency (DTRA) program, 76
high power, 75
military missions, 75
self-sufficiency, 75
solar energy flux, 74, 74f
survivability, 75
Power dissipation, 198, 200f
Prandtl number, 357
Pressure Controlled Heat Pipe (PCHP), 227, 233f, 236f
applications, 234
Non-Condensable Gas (NCG), 233–234, 235f
temperature control, 234
volume modulated, 235–236, 235f
Production tools, 169
Project Prometheus 2003, 97–99
Pulsating Heat Pipe (PHP), 196–197, 197f–198f, 216–217
Pumped Hydro-Power Storage (PHPS), 124

R

Radar Ocean Reconnaissance Satellites (RORSATs), 68
Radiative coupling, 71–72
Radiator Heat Pipes (RHPs), 222
Radioactive Heater Units (RHUs), 66
Radioisotope systems
Cassini–Huygens mission, 64
Galileo, 63–64
Multi-Mission RTG (MMRTG), 62
New Horizons, 64–68, 66f–67f
Radioisotope Thermoelectric Generators (RTGs), 62, 63f
Ulysses, 62–63
Radioisotope Thermoelectric Generators (RTGs), 62, 66
Reactor sitting, 94–95
Remote Heat Exchanger (RHE), 110–111
Residential building, 121–122, 122f
Resistance thermometer, 211
Reynolds number, 356
Rotating and revolving heat pipes, 209f–210f
common components, 210, 212f
compact air-conditioning unit, 209, 209f
copper-methanol rotating heat pipe, 210, 211f
cross-section, 210, 212f
resistance thermometer, 211
Rutherford High Energy Laboratory (RHEL), 213

S

SAFE-400 (Safe Affordable Fission Engine) space fission reactor, 77
Screen mesh, 21–23, 22f–23f
Sensible Thermal Energy Storage Systems (STESs), 126
Silicon-controlled rectifiers (SCRs), 108–109
Sintered powder, 21
Small Business Innovation Research (SBIR), 153–154, 157
Snow melting, 319
Sodium, 325
Sodium/molybdenum heat pipes, 7, 10f
Solar energy flux, 74, 74f
Solar house, 159, 160f
Solar thermal water heaters, 49
Sonic limitation, 6–7, 28, 277–278
Spacecraft, 53–54, 186
aluminum/ammonia, 189
Apollo spacecraft, 78
applications, 98–99
Cassini spacecraft, 62, 67
interplanetary, 74
nuclear power cells, 48
thermal radiators, 312–313
Voyager spacecraft, 62
Space electronics, 113–114
Space Nuclear Auxiliary Power (SNAP), 93
Space nuclear systems, 11–12
Space Power 100 (SP-100), 83, 83f
cross-section, 87, 87f
initial design, 87, 88f
Space Power Reactors (SPRs), 66
Space reactor power systems, 79t
heat pipe and fuel pins configuration, 82–85, 82f–83f, 84t, 85f
Heat pipe Operated Mars Exploration Reactor-15 (HOMER-15), 80–82, 80f, 82t
Heat Pipe Operated Mars Exploration Reactor (HOMER), 78–80
Space shuttle orbiter heat pipe applications, 104–107, 106f
Space systems
fission systems
heat, 68–69
propulsion, 69

heat pipe design, 86, 86f
Heat Pipe Power System (HPS), 77–78
Kilopower Reactor Using Stirling Technology (KRUSTY), 101–104
Mars One mission, 99–101
material choices, 92
neutron shielding, 93–94
Nuclear Energy Propulsion of Aircraft (NEPA), 95–97
nuclear reactor power system, 87–92
nuclear thermionic technology development, 69–74
potential space nuclear thermionic missions, 74–77
Project Prometheus 2003, 97–99
radioisotope systems, 62–68
reactor control, 92–93
reactor sitting, 94–95
safety considerations, 92
space reactor power systems, 78–85
Stirling engine system, 85–86
Space Technology 8 (ST8), 7
Space Technology Mission Directorate (STMD), 102
Space thermionic advanced reactor compact (Star-C) concept, 73
SpaceX, 100
Spiral Heat Exchanger (SHE), 313–314
Spirit Rover, 67, 67f
Standard heat pipe configuration, 184, 184f
StarFin, 50
Stirling engine system, 85–86
Stirling Radioisotope Generator (SRG), 65–66
Superalloy, 189
Super-conductors, 5
Super-thin Micro-Heat Pipe, 200, 201f
Surface energy balance, 142
Surface tension, 30–31, 30f
Surface tension effects, 196–197
Surging (chugging) instability, 333–334
System for Nuclear Auxiliary Power (SNAP), 61

T

Temperature control, 49, 53
Temperature flattening, 52
Thermacore®, 56, 280–281
Thermal Energy Storage (TES), 128, 159–160
Concentrated Solar Power (CSP), 123, 123f
energy storage methods, 124–126
Latent Heat Thermal Energy Storage (LHTES), 126–127, 134–152
latent heat thermal storage materials
chemical properties, 128
physical properties, 127
thermal properties, 127
Phase Change Materials (PCMs), 128–134
phase change medium, 123–124
Thermal insulation, 52
Thermionic Fuel Element (TFE), 71
Thermionic reactor concept, 52, 52f
Thermodynamic analysis
Eolic energy, 345b–346b
general model and flooding, 347–350, 348f–349f
operation, 346–347
two-dimensional model, 347
two-phase thermosyphon thermodynamic analysis, 350–352, 350f–351f
Thermodynamic working fluid, 5–6
Thermosyphons (TSs), 6, 195, 196f, 275
alkaline metal, 329–332
application
automobile industry, 319
cold regions, 317–319, 318f
electrical industry, 319–320
electronics, 320
heating and cooling, 320
passive decay heat removal system, 320
railroad industry, 319
space systems, 317
characteristics, 316
design, 321–325
fundamental dimensions, 356
geometric parameter, 357
geometry, 321f, 322–323
heat transport limitations
sonic limit, 327–328
viscous limit, 328–329, 330f
historical development and background
adjustable heat transfer performance, 312–313
evaporator–controlled two-accumulator PPM systems, 313, 314f
heat transfer performance, 311
liquid-vapor interfacial shear stress, 313
natural convection, 309
Next Generation Nuclear Plant (NGNP), 313–314
operating and limiting mechanisms, 310–311
Perkins boiler, 309–310, 310f
two-phase closed thermosyphon, 310–311, 311f
two-phase closed U tube thermosyphon loop, 312–313, 312f

Thermosyphons (TSs) (*Continued*)
inclination effects, 338f
experimental setup, 338–339
gravitational influence, 337–338
inclination angle, 339–340, 339f
interactive influence, 338–339
mass flow rate and sonic velocity analysis, 326–327
Next Generation Nuclear Plant (NGNP), 355–356
nucleation sites, 334–337
physical quantity, 356
Prandtl number, 357
repeating variables, 356
Reynolds number, 356
simple controllable, 321, 321f
sonic velocity, 356–357
startup, 332, 333f
thermodynamic analysis. *See* Thermodynamic analysis
two-phase instabilities
fluid superheating, 334
surging and geysering instability, 333–334
thermosyphon evaporator instability, 334
vapor viscosity, 357
working fluids
alkaline metals, 323–324, 324t
cesium, 325
lithium, 324–325
potassium, 325
sodium, 325
Titanium-Zirconium-Molybdenum (TZM) alloy, 283–285
Topaz-1, 68–69
Traditional heat pipe, 7, 7f
Trans-Alaska Pipeline System (TAPS), 49, 50f, 318
Transportation systems, 173–174
TrueLeaf, 50
Tungsten inert gas (TIC) welding, 253
Two-phase closed thermosyphon (TPCT), 281, 310–311, 311f
Two-phase instabilities
fluid superheating, 334
surging and geysering instability, 333–334
thermosyphon evaporator instability, 334
Two-phase thermosyphon thermodynamic analysis, 350–352, 350f–351f

U

Ultra-high temperature heat pipes, 283, 285f
Ulysses, 62–63

V

Vapor flow control, 206, 207f
Vaporization, 3–4
Vapor Trap Diode, 218f–219f
Advanced Cooling Technologies (ACTs), 219
components, 221
General Purpose Heat Source (GPHS), 220–221
High Temperature Thermal Management System (HTTMS), 221
Non-Condensable Gas (NCG), 219
steady state temperature profiles, 219–220, 220f
Variable Conductance Heat Pipes (VCHPs), 14, 29–30, 158–159, 158f, 221, 233–234
cross section, 298, 299f
heat-radiation characteristics, 17–18, 18f
heat transfer, 15–16, 16f
measure outlet hydrogen temperature plot, 299, 300f
non-condensable gas, 298
structural comparison, 17, 17f
testing condition, 299, 299f
Variable Specific Impulse Magnetoplasma Rocket (VASIMR), 69
Vertical Direct-contact Heat Exchanger (VDHX), 294f
air flow rate cooling capacity, 297f
Coefficient of Performance (COP), 297–298
demonstration test bed schematic, 296, 296f
Energy Efficiency Ratio (EER), 297–298
evaluation data, 296, 298t
top-view, 296, 297f
Viscous limit, 28
Voyager spacecraft, 62

W

Warm Electronic Box (WEB), 158–159
Wick, 5–6, 20–21, 244, 315
forming and insertion, 253
structures, 26–27, 27f, 258b–259b
Wickless heat pipe, 313–314
Wrap-around heat pipe (WAHP), 214

Kolibri 360 GmbH
Ferdinand-Jühlke-Straße 7
99095 Erfurt, Deutschland
produktsicherheit@kolibri360.de